ORGANOTIN COMPOUNDS

VOLUME 2

ORGANOTIN COMPOUNDS

IN THREE VOLUMES

Edited by ALBERT K. SAWYER

Department of Chemistry
University of New Hampshire
Durham, New Hampshire

Volume 2

MARCEL DEKKER, INC., New York 1971

MARCEL DEKKER, INC.
95 Madison Avenue, New York, New York 10016

LIBRARY OF CONGRESS CATALOG CARD NUMBER 71-142895

ISBN 0-8247-1597-7

PRINTED IN THE UNITED STATES OF AMERICA

DEDICATED TO HENRY G. KUIVILA

for whom I have the greatest respect, both as an individual and as a chemist, and who is responsible for my first contact with organotin chemistry.

PREFACE

There has been a marked increase of activity in organotin chemistry in recent years. This work was undertaken to provide up-to-date, comprehensive coverage of this field, prepared by individuals who are well informed in their specialized areas. As editor I have been most fortunate to have such well-qualified authors for the individual chapters.

It is hoped that this book will be of value not only to active workers in organotin chemistry and related areas, but also to new workers in providing the present state of knowledge in the field. In addition to chapters on the chemistry of compounds containing tin bonded to the main group elements, there are chapters on tin- other element bonds, and on such specialized topics as organotin polymers, applications, biological effects and analyses. Complete referencing has not been attempted although such referencing has been attempted for publications since the comprehensive review article by Ingham, Rosenberg, and Gilman in *Chemical Reviews*, October, 1960.

I am particularly indebted to Drs. William Considine and Gerald Reifenberg for critical reviews of several chapters. In addition I wish to thank Dr. Paul Jones for translating the chapter on "Organotin Compounds with Sn–S, Sn–Se, and Sn–Te Bonds," and Mr. Ingo Hartmann for translating the chapter on "Organotin Compounds with Sn–P, Sn–As, Sn–Sb, and Sn–Bi Bonds." I wish also to express appreciation for the help, advice, and encouragement given by many friends too numerous to mention, but particularly for the reading of individual chapters by Drs. Henry Kuivila, Paul Jones, and John Uebel.

Durham, New Hampshire A. K. SAWYER

FOREWORD

There can be little question of the usefulness of a book that brings together up-to-date material in an active area of organometallic chemistry. Professor Sawyer has done this with *Organotin Compounds*. Not only is there included his continuing, extensive studies but also the cooperative efforts of a group of eminent chemists distinguished for the breadth and depth of their researches in different aspects of this area.

Where does organotin chemistry stand relative to organometallic chemistry as a whole? This is a query that carries with it a relatively high degree of subjectivity. Organotin compounds were among the first organometallic species to be investigated. They have an increasing importance as synthetic agents. However, for a long period no organometallic group equaled the versatility of Grignard reagents as synthetic tools. What of industrial applications? On a tonnage basis, tetraethyllead and tetramethyllead (used primarily as antiknock agents) for many years have exceeded other organometallic groups. In this connection we wrote only about 35 years ago: "Undoubtedly the greatest value of organometallic compounds is their laboratory use for synthesis. It is doubtful that any other group of organic compounds combines at the same time an astonishingly high utility in the laboratory with an equally low usefulness in industry." However, the industrial picture of organometallic chemistry has changed markedly in the last two or three decades.

What shall be said of the use of organotin compounds in the investigation of bonding? Here studies with every type of organometallic compound are important with this most fundamental of all concepts. Some organometallic compounds currently lend themselves to a greater extent than others to the development and testing of new principles by a variety of kinetic and spectroscopic measurements and techniques. One that comes to mind with organotin compounds is the Mössbauer effect.

What of biological or physiological properties? One is aware of the expanding applications of organotin compounds as biocides. But how is one to compare tonnage use of some materials with the extraordinary importance of trace quantities of metal combinations in basically vital processes such as animal enzymatic transformations and photosynthesis? In an exercise of this kind, involved with some comparisons or correlations of similarities and differences, how shall one evaluate the relative impact of organometallic types in the effective interdisciplinary bridging of different branches of chemistry as well as currently disparate areas of science generally.

How much consideration should be given to the recent and current rate of growth of research interest, both academic and industrial? Here organotin chemistry has a highly distinguished record. Might this be ephemeral, or may there be great pauses or interruptions in activity? It is known that more than 100 years passed before the problem of dialkyltins was clarified, and that more than 40 years went by between the times when the structure of tetrakis-(triphenylstannyl)tin was suggested and confirmed. But interruptions in some developments are not atypical. With greater activity in the broad domain of organometallic chemistry, the prospects of fewer discontinuities will improve.

Whatever the prognosis may be, it is reasonably certain that the significant development of organotin chemistry will continue. This book will assist by providing a most helpful background; by suggesting new avenues of approach; by indicating useful correlations, particularly because of the special situation of tin in the periodic arrangement of the elements; and, by setting up research targets for those who hunt intentionally or unknowingly.

Iowa State University
Ames, Iowa

HENRY GILMAN

CONTRIBUTORS TO THIS VOLUME

K. JONES, Department of Chemistry, University of Manchester, Institute of Science and Technology, Manchester, England

M. F. LAPPERT, School of Molecular Sciences, University of Sussex, Brighton, England

M. OHARA, Osaka University, Osaka, Japan

R. OKAWARA, Osaka University, Osaka, Japan

MAX SCHMIDT, Institute for Inorganic Chemistry of the University of Würzburg, Würzburg, Germany

HERBERT SCHUMANN, Institute for Inorganic and Analytical Chemistry of the Technical University of Berlin, Berlin, Germany

INGEBORG SCHUMANN-RUIDISCH, Institute for Inorganic and Analytical Chemistry of the Technical University of Berlin, Berlin, Germany

Volume 3 contains the Author and Subject Indexes

CONTENTS

CONTENTS OF OTHER VOLUMES

VOLUME 1

VOLUME 3

ORGANOTIN COMPOUNDS

VOLUME 2

5. ORGANOTIN COMPOUNDS WITH Sn—O BONDS

Organotin Carboxylates, Salts, and Complexes

R. OKAWARA AND M. OHARA

Osaka University
Osaka, Japan

I. Organotin Carboxylates

A. General Preparative Methods

The preparative methods for simple organotin carboxylates are only outlined here. For details refer to Ref. (*42*).

The oldest and the most general procedure is the reaction of organotin oxides, hydroxides, or halides with carboxylic acids or acid anhydrides:

$$(R_3Sn)_2O + 2\,R'COOH \longrightarrow 2\,R_3SnO_2CR' + H_2O$$

$$R_2SnO + (R'CO)_2O \longrightarrow R_2Sn(O_2CR')_2$$

Another general method includes the reaction of organotin halides with metal salts of carboxylic acids (*8*, *131*) or organotin sulfides with silver salts of carboxylic acids (*8a*):

$$R_{4-n}SnX_n + m\,M(O_2CR')_q \longrightarrow R_{4-n}Sn(O_2CR')_n + m\,MX_q$$

$$(M = Na, K, Ag, Pb)$$

Sometimes the direct reaction of tetraorganotins with metal salts of carboxylic acids is used (*7*) as well as the cleavage of one or more organic groups from tetraorganotin compounds by carboxylic acids (*7*, *37*, *126*):

$$R_4Sn + R'COOH \xrightarrow{\Delta} R_3SnO_2CR' + RH$$

Vinyl groups are cleaved more readily than normal alkyl groups but less readily than phenyl groups. Many other procedures have been reported, but are used less extensively.

B. Triorganotin Carboxylates

The structure and bonding of triorganotin carboxylates have been most extensively investigated and discussed (*83*). The ionic nature of the bonding was postulated by Freeman (*33*), and the ionic structure with a planar trimethyltin cation and a formate anion was proposed by Okawara et al. (*84*), but the possibility of chelating or bridging through weak coordination of oxygen atoms to tin atoms was pointed out by Beattie and Gilson (*11*).

From the different features of the infrared spectra of various trialkyltin carboxylates in the solid state and in carbon tetrachloride solution as shown in Table 1, and from the result of the temperature dependence of the absorption intensities, Janssen et al. (*44*) and Cummins and Dunn (*25*) have concluded that trialkyltin carboxylates are polymeric in the solid state with planar trialkyltin groups and bridging acyloxy groups, and are ester-like monomers in dilute solution (see Fig. 1).

TABLE 1

INFRARED FREQUENCIES OF TRIALKYLTIN CARBOXYLATES (IN cm^{-1})

	CO_2 frequencies		SnC_3 frequencies	
	Solid state	Solution	Solid state	Solution
$(CH_3)_3SnO_2CCH_3$	1565, 1412	—	547, —	
$(C_2H_5)_3SnO_2CCH_3$	1572, 1412	1655, 1302		
$(n\text{-}C_4H_9)_3SnO_2CCH_3$	1572, 1410	1647, 1300		
$(C_6H_{13})_3SnO_2CCH_3$	1570, 1408	1650, 1304		
$(CH_3)_3SnO_2CC_{11}H_{23}$	1567, 1410	1642, 1302	548, —	548, 516

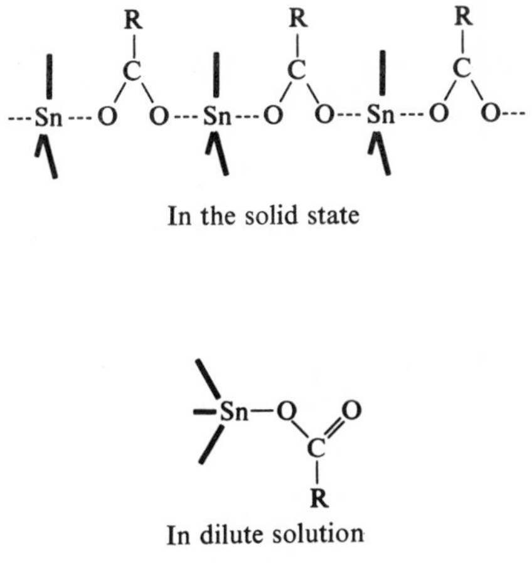

In the solid state

In dilute solution

Fig. 1. Structure of trialkyltin carboxylates.

Okawara and Ohara (*79, 80*) examined the infrared and far-infrared spectra of triethyltin and tripropyltin formates in various states. From the spectroscopic data shown in Table 2, and molecular weights determined cryoscopically in cyclohexane (Fig. 2), they concluded that trialkyltin formates form infinite linear polymers through bridging of formoxy groups. This was effected by the coordination of oxygen to tin atoms in the solid state or in neat liquid. Conversely, in solution the coordination bonds are broken at random and infinite polymers become low polymers having both terminal and bridging formoxy groups, as shown in Fig. 3. They assigned the broad

TABLE 2

CHARACTERISTIC FREQUENCIES OF R_3SnO_2CH (IN cm^{-1})

$R = C_2H_5$		$R = n\text{-}C_3H_7$		
Nujol mull	Cyclohexane solution	Liquid film	Cyclohexane solution	Assignment
778	778	779	777	CO_2 deform.
1366	1242 1359	1362	1245 1359	CO_2 sym. str.
1592	1585 1658	1587	1587 1661	CO_2 asym. str.

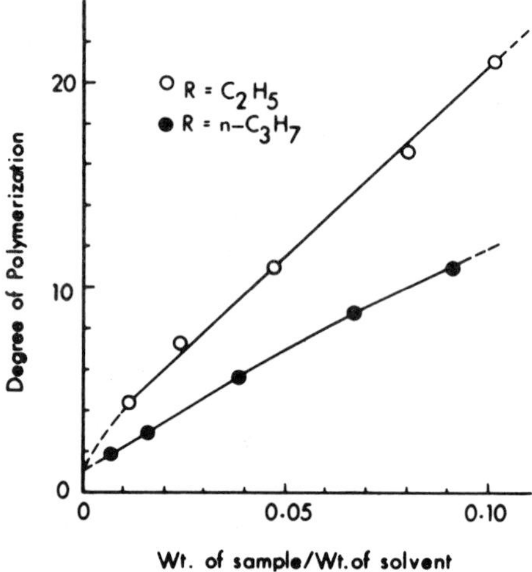

Fig. 2. Molecular weights of R_3SnO_2CH($R = C_2H_5$ and $n\text{-}C_3H_7$) in cyclohexane.

absorption around 290 cm^{-1} to the coordination band, referring to the Sn—O coordination band at 300 cm^{-1} in methyltin formates (*95*). However, Lohmann (*64*) measured the far lower region and proposed different results.

Usually trimethyltin formate and acetate are very insoluble in nonpolar solvents. However, Simons and Graham (*105*) prepared trimethyltin formate and acetate in a soluble form by heating the compounds with cyclohexane in a sealed tube at 90°C. The infrared spectra of the soluble compounds are closely parallel to the preceding observations. The molecular weight determination in methylene dibromide by a vapor-pressure osmometer shows that

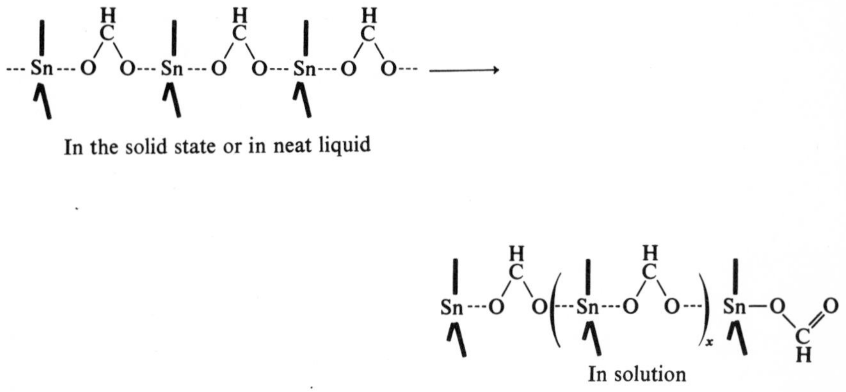

In the solid state or in neat liquid

In solution

Fig. 3. Structure of trialkyltin formates.

trimethyltin formate is associated, and that the acetate is monomeric. From the infrared spectra in solution, and from the coupling constants the soluble trimethyltin formate is postulated to exist in solution as a cyclic form in equilibrium with free monomers.

In the sterically hindered trineophyltin formate, acetate, and perfluoroacetate, CO_2 stretching bands in the solid state appear at the ester type position and tetravalent tin atoms are suggested (*89*, *135*).

Trialkyltin chloro- and dichloroacetates (*26*, *106*, *120*) are reported to be almost monomeric in solution. Splitting in the carbonyl absorption bands in solution suggests the occurrence of rotational isomers (*106*). Mössbauer effects have been studied by several workers (*2*, *14*, *89*, *135*) and the quadrupole splitting is found to appear as shown in Table 3.

Among the reactions of triorganotin carboxylates decarboxylation and disproportionation have been most extensively investigated. Triphenylcyanomethyltin was prepared by heating triphenyltin cyanoacetate under vacuum (*118*):

$$(C_6H_5)_3SnO_2CCH_2CN \xrightarrow{\Delta} (C_6H_5)_3SnCH_2CN \quad (50\%)$$

In the aliphatic series, the reaction proceeds with more difficulty than in the aromatic series. Triethyltin cyanoacetate gives only tetraethyltin, whereas the tri-*n*-butyltin compound yields about 20% tri-*n*-butylcyanomethyltin.

TABLE 3

MÖSSBAUER EFFECTS OF ORGANOTIN CARBOXYLATES

Compound	Isomer shift[a] (mm/sec)	Quadrupole splitting (mm/sec)
$(Neophyl)_3SnO_2CCH_3$[b]	1.35	2.447
$(C_2H_5)_3SnO_2CCH_3$	1.60	3.20
$(C_6H_5)_3SnO_2CC(CH_3){=}CH_2$	1.15	2.10
$(C_4H_9)_2Sn(O_2CCH_3)_2$	1.40	3.45
$(C_4H_9)_2Sn(O_2CC_7H_{15})_2$	1.45	3.50
$(C_4H_9)_2Sn(O_2CC_{17}H_{35})_2$	1.45	3.30
$(C_4H_9)_2Sn[O_2CC(CH_3){=}CH_2]_2$	1.40	3.50
$(C_4H_9)_2Sn(O_2CCH_2Cl)_2$	1.60	3.65
$(C_4H_9)_2Sn(O_2CCCl_3)_2$	1.65	3.80

[a] Relative to SnO_2.
[b] Neophyl = $C_6H_5C(CH_3)_2CH_2$—.

By heating triorganotin acetylene dicarboxylates, propiolates, and phenyl propiolates under vacuum (*65*), various acetylene derivatives with triorganotin groups were obtained.

$$R_3SnO_2CC{\equiv}CCO_2SnR_3 \xrightarrow{\Delta} R_3SnC{\equiv}CSnR_3$$

$$(R = C_2H_5,\ n\text{-}C_3H_7,\ \text{and}\ n\text{-}C_4H_9)$$

Bis(trimethyltin)acetylenedicarboxylate cannot be decarboxylated in the same manner. Triphenyltin propiolate, after decarboxylation, yields bis(triphenylstannyl)acetylene. Triphenyltin and tri-*n*-butyltin phenylpropiolates give triphenylstannyl and tri-*n*-butylstannylphenylacetylenes, respectively.

Trialkyltin hydrides were prepared by thermal decomposition of the corresponding formates under reduced pressure (*76*):

$$R_3SnO_2CH \xrightarrow{\Delta} R_3SnH + CO_2$$

$$[R = n\text{-}C_3H_7\ (25\%)\ \text{and}\ n\text{-}C_4H_9\ (60\%)]$$

Triorganotin trihalogenoacetates were used as CX_2 transfer agents by refluxing them with cyclooctene neat or in solvents (*103*):

$$R_3SnO_2CCX_3 + \text{cyclooctene} \longrightarrow R_3SnX + \text{(bicyclo[6.1.0]nonane)}{>}X_2 + CO_2$$

$$(R = CH_3\ \text{and}\ C_6H_5)$$
$$[X = Cl\ (56\%)\ \text{and}\ Br\ (36\%)]$$

It is not clear whether R_3SnCCl_3 or R_3SnCBr_3 are intermediates in this reaction, or whether decarboxylation and CX_2 transfer to olefins occur simultaneously in a concerted fashion. Trialkyltin moieties in amino acid derivatives are available as masking agents for the carboxyl group in peptide synthesis (*32*).

C. Diorganotin Dicarboxylates

The structure of diorganotin dicarboxylates was first suggested for dimethyltin diformate by Okawara et al. (*84*), which included a linear dimethyltin cation and a formate anion.

The symmetrically chelated structure in Fig. 4 was, however, proposed

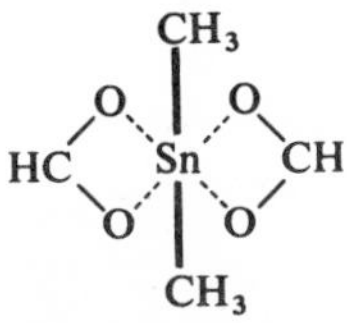

Fig. 4. Symmetrically chelated structure of dimethyltin diformate.

from infrared and far-infrared spectra (*95*), the absorption band around 300 cm^{-1} being assigned to the Sn—O coordination band. Further studies have been carried out on dialkyltin diacetates (*66, 67*). From the infrared spectral data as shown in Table 4, and monomeric values of molecular weights in benzene, a nonsymmetrically chelated configuration in Fig. 5(a) has been proposed for dialkyltin diacetates in solution.

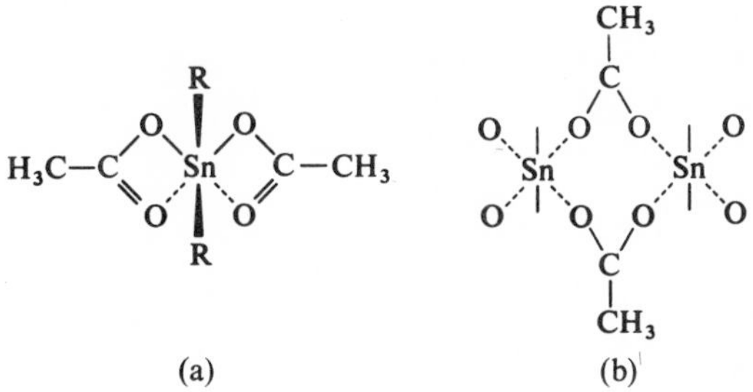

(a)

(b)

Fig. 5. Structure of dialkyltin diacetates.

However, an additional band at 1560–1570 cm^{-1} and an increase in intensity of the band at 1400–1440 cm^{-1} in neat liquid or in the crystalline state have been attributed to the partial bridging of acetoxy groups, as shown in Fig. 5(b), from the similarity in the position of CO_2 stretching bands in trialkyltin acetates where the acetoxy groups are bridging between two tin atoms.

TABLE 4

RELEVANT INFRARED FREQUENCIES OF $R_2Sn(O_2CCH_3)_2$ (IN cm^{-1})

$R = CH_3$		$R = C_2H_5$		$R = n\text{-}C_3H_7$		$R = n\text{-}C_4H_9$		
Crystal film	C_6H_{12} 7% soln.	Neat liquid	C_6H_{12} 5% soln.	Neat liquid	C_6H_{12} 5% soln.	Neat liquid	C_6H_{12} 5% soln.	Assignments
1600 s	1607 s	1600 s	1607 s	1605 s	1609 s	1605 s	1609 s	CO_2 asym. str.
1560 s		1570 s		1570 s		1570 s		
1438 sh	1433 sh	1422 sh	1425 sh	1432 sh	1425 sh	1425 s	1425 sh	
	1405 sh		1400 sh		1400 sh		1400 sh	CO_2 sym. str.[a]
1374 s	1380 s	1376 s	1378 s	1378 s	1377 s	1380 s	1377 s	
1334 s	1331 s	1330 s	1331 s	1332 s	1330 s	1333 m	1333 s	CH_3 deform.[a]
698 s	698 s	679 s	697 s	693 s	695 s	690 s	694 s	
		667 s	685 s	666 s				CO_2 scissor.
619 m	622 m	622 m	622 m	622 m	622 m	623 m	622 m	CO_2 out-of-plane bend.
571 m	574 m	542 m	542 m					SnC_2 antisym. str.
526 m	528 m	501 m	501 w					SnC_2 sym. str.
492 w	493 w	492 sh	492 w	492 w	491 w	491 w	490 w	CO_2 rock. (in plane)
304 s	305 s	302 s	304 s	303 s	304 s	302 s	303 s	SnO str.
280 sh		280 sh		281 sh				

[a] The assignments were tentatively carried out considering the intensities.

Diorganotin derivatives of dicarboxylic acids are usually polymeric, but in several cases, discrete species have been reported. Dibutyltin succinate is found to be a cyclic tetramer from its molecular weight and physical properties (*9*). Dimeric and cyclic structures have been proposed for diethyltin derivatives based on ebullioscopic molecular weight determinations (*108*). In the case of dibutyltin maleate, the monomeric monohydrate, and the anhydrous trimer and tetramer have been isolated, respectively (*73*).

Diorganotin dicarboxylates are reactive and several reactions have been reported. They are hydrolyzed into $[R_2(R'CO_2)SnOSnR_2(O_2CR')]$ and the hydrolysis goes to a further stage with or without bases to give$[R_2(R'CO_2)\text{-}SnOSnR_2(OH)]_2$. The same compounds are prepared by the reaction of diorganotin dicarboxylates and oxides with the latter compound requiring the presence of advantitious moisture.

Diorganotin dicarboxylates react easily with dihydrides, dihalides, and dialkoxides to give compounds of the type $R_2SnX(O_2CR')$:

$$R_2Sn(O_2CR')_2 + R_2SnX_2 \longrightarrow 2\ R_2SnX(O_2CR')$$
$$(X = H, Cl, Br, I, OMe)$$

Heating dibutyltin diacetate with dibutyltin dibutoxide results in the splitting out of butyl acetate giving a polystannoxane compound terminated with butoxy and acetoxy groups (*134*).

D. Organotin Tricarboxylates

Organotin tricarboxylates are expected to be very reactive and from the structural aspects they seem to be interesting. However, these compounds have only been reported with ambiguities or discretely (*87*). Recently *n*-butyltin and phenyltin tricarboxylates have been prepared by the reaction of

TABLE 5

Organotin Tricarboxylates

Compound	bp, °C/mmHg	mp, °C	d_4^{20}	n_D^{20}
$C_2H_5Sn(O_2CC_6H_5)_3$ (*87*)		185–8		
$n\text{-}C_4H_9Sn(O_2CCH_3)_3$	117–9/1	46	1.474	1.476
$n\text{-}C_4H_9Sn(O_2CC_2H_5)_3$	133–5/1		1.360	1.470
$n\text{-}C_4H_9Sn(O_2C\text{-}n\text{-}C_3H_7)_3$	156–8/1		1.281	1.463
$n\text{-}C_4H_9Sn(O_2C\text{-}i\text{-}C_3H_7)_3$	139–141/1		1.262	1.4556
$n\text{-}C_4H_9Sn(O_2C\text{-}n\text{-}C_4H_9)_3$	182–4/1		1.213	1.4627
$C_6H_5Sn(O_2CCH_3)_3$	~180/1	76		
$C_6H_5Sn(O_2CC_2H_5)_3$	~195/1	67.5		
$C_6H_5Sn(O_2C\text{-}n\text{-}C_3H_7)_3$	193–6/1		1.355	1.5105
$C_6H_5Sn(O_2C\text{-}i\text{-}C_3H_7)_3$	171–3/1	50.5	1.336	1.4925

trichlorides with an excess of the silver salts of carboxylic acids in carbon tetrachloride (*8*).

In camphor solution, *n*-butyltin triacetate and tripropionate are monomeric. The organotin tricarboxylates are easily hydrolyzed in 97% ethanol to give polymeric organotin oxycarboxylates:

$$RSn(O_2CR')_3 + H_2O \longrightarrow [RSn(O)(O_2CR')]_n + 2\,R'COOH$$

An excess of boiling carboxylic acids gives pure alkyltin tricarboxylates from alkyltin oxycarboxylates but tin tetracarboxylate from phenyltin oxycarboxylate. The infrared spectra of these tricarboxylates in carbon tetrachloride solution show coordinating CO_2 stretching bands around 1570 cm^{-1} and 1420 cm^{-1} with weak to medium carbonyl absorption near 1700 cm^{-1} which has been attributed to the occurrence of partial hydrolysis.

E. Diorganotin Monocarboxylates, $R_2SnX(O_2CR')$

The problems of the structure, or configuration of compounds of this type have not been completely settled and the molecular weights have not been determined, except in a few cases. However, the formula $R_2SnX(O_2CR')$ is used conventionally.

The compounds of this type were first prepared by the following reactions (*81*):

$$(CH_3)_2SnCl_2 + NaO_2CH \longrightarrow (CH_3)_2SnCl(O_2CH) + NaCl$$

$$[Cl(CH_3)_2Sn]_2O + 2\,HCO_2H \longrightarrow (CH_3)_2SnCl(O_2CH) + H_2O$$

$$(CH_3)_2SnCl_2 + (CH_3CO)_2O \longrightarrow (CH_3)_2SnCl(O_2CCH_3) + CH_3COCl$$

Dimethylchlorotin acetate was prepared also by the direct reaction of dimethyltin oxide and acetyl chloride (*88*).

The direct reactions of dibutyltin dihalides and diacetate have been used conveniently to obtain dibutylhalogenotin acetates (*3, 27, 97*):

$$(n\text{-}C_4H_9)_2SnX_2 + (n\text{-}C_4H_9)_2Sn(O_2CCH_3)_2 \longrightarrow 2\,(n\text{-}C_4H_9)_2SnX(O_2CCH_3)$$

$$(X = Cl, Br, I)$$

Dialkylchlorotin formates and acetates have also been prepared by the reactions of $[ClR_2Sn]_2O$ with acids (*124*). From infrared spectra and cryoscopic molecular weight determinations in benzene, a monomeric and pentavalent structure with a chelated carboxyl group, as shown in Fig. 6, has been

```
     R  O
     | ⋰  \
Cl—Sn      C—R
     | ⋱  /
     R  O
```

Fig. 6. Pentavalent configuration of dialkylchlorotin carboxylates.

TABLE 6

R_2SnXY COMPOUNDS

R	X	Y	mp, °C	References
n-C_4H_9	H	O_2CCH_3	17–20	(*96*, *99*)
CH_3	Cl	O_2CH	143	(*81*)
CH_3	Cl	O_2CCH_3	189, 184–186	(*81*, *88*)
CH_3	Cl	$O_2CC_2H_5$	104–106	(*68*)
CH_3	Br	O_2CCH_3	190–191	(*68*)
CH_3	I	O_2CCH_3	172–173	(*68*)
C_2H_5	Cl	O_2CH[a]	82	(*124*)
C_2H_5	Cl	O_2CCH_3	94	(*124*)
n-C_3H_7	Cl	O_2CH[a]	81	(*124*)
n-C_3H_7	Cl	O_2CCH_3	73	(*124*)
n-C_4H_9	Cl	O_2CH[a]	69	(*124*)
n-C_4H_9	Cl	O_2CCH_3	56.5–57.5, 61, 63–65	(*27*, *124*, *97*)
n-C_4H_9	Br	O_2CCH_3	66–67, 67.0–68.5	(*3*, *98*)
n-C_4H_9	I	O_2CCH_3	59–60	(*27*)
n-C_4H_9	Cl	methyl maleyloxy	41–42	(*73*)
n-C_4H_9	OCH_3	O_2CCH_2Cl[b]	95–96	(*133*)
n-C_4H_9	OCH_3	O_2CCH_3	94	(*27*)
n-C_4H_9	OCH_3	$O_2CC_{11}H_{23}$	oil	(*27*)

[a] Each compound has one molecule of water.
[b] Denoted as $(n\text{-}C_4H_9)_2Sn(OCH_3)_2 \cdot (n\text{-}C_4H_9)_2Sn(O_2CCH_2Cl)_2$ in the original paper.

TABLE 7

HEATS OF FORMATION OF R_2SnXY[a] COMPOUNDS

R_2SnX_2	R_2SnY_2	ΔH, kcal/mole
$(n\text{-}C_4H_9)_2SnCl_2$	$(n\text{-}C_4H_9)_2Sn(O_2CCH_3)_2$	1.5
$(n\text{-}C_4H_9)_2Sn(OCH_3)_2$	$(n\text{-}C_4H_9)_2Sn(O_2CCH_3)_2$	10.3
$(n\text{-}C_4H_9)_2Sn(O\text{-}n\text{-}C_4H_9)_2$	$(n\text{-}C_4H_9)_2Sn(O_2CCH_3)_2$	10.1
$(n\text{-}C_4H_9)_2Sn(OCH_3)_2$	$(n\text{-}C_4H_9)_2Sn(O_2CC_{11}H_{23})_2$	7.8
$(n\text{-}C_4H_9)_2Sn(OCH_3)_2$	$(n\text{-}C_4H_9)_2Sn(O_2CCH_2Cl)_2$	7.8

[a] The products were denoted as $R_2SnX_2 \cdot R_2SnY_2$ in the original paper except for $(n\text{-}C_4H_9)_2SnCl(O_2CCH_3)$.

proposed for these compounds. Interestingly each dialkylchlorotin formate has one molecule of water.

The occurrence of an equilibrium has been pointed out in the case of the direct reaction of dibutyltin dihydride and diacetate, and dibutylhydridotin acetate has successfully been isolated (*96, 99*):

$$(n\text{-}C_4H_9)_2SnH_2 + (n\text{-}C_4H_9)_2Sn(O_2CCH_3)_2 \rightleftharpoons 2\,(n\text{-}C_4H_9)_2Sn(H)(O_2CCH_3)$$

Diethylethoxytin benzoate and acetate have been reported as the products in the reaction of triethyltin alkoxide with benzoyl peroxide or acetyl benzoyl peroxide (*121*). Dibutylmethoxytin acetate has been prepared by the reaction of the dimethoxide with the diacetate (*27*):

$$(C_4H_9)_2Sn(OCH_3)_2 + (C_4H_9)_2Sn(O_2CCH_3)_2 \longrightarrow 2\,(C_4H_9)_2Sn(OCH_3)(O_2CCH_3)$$

The heats of formation have been measured upon mixing equimolar amounts of both compounds (*133*).

F. Compounds with Sn—O—Sn Bonds

Distannoxane derivatives

$$R_2(R'CO_2)SnOSnR_2(O_2CR') \quad \text{and} \quad R_2(R'CO_2)SnOSnR_2(OH)$$

have been prepared by the same methods as for other distannoxane compounds (*3, 4, 27, 36, 55, 77, 81, 88, 121*), and the analogous dimeric structures (a) and (b) in Fig. 7 have been suggested.

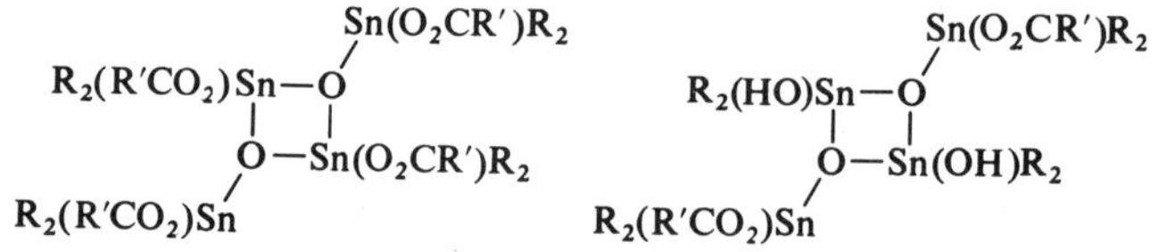

```
                        Sn(O2CR')R2                              Sn(O2CR')R2
                       /                                        /
      R2(R'CO2)Sn—O                            R2(HO)Sn—O
                  |     |                                  |     |
                  O—Sn(O2CR')R2                            O—Sn(OH)R2
                 /                                        /
  R2(R'CO2)Sn                              R2(R'CO2)Sn

                  (a)                                      (b)
```

Fig. 7. Dimeric structure of distannoxane derivatives.

From the molecular weights and infrared spectra in solutions, the occurrence of the following equilibrium has been suggested for $[R_2(CH_3CO_2)Sn]_2O$:

$$[R_2(CH_3CO_2)SnOSnR_2(O_2CCH_3)]_2 \rightleftharpoons 2\,R_2(CH_3CO_2)SnOSnR_2(O_2CCH_3)$$

and the association of dimeric species through acetoxy bridging for $[R_2(CH_3CO_2)SnOSnR_2(OH)]_2$ (*67*). These distannoxanes are several times as effective as the usual catalysts, such as dibutyltin dilaurate and tertiary amines in polyurethane formation (*130*). Polystannoxane derivatives terminated with functional groups have been prepared from distannoxanes (*132*), or the diacetate and the dibutoxide (*134*):

$$[Bu_2Sn(O_2CCH_3)]_2O + CH_2N_2 \xrightarrow{H_2O} (CH_3CO_2)Bu_2SnO(SnBu_2O)_nSnBu_2(O_2CCH_3)$$
$$(n = 2, 6, 14)$$

$$Bu_2Sn(OBu)_2 + Bu_2Sn(O_2CCH_3)_2 \xrightarrow[16\text{-}18h]{\Delta}$$
$$(BuO)Bu_2SnO(SnBu_2O)_nSnBu_2(O_2CCH_3) + CH_3COOBu$$
$$(\text{M.W.} \sim 4{,}000)$$

G. Organotin Inner Salts

Organotin inner salts have been prepared by the hydrolysis of bromo(2-carboxyethyl)tin compounds (*119*) or by the thermal decomposition of triorgano(2-carboxyethyl)tin. Triphenyl(2-carboxyethyl)tin from triphenyltin hydride and acrylic acid is unstable even at room temperature to give an inner salt:

$$R_2Sn(Br)CH_2CH_2COOH \xrightarrow{OH^-} R_2SnCH_2CH_2CO_2$$

$$R_3SnCH_2CH_2COOH \xrightarrow{\Delta} R_2SnCH_2CH_2CO_2 + RH$$

Refluxing tetra(2-carboxyethyl)tin with water gives bis(2-carboxyethyl)(2-carboxylatoethyl)tin and bis(2-carboxylatoethyl)tin (*90, 45*):

$$Sn(CH_2CH_2COOH)_4 \xrightarrow[H_2O]{\Delta}$$
$$(HOOCCH_2CH_2)_2SnCH_2CH_2COO\ (90\%) + Sn(CH_2CH_2COO)_2\ (5\%)$$

H. Miscellaneous

1. *Tin Tetracarboxylates* $Sn(O_2CR)_4$

Tin tetraacetate (*101*) and tetraisobutyrate (*8*) have been prepared but little studied. However, various tin tetracarboxylates have recently been prepared as colorless needles by heating tetravinyltin and carboxylic acids in a sealed tube at 110°C for a period of up to 40 h (*37*). Shorter reaction time or lower reaction temperature normally gave incomplete reaction and no intermediates, except with formic acid, which gave divinyltin diformate.

Cryoscopic molecular weight determination in benzene showed a fall over 30 min to a monomeric value after mixing the solute and the solvent, suggesting association in the solid and its dissociation in solution.

The carbonyl absorptions of tin tetraacetate and tin tetrapropionate in the solid state appear at 1262 and 1704 cm^{-1}. These show an increase of intensity in carbon tetrachloride solution, whereas two additional bands at 1440 and 1560 cm^{-1}, found in the solid state, disappear completely in solution.

With tin tetrastearate, the latter two absorptions decrease but do not disappear in solution. The tetrastearate is sufficiently resistant to hydrolysis to be recrystallized from benzene.

Tin tetrapropionate reacts with dimethylamine at $-20°C$ and with hydrogen chloride at $-78°C$:

$$Sn(O_2CC_2H_5)_4 + 4\,(CH_3)_2NH \xrightarrow{-20°C} 2\,H_2O + 2\,C_2H_5CON(CH_3)_2 + [(CH_3)_2N]_2Sn(O_2CC_2H_5)_2$$

$$Sn(O_2CC_2H_5)_4 + 6\,HCl \xrightarrow{-78°C} Cl_3Sn(O_2CC_2H_5) + 3\,(C_2H_5CO_2H_2)^+Cl^-$$

TABLE 8

TIN TETRACARBOXYLATES
$Sn(O_2CR)_4$

R	mp, °C
H	250
CH_3	255–256
C_2H_5	146
i-C_4H_9	116
n-$C_{17}H_{35}$	65–66

2. *Hexaacetoxyditin* $Sn_2(O_2CCH_3)_6$ (*126*)

This compound has been prepared as white, fine crystals by heating hexaphenylditin with glacial acetic acid at 120°C:

$$(C_6H_5)_3Sn—Sn(C_6H_5)_3 + 6\,CH_3COOH \xrightarrow{120°C} (CH_3CO_2)_3Sn—Sn(O_2CCH_3)_3 + 6\,C_6H_6$$

Bromine in glacial acetic acid causes cleavage of the tin-tin bond at room temperature to give bromotriacetoxytin, $BrSn(O_2CCH_3)_3$. At $-100°C$, with hydrogen chloride in ether or liquid hydrogen chloride, only the displacement of acetoxy groups with chlorine occurs:

$$(CH_3CO_2)_3Sn—Sn(O_2CCH_3)_3 + 4\,HCl \xrightarrow[\text{ether}]{-100°C} Sn_2Cl_4(O_2CCH_3)_2 + 4\,CH_3CO_2H$$

$$(CH_3CO_2)_3Sn—Sn(O_2CCH_3)_3 + 6\,HCl \xrightarrow{-100°C} Sn_2Cl_6 + 6\,CH_3CO_2H$$

The tin-tin bond in $Cl_4Sn_2(O_2CCH_3)_2$ is also cleaved by chlorine in carbon tetrachloride at room temperature:

$$Cl_4Sn_2(O_2CCH_3)_2 + Cl_2 \longrightarrow 2\,Cl_3Sn(O_2CCH_3)$$

From the infrared spectra and the stability of the tin-tin bond, the acetoxy-bridged structures shown in Fig. 8 have been suggested.

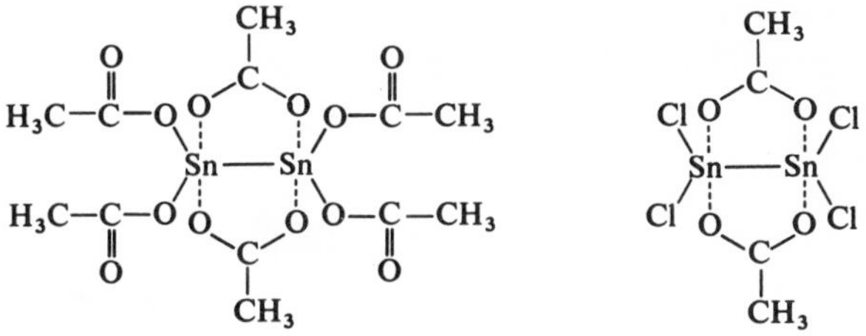

Fig. 8. Examples of the acetoxy-bridged structures.

3. 1,1,2,2,-*Tetraorgano*-1,2-*diacyloxyditins* $[R_2Sn(O_2CR')]_2$

The title compounds have been prepared by the reaction of diorganotin dihydrides with carboxylic acids or by mixing equimolar amounts of diorganotin dihydrides and diorganotin dicarboxylates (*96*, *98*):

$$2\,R_2SnH_2 + 2\,R'COOH \longrightarrow [R_2Sn(O_2CR')]_2 + 3\,H_2$$

$$R_2SnH_2 + R_2Sn(O_2CR')_2 \longrightarrow [R_2Sn(O_2CR')]_2 + H_2$$

The mechanism of the reaction, investigated on the dibutyltin dihydride and acetic acid system by infrared spectra, is shown in Fig. 9 (*99*).

TABLE 9

$[R_2Sn(O_2CR')]_2$

R	R′	mp, °C
C_6H_5	H	153[a]
C_6H_5	CH_3	152
C_6H_5	CH_2Cl	150
C_6H_5	$CHCl_2$	169
C_6H_5	CCl_3	170
C_6H_5	CF_3	165
C_6H_5	C_6H_5	185
C_6H_5	*o*—Cl—C_6H_4	161
C_6H_5	*o*—HO—C_6H_4	197
C_6H_5	C_5H_{11}	85–87
C_6H_5	C_7H_{15}	86–88
C_4H_9	CH_3	−4.0 to −7.0
C_4H_9	C_6H_5	31.5–32.5
C_4H_9	*o*—Cl—C_6H_4	65–66.5
C_4H_9	*p*—Cl—C_6H_4	75–77

[a] Product is not analytically pure.

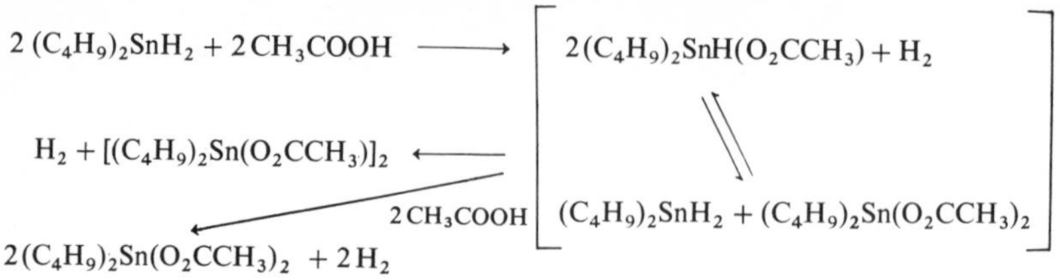

$$2\,(C_4H_9)_2SnH_2 + 2\,CH_3COOH \longrightarrow \left[\,2(C_4H_9)_2SnH(O_2CCH_3) + H_2 \rightleftharpoons (C_4H_9)_2SnH_2 + (C_4H_9)_2Sn(O_2CCH_3)_2\,\right]$$

$$\left[\,\ldots\,\right] \longrightarrow H_2 + [(C_4H_9)_2Sn(O_2CCH_3)]_2$$

$$\left[\,\ldots\,\right] \xrightarrow{2\,CH_3COOH} 2(C_4H_9)_2Sn(O_2CCH_3)_2 + 2\,H_2$$

Fig. 9. The reaction scheme of dibutyltin dihydride-acetic acid system.

II. Organotin Salts

A. Organotin Nitrates

1. *Triorganotin Nitrates*

Several inconsistent results have been published on the structure of the simplest compound, trimethyltin nitrate (*21*, *78*), but it is understood that they are derived from careless treatment of hygroscopic anhydrous trimethyltin nitrate (*22*, *128*). Reaction of trimethyltin hydroxide with nitric acid solution has been used to obtain crystalline trimethyltin nitrate monohydrate, $(CH_3)_3SnNO_3 \cdot H_2O$, which is soluble in water, methanol, or acetone (*30*). Sublimation of the monohydrate under vacuum gives anhydrous trimethyltin nitrate. The anhydrous trimethyltin nitrate is fairly hygroscopic, and on standing for a few days in air it absorbs one mole of water to give the monohydrate. The symmetry of the NO_3 group in trimethyltin nitrate and its monohydrate, assumed from the infrared spectra, is C_{2v} or lower. However, an absorption band due to the coordination of oxygen of the NO_3 group with tin in these compounds is not found down to 300 cm^{-1}, and it is assumed that the Sn—O bond is weak. In trimethyltin nitrate, only one absorption band is observed in the KBr region at 552 cm^{-1}, which can be assigned to the Sn—C degenerate stretching vibration band. This fact suggests the structure given in Fig. 10, in which the planar trimethyltin groups are bridged

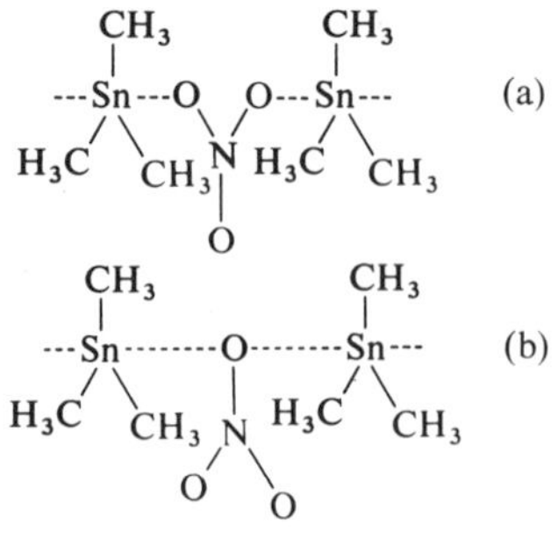

```
        CH3           CH3
         |             |
    ---Sn---O    O---Sn---        (a)
       / \     \ /    / \
    H3C   CH3   N  H3C   CH3
                |
                O

        CH3            CH3
         |              |
    ---Sn--------O--------Sn---   (b)
       / \       |       / \
    H3C   CH3    N    H3C   CH3
                / \
               O   O
```

Fig. 10. Possible structures of trimethyltin nitrate.

by two oxygen atoms of the NO_3 group. A Stuart's model shows that bridging by a single oxygen atom of the NO_3 group (b) is also possible.

The structures shown in Fig. 11 have been suggested for trimethyltin

```
H2O     CH3          H2O     CH3
   \   /                \   /
    Sn--------O    O-------Sn                (a)
   /  \        \  /       /  \
H3C    CH3      N     H3C     CH3
                |
                O

   H2O     CH3    H2O     CH3
      \   /          \   /
    ---Sn--------O--------Sn---              (b)
      /  \       |       /  \
   H3C    CH3    |   H3C     CH3
                 N
                / \
               O   O
```

Fig. 11. Possible structures of trimethyltin nitrate monohydrate.

nitrate monohydrate due to the three following facts: (i) the positions of the bands associated with the NO_3 group in the anhydrate and in the monohydrate are almost identical, (ii) two bands due to the Sn—C stretching vibration are observed in the monohydrate and (iii) the bands due to the water molecule, which coordinates weakly to the tin atom, are observed.

Triphenyltin nitrate has been prepared by the reaction of a triphenyltin halide with silver nitrate in dry methanol (*19*), acetone (*19*), benzene (*109*), or acetonitrile (*30*). This compound can be obtained when bis(triphenyltin) oxide in 95% ethanol is neutralized by dilute nitric acid using methyl orange as an indicator (*109*). There were inconsistent reports (*104*, *107*) on the stability of this compound; however, it is believed that the pure compound is stable (*19*, *30*, *109*, *116*). Antimicrobial activity of this compound is described (*109*).

Trineophyltin nitrate (*89*), $[C_6H_5C(CH_3)_2CH_2]_3SnNO_3$ is prepared from trineophyltin hydroxide and concentrated nitric acid in heptane. The ir, nmr and Mössbauer (*38*) spectra of this compound are recorded. Molecular weights in dilute solution indicate a monomeric species and it is assumed that the nitrate is a bidentate ligand attached to a five-coordinated tin atom in the monomer.

2. *Diorganotin Dinitrates* (**5**)

The dialkyltin dinitrates are very deliquescent compounds which are soluble in polar organic solvents. The nitrates undergo decomposition with the liberation of oxides of nitrogen at room temperature even when stored in a sealed tube, but they can be stored at low temperature (−18°C) without appreciable decomposition. The 1,10-phenanthroline complexes, on the other hand, are perfectly stable at room temperature and are nondeliquescent.

There are three types of compounds among the partial hydrolysis products of dialkyltin dinitrates (*129*). The relations of these compounds are shown in Fig. 12.

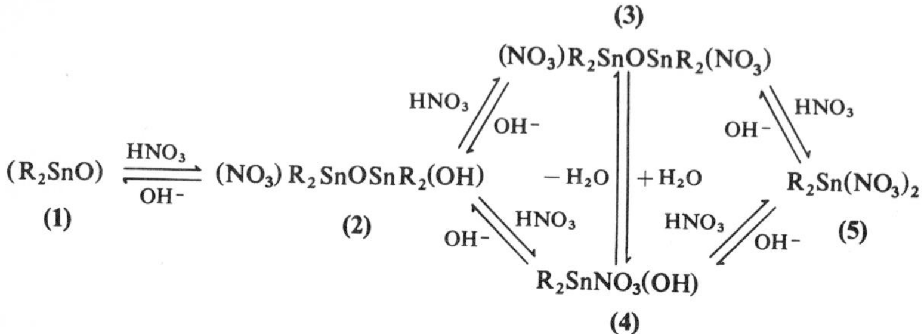

Fig. 12. Partial hydrolysis products of diorganotin dinitrates.

Hygroscopic dimethyltin dinitrate was first prepared by Addison (*1*). Anhydrous dimethyltin dinitrate is obtained by the reaction of tetramethyltin with dinitrogen tetroxide in dry ethyl acetate. It is a white crystalline solid, is highly deliquescent, and on exposure to the atmosphere is converted to a liquid within one minute. On heating, some nitrogen dioxide is evolved, and the compound explodes. On heating dimethyltin dinitrate at 85°C/10^{-2} mm, a crystalline sublimate is obtained. It is soluble in water and many polar organic solvents (ethyl alcohol, ethyl acetate, methyl cyanide, dimethyl sulfoxide), giving clear solutions. It has a slight solubility in chloroform, from which it crystallizes readily, but is insoluble in carbon tetrachloride, benzene, and nitrobenzene. No solvent suitable for molecular-weight measurements has yet been found, since solution appears to be accompanied by ionic dissociation:

$$(CH_3)_2Sn(NO_3)_2 \xrightarrow{(CH_3)_2SO} (CH_3)_2Sn[(CH_3)_2SO]_n^{2+} + 2\,NO_3^-$$

In the solid it has a structure involving tetrahedral tin with monodentate nitrate substituents. This compound can also be prepared by the silver nitrate method (*35*):

$$(CH_3)_2SnCl_2 + 2\,AgNO_3 \longrightarrow (CH_3)_2Sn(NO_3)_2 + 2\,AgCl$$

The Raman and infrared spectra of aqueous solutions of dimethyltin dinitrate (*70*) (and perchlorate) have been determined, and a complete vibrational assignment has been made on the basis of a linear C—Sn—C skeleton with an effective point group of D_{3d}. No lines attributable to tin-oxygen stretching frequencies are observed, and it is concluded that four water molecules are probably coordinated about the cation in the equatorial plane by highly polar bonds.

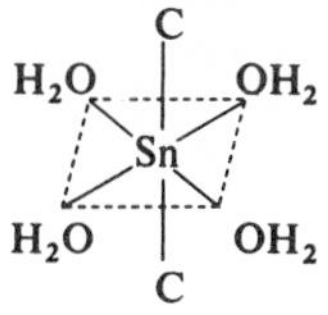

C

H_2O OH_2

Sn

H_2O OH_2

C

The preparation of diethyltin dinitrate is claimed in the early literature [A. Cahours, *Ann. Chem.*, **114**, 354 (1860); C. Lowig, *Ann. Chem.*, **84**, 308 (1852)].

Dimethyltin dinitrate (*35*), di-*n*-propyltin dinitrate (*35*), and di-*n*-butyltin dinitrate (*129*) have been obtained by the reactions of the corresponding dialkyltin dihalides with silver nitrate in methanol. By mixing equimolar proportions of 1,10-phenanthroline monohydrate and di-*n*-propyltin dinitrate in absolute ethanol, the di-*n*-propyltin dinitrate-1,10-phenanthroline complex precipitates immediately, which may be recrystallized from the absolute ethanol, mp 205–206°C (decomp.).

The dihydrate of di-*n*-butyltin dinitrate (*129*) is obtained by dissolving dibutyltin oxide in acetone containing two equivalents of nitric acid. On heating this hydrate at reduced pressure, $(n\text{-}C_4H_9)_2Sn(NO_3)_2$ is obtained. Di-*n*-butyltin dinitrate-1,10-phenanthroline melts at 209–212°C (decomp.).

Diphenyltin dinitrate (*30*) can be obtained by the reaction of diphenyltin dichloride with silver nitrate in acetonitrile. It is soluble in polar organic solvents and slowly decomposes at room temperature. However, it may be refrigerated without decomposition for several weeks and is stable as its 1:1 reaction product with 1,10-phenanthroline (mp 276–8°C, decomp.). Whether diphenyltin dinitrate can be assumed as an intermediate in the thermal decomposition of triphenyltin nitrate (*104*, *116*) or not (*30*, *107*), has been considered.

3. *Diorganotin Hydroxide Nitrates*, $R_2SnNO_3(OH)$ (**4**)

Dimethyl or diethyltin hydroxide nitrates (*128*) may be obtained by reactions of the corresponding oxides with 2.5 *M* nitric acid solution. The crystalline compounds are soluble in water or methanol, have high melting points and are insoluble in nonpolar solvents.

Di-*n*-propyltin hydroxide nitrate monohydrate (*129*) is obtained by evaporating a solution from the reaction of di-*n*-propyltin oxide with a slight excess of nitric acid in acetone. The compound may be recrystallized from acetone containing a small amount of nitric acid. The anhydrous salt is obtained by dehydration of the monohydrate in a desiccator over calcium chloride for several days.

Di-*n*-butyltin hydroxide nitrate (*129*) results from the reaction of di-*n*-butyltin oxide with an equivalent amount of nitric acid in benzene.

From the results of infrared spectra, the structures of the methyl and ethyl compounds have been suggested as shown in Fig. 13, which indicate that the

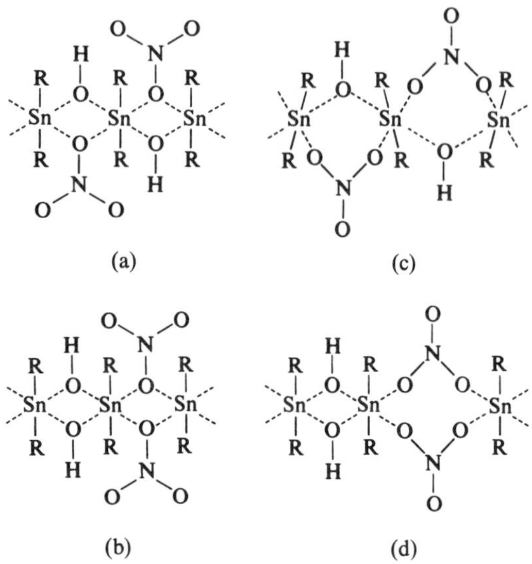

Fig. 13. Some possible configurations of dialkyltin hydroxide nitrate.

oxygen atoms of the bridging OH and NO_3 groups coordinate strongly with the tin atoms.

4. *Compounds with* Sn—O—Sn *Bonds*

Tetraalkyl-1,3-dinitratodistannoxanes, $(NO_3)R_2SnOSnR_2(NO_3)$ (**3**) (*129*) can, in general, be obtained by dehydrating compounds (**4**) on heating under reduced pressure. The *n*-propyl and *n*-butyl compounds are more easily obtained, compared with the methyl and the ethyl compounds. All of these compounds are easily hydrolyzed by atmospheric moisture. The propyl compound is dimeric in boiling benzene.

Tetraphenyl-1,3-dinitratodistannoxane, $(NO_3)(C_6H_5)_2SnOSn(C_6H_5)_2NO_3$ (*30*), is obtained when a suspension of diphenyltin dinitrate is stirred overnight in a flask open to the atmosphere. It is soluble in polar organic solvents and is monomeric in acetone. Triphenyltin nitrate also yields this compound when warmed in chloroform with access to air.

The structure shown below is suggested for these distanoxanes, for the infrared spectra show a broad band near 600 cm^{-1}, characteristic of the Sn—O—Sn stretching vibration of dimeric distannoxanes and the patterns due to only one kind of NO_3 group are observed.

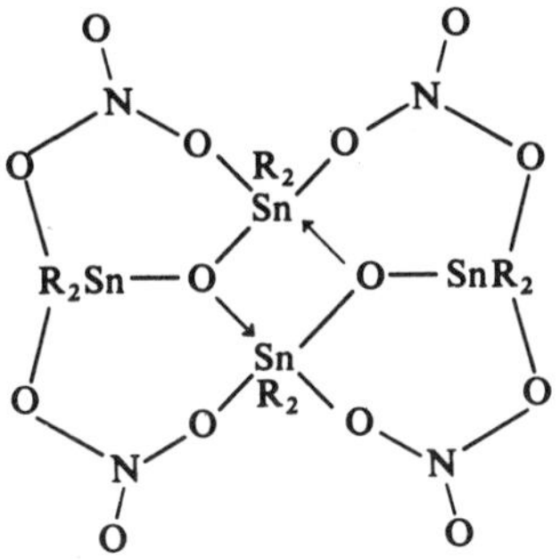

TABLE 10

ORGANOTIN NITRATES

	mp, °C	References
$(CH_3)_3SnNO_3$	127–128 (sealed tube) 140	(*21, 128*)
$(CH_3)_3SnNO_3 \cdot H_2O$	98–99	(*128*)
$(n\text{-}C_4H_9)_3SnNO_3$	bp 156–7/1 mm $n_D^{20} = 1.4971$	(*28*)
$(C_6H_5)_3SnNO_3$	170–170.5, 180–182, 182–184, 184–186	(*107, 109, 116, 30*)
$[C_6H_5C(CH_3)_2CH_2]_3SnNO_3$	122–123	(*89*)
$(CH_3)_2Sn(NO_3)_2$	exploded	(*1*)
$(n\text{-}C_3H_7)_2Sn(NO_3)_2$	137–138	(*35*)
$(n\text{-}C_4H_9)_2Sn(NO_3)_2$	83–87, 103.5–104.5	(*129, 35*)
$(C_6H_5)_2Sn(NO_3)_2$	195–197 (decomp.)	(*30*)
$(n\text{-}C_4H_9)_2Sn(NO_3)_2 \cdot 2H_2O$	83–87	(*129*)
$(CH_3)_2Sn(NO_3)(OH)$	>250	(*128*)
$(C_2H_5)_2Sn(NO_3)(OH)$	214 (decomp.)	(*128*)
$(n\text{-}C_3H_7)_2Sn(NO_3)(OH)$	183 (decomp.)	(*129*)
$(n\text{-}C_3H_7)_2Sn(NO_3)(OH) \cdot H_2O$	183 (decomp.)	(*129*)
$(n\text{-}C_4H_9)_2Sn(NO_3)(OH)$	92.5–95	(*129*)
$[(NO_3)(CH_3)_2Sn]_2O$	>250	(*129*)
$[(NO_3)(C_2H_5)_2Sn]_2O$	214 (decomp.)	(*129*)
$[(NO_3)(n\text{-}C_3H_7)_2Sn]_2O$	183 (decomp.)	(*129*)
$[(NO_3)(n\text{-}C_4H_9)_2Sn]_2O$	92.5–95.0	(*129*)
$[(NO_3)(C_6H_5)_2Sn]_2O$	287–289	(*30*)
$(NO_3)(n\text{-}C_3H_7)_2SnOSn(n\text{-}C_3H_7)_2(OH)$	221–222.5 (decomp.)	(*129*)
$(NO_3)(n\text{-}C_4H_9)_2SnOSn(n\text{-}C_4H_9)_2(OH)$	210–213	(*129*)

Tetraalkyl-1-nitrato-3-hydroxy distannoxanes $(NO_3)R_2SnOSnR_2(OH)$ (**2**), ($R = n\text{-}C_3H_7$, $n\text{-}C_4H_9$) (*129*) are obtained when a slight excess of NaOH solution equivalent to a half mole of (**4**) is added to an aqueous acetone solution of (**4**) to give a small amount of white precipitate. The filtrates, on evaporation, give precipitates of (**2**).

B. Organotin Carbonates

Bis(trimethyltin) carbonate and bis(triethyltin) carbonate are obtained by the reaction of the corresponding trialkyltin hydroxides with carbon dioxide. Bis(tri-*n*-butyltin) carbonate is obtained by the reaction of bis(tri-*n*-butyltin) oxide with carbon dioxide (*28*). The methyl compound is insoluble in organic solvents whereas the ethyl compound is soluble in carbon tetrachloride and carbon disulfide.

From its infrared spectra, the methyl compound is expected to have the following configuration with nonplanar C_3Sn and the CO_3 group having lower symmetry than D_{3h}.

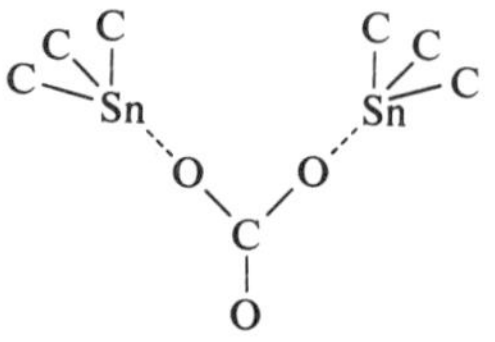

The absorption due to the weak Sn···O was found at about 380 cm^{-1}.

TABLE 11

Organotin Carbonates

	mp, °C	References
$[(CH_3)_3Sn]_2CO_3$	>200	(*94*)
$[(C_2H_5)_3Sn]_2CO_3$	119–142	(*94*)
$[(n\text{-}C_4H_9)_3Sn]_2CO_3$	waxy solid	(*28*)

C. Organotin Perchlorates

Anhydrous silver perchlorate reacts with trimethyltin bromide in a 1:1 mole ratio in dry methanol to give trimethyltin perchlorate. Upon removal of methanol it sublimes from the flask as a white crystalline solid. It is extremely soluble in methanol and other polar solvents. It is very hygroscopic, being converted to a liquid in less than a minute on exposure to moist air.

Triphenyltin perchlorate (*19*) has likewise been obtained when silver perchlorate and triphenyltin chloride are shaken in anhydrous ether for three days. It is also very soluble in both methanol and ether and is hydrolyzed in moist air. Trineophyltin perchlorate (*89*), $[C_6H_5C(CH_3)_2CH_2]_3SnClO_4$, is prepared by the reaction of 70% perchloric acid with trineophyltin hydroxide in ethanol. Dimethyltin diperchlorate (*115*), $(CH_3)_2Sn(ClO_4)_2$, has been obtained as needlelike crystals by the storage of concentrated solutions of it in a vacuum desiccator over phosphoric oxide. It is potentially explosive.

The results of infrared spectra of trimethyltin perchlorate show that the planar trimethyltin groups are bridged by ClO_4 groups as shown below (*21*, *128*).

$$-\mathrm{Sn}-\mathrm{O}\diagdown\ \mathrm{Cl}(\mathrm{O})(\mathrm{O})\ \diagup\mathrm{O}-\mathrm{Sn}-$$

The diammonia adduct is formed when anhydrous ammonia is condensed into trimethyltin perchlorate and excess ammonia removed *in vacuo*. This adduct has low solubility in organic solvents and is stable in air. The results of infrared spectra in the solid state indicate the presence of a free ClO_4^- ion and a planar $(CH_3)_3Sn$ group coordinated by two NH_3 groups forming a trigonal bipyramidal configuration as shown below.

$$(H_3N-Sn-NH_3)^+\,ClO_4^-$$

The ir, nmr and Mössbauer spectra (*38*) of trineophyltin perchlorate, which has a sterically hindered tin atom, are recorded. Molecular weights in dilute solution indicate a monomeric species and it is assumed that the perchlorate is a bidentate ligand attached to a five-coordinated tin atom in the monomer.

TABLE 12

ORGANOTIN PERCHLORATES

	mp, °C	References
$(CH_3)_3SnClO_4$	125–127 (sealed tube)	(*21*)
$(C_6H_5)_3SnClO_4$		(*19*)
$[C_6H_5C(CH_3)_2CH_2]_3SnClO_4$	162–163	(*89*)
$(CH_3)_2Sn(ClO_4)_2$	explosive	(*115*)

D. Organotin Phosphates and Phosphinates

Triorganotin diphenylphosphinates $R_3SnOP(O)(C_6H_5)_2$ (*16*) have been prepared in excellent yield by the condensation of diphenylphosphinic acid with trialkyltin hydroxides or oxides. Oxidation of stannylphosphines or the interaction of sodium diphenylphosphinite $(C_6H_5)_2PONa$ with a trialkyltin bromide or the reaction of diphenylphosphine oxide $R_2P(O)H$ with hexa-organodistannoxanes (*43*) also give these compounds. Diethyltin bisdiphenyl-phosphinate $(C_2H_5)_2Sn[OP(O)(C_6H_5)_2]_2$ (*16*) is prepared by the reaction of 2 moles of sodium diphenylphosphinate and 1 mole of diethyltin dibromide in dry benzene under refluxing. Various organotin salts of phosphoric acid esters have been prepared, and their biological activities have been studied (*60*).

The triorganotin diphenylphosphinates have been found to be dimeric in benzene, with the exception of the trimethylstannyl derivatives which are tetrameric. Examination of the infrared spectra of these esters shows a lowering of the P=O stretching frequency, indicating that the dimers are formed by coordination of the phosphoryl oxygen to the tin, which therefore becomes five-coordinate in the eight membered ring.

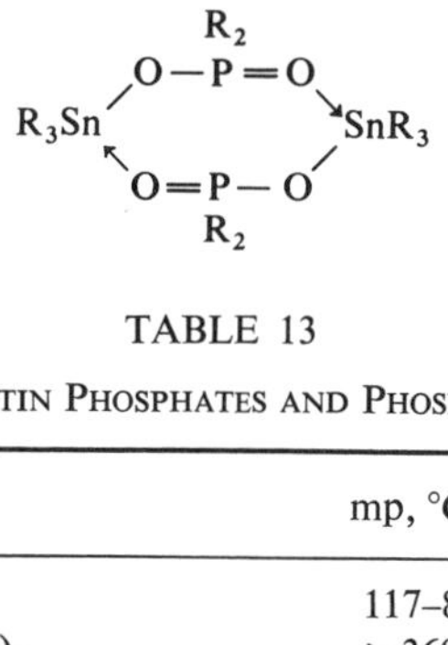

```
          R2
       O—P=O
      /      ↘
  R3Sn        SnR3
      ↖      /
       O=P—O
          R2
```

TABLE 13

Organotin Phosphates and Phosphinates

	mp, °C	References
$[(C_6H_5)_3Sn]_3PO_4$	117–8	(*109*)
$(CH_3)_3SnOP(O)(C_6H_5)_2$	>360	(*16*)
$(C_2H_5)_3SnOP(O)(C_6H_5)_2$	246–7	(*16, 22*)
$(n\text{-}C_3H_7)_3SnOP(O)(C_6H_5)_2$	226, 218	(*16*)
$(n\text{-}C_4H_9)_3SnOP(O)(C_6H_5)_2$	215, 217–8, 209	(*16, 22*)
$(C_6H_5)_3SnOP(O)(C_6H_5)_2$	>360	(*16*)
$(C_2H_5)_2Sn[OP(O)(C_6H_5)_2]_2$	370–372	(*16*)
$(C_6H_5)_3SnOP(O)(OC_2H_5)_2$	193	(*60*)
$(n\text{-}C_4H_9)_3SnOP(O)(OC_2H_5)_2$	liq.	(*60*)
$(C_6H_5)_3SnOP(O)(OC_6H_5)_2$	170	(*60*)
$(C_6H_5)_3SnOP(O)[N(CH_3)_2]_2$	>250	(*60*)
$(n\text{-}C_4H_9)_3SnOP(O)[N(C_2H_5)_2]_2$	151–3	(*60*)
$(n\text{-}C_4H_9)_3SnOP(O)[N(CH_3)_2]_2$	158	(*60*)
$(C_6H_5)_3SnOP(O)(NHC_2H_5)_2$	218–221	(*60*)
$(C_6H_5)_3SnOP(O)[NH(n\text{-}C_3H_7)]_2$	153	(*60*)
$(C_6H_5)_3SnOP(O)[NH(C_6H_5)]_2$	187–9	(*60*)

E. Organotin Sulfates and Sulfonates

Bis(trimethyltin) sulfate can be prepared by the reaction of trimethyltin bromide and silver sulfate in dry methanol. Upon removal of the methanol under vacuum at 25°C, white crystals of $[(CH_3)_3Sn]_2SO_4(CH_3OH)_2$ (*78*) are obtained. This product is readily dissolved in water and methanol and is rapidly hydrolyzed on exposure to air. When heated under vacuum at 100°C for 4 h, the methanol is removed, to give white $[(CH_3)_3Sn]_2SO_4$ (*19*). This product is also rapidly hydrolyzed on exposure to air.

The reaction of sulfuric acid with trimethyltin hydroxide gives bis(trimethyltin) sulfate dihydrate $[(CH_3)_3Sn]_2SO_4 \cdot 2\,H_2O$. The anhydrous compound $[(CH_3)_3Sn]_2SO_4$ (*21*) is obtained by keeping the hydrate above 80°C *in vacuo*. The results of the infrared spectra for the hydrate show planar C_3Sn and SO_4 group (*19, 82*) with a symmetry lower than tetrahedral; whereas for the anhydrous compound the results show planar C_3Sn and tetrahedral (*19*), or nearly tetrahedral SO_4 (*82*).

It remains incorrect to assume the configuration of these compounds only through the results of the infrared spectra. However, it is important that the band due to the Sn···O weak coordination cannot be found down to 200 cm^{-1} in the dehydrated and the hydrated compound (*82*).

Bis(triethyltin) sulfate (*82*) is obtained by the reaction of triethyltin hydroxide or hexaethyldistannoxane with sulfuric acid. Its infrared spectra have also been studied by Lohman (*64*). The toxity of this compound has been

TABLE 14

Organotin Sulfates and Sulfonates

	mp, °C	References
$[(CH_3)_3Sn]_2SO_4$	>250	(*19, 82*)
$[(CH_3)_3Sn]_2SO_4 \cdot 2\,H_2O$	>250	(*82*)
$[(C_2H_5)_3Sn]_2SO_4$		(*64, 82*)
$[(n\text{-}C_4H_9)_2Sn]_2SO_4$		(*2*)
$[(C_6H_5)_3Sn]_2SO_4$	>290	(*13, 109*)
$(CH_3)_2Sn(OSO_2CH_3)_2$	325	(*6*)
$(C_2H_5)_2Sn(OSO_2CH_3)_2$	334	(*6*)
$(C_2H_5)_2Sn(OSO_2C_2H_5)_2$	309	(*6*)
$(n\text{-}C_3H_7)_2Sn(OSO_2CH_3)_2$	307	(*6*)
$(n\text{-}C_3H_7)_2Sn(OSO_2C_2H_5)_2$	298	(*6*)
$(i\text{-}C_3H_7)_2Sn(OSO_2CH_3)_2$	275	(*6*)
$(n\text{-}C_4H_9)_2Sn(OSO_2CH_3)_2$	312	(*6*)
$(n\text{-}C_4H_9)_2Sn(OSO_2C_6H_4\text{-}p\text{-}CH_3)$	320	(*102*)
$(i\text{-}C_3H_7)_3Sn(OSO_2CH_3)$	244	(*6*)
$(n\text{-}C_4H_9)_3Sn(OSO_2\text{-}2\text{-}C_{10}H_7)$	waxy solid	(*28*)

studied (*114*). Di-*n*-butyltin sulfate has been reported to change into di-*n*-butyltin sulfite upon irradiation with gamma-rays (*2*). Bis(triphenyltin) sulfate has been prepared and its antimicrobial activities have been studied (*114*).

Di-*n*-butyltin bis(*p*-toluenesulfonate) (*102*) is a white solid, melting above 320°C, made from di-*n*-butyltin oxide and *p*-toluenesulfonic acid monohydrate in methanol, evaporation to a small volume, and then extraction with ether. Alkyltin sulfonates are best prepared in a similar way from an excess of alkanesulfonic acid and a dialkyltin oxide, or a bis(trialkyltin) oxide (*6*). These compounds are soluble in the polar solvents, water, dimethylformamide, dimethyl sulfoxide, and lower alcohols. These have infrared absorptions for asymmetric and symmetric sulfonate stretching at 1188–1195 and 1050–1066 cm^{-1} showing a partly ionic nature of the tin-sulfonate bondings.

F. Miscellaneous Salts

Reaction of trimethyltin bromide and silver chromate in dry methanol gives yellow, crystalline $[(CH_3)_3Sn]_2CrO_4$. Trimethyltin chromate is soluble in water and methanol, but decomposes slowly in solution; it is only slightly soluble in acetone and insoluble in ether. It is not hydrolyzed in moist air; samples exposed to air for 3 days showed no change spectroscopically. The results of infrared spectra show that the $(CH_3)_3Sn$ group is planar and that the CrO_4 group has a tetrahedral symmetry. Dimethyltin hydrogen arsenate, molybdate, and tungstate are prepared as precipitates in aqueous solution by the reaction of the dichloride with corresponding acids or sodium salts (*91*). Triphenyltin arsenate and phenyl arsenate are prepared by the reaction

TABLE 15

Miscellaneous Salts

	mp, °C	References
$[(CH_3)_3Sn]_2CrO_4$		(*19*)
$(CH_3)_3SnBF_4$		(*18*)
$(CH_3)_3SnAsF_6$		(*18*)
$(CH_3)_3SnSbF_6$		(*18*)
$(CH_3)_2SnHAsO_4$	darken at 350	(*91*)
$(CH_3)_2SnMoO_4$	343(decomp.)	(*91*)
$(CH_3)_2SnWO_4$	darken at 330	(*91*)
$[(C_6H_5)_3Sn]_3AsO_4$	227–230, 195–6	(*17*, *109*)
$[(C_6H_5)_3Sn]_2O_3AsC_6H_5$	245–247	(*17*)

of the hydroxides with acid in methanol or acetone (*17*). Trimethyltin tetrafluoroborate, hexafluoroarsenate, and hexafluoroantimonate are prepared by the reaction of trimethyltin chloride with silver salts in sulfur dioxide (*18*). These compounds are bridged by two fluorine atoms: AsF_6 groups form *trans*-fluorine bridges but SbF_6 groups form *cis*-bridges.

III. Organotin Complexes

A. Organotin Oxinates

1. *Triorganotin Oxinates*, $R_3Sn(OX)$ (*47*) ($OX = C_9H_6NO$)

To a cyclohexane solution of 8-hydroxyquinoline (oxine) mixed with sodium methoxide solution in methanol is added trimethyltin chloride in methanol to give a product which, when concentrated and distilled under reduced pressure, gives a clear, yellow distillate of $(CH_3)_3Sn(OX)$. The large amount of residue in the still is dimethyltin bisoxinate. The following disproportionation might occur during distillation.

$$2\,(CH_3)_3Sn(OX) \xrightarrow{\Delta} (CH_3)_2Sn(OX)_2 + (CH_3)_4Sn$$

Other triorganotin oxinates ($R = C_2H_5$, $n\text{-}C_4H_9$ and C_6H_5) are prepared similarly. These compounds are listed in Table 18.

From the electronic spectra (*10, 92, 93, 123*), it has been concluded that in cyclohexane and in benzene a chelated triphenyltin oxinate exists while in 95% aqueous ethanol and even in dry methanol breakage of the tin-nitrogen bond takes place.

2. *Diorganotin Bisoxinates*, $R_2Sn(OX)_2$ (*12, 34, 113*)

An ethanol solution of dialkyl- or diphenyltin dichloride is mixed with oxine in ethanol (mole ratio 1:2) followed by neutralization with aqueous ammonia. The yellow precipitates formed are recrystallized from benzene or ligroin. All of these bright yellow, crystalline compounds are stable in air and have the composition $R_2Sn(OX)_2$. They are soluble in common organic solvents, but insoluble in water. Bis(pentafluorophenyl)tin bisoxinate is prepared from a mixture of tetrakis(pentafluorophenyl)tin or tris(pentafluorophenyl)tin chloride and an excess of oxine in ethanol under reflux (*40*).

Dimethyltin bisoxinate has been studied by spectroscopic techniques (*10, 47, 54, 71, 75, 113, 125*) and at least two aspects of the structure have been the subject of conflicting interpretation. The oxinate groups are chelating, as has been predicted from consideration of the visible-ultraviolet spectrum (*10*) and the infrared spectrum (*47, 113*), and by the failure to coordinate dimethyl sulfoxide in solution (*54*). As predicted by McGrady and Tobias (*71*), an x-ray result (*100*) shows that the Sn—O bonds are shorter (2.11 Å) and, presumably, stronger than the Sn—N bonds (2.35 Å). The bond angles around the tin atom are shown in Fig. 14 and the molecular structure is shown in Fig. 15. The observed C—Sn—C angle (110.7 ± 0.8°) is remarkably

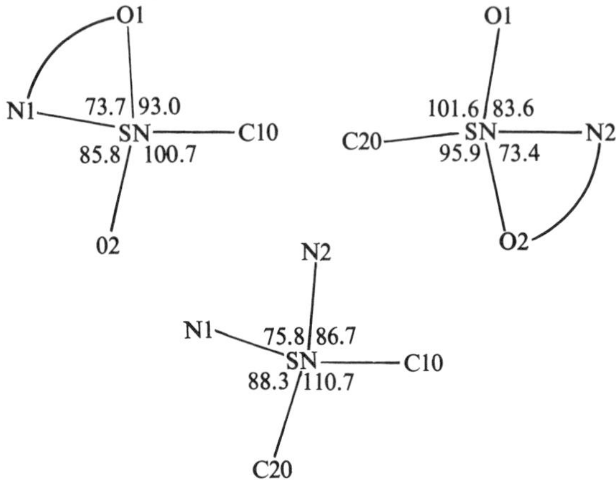

Fig. 14. Bond angles around the tin atom in $(CH_3)_2Sn(C_9H_6NO)_2$.

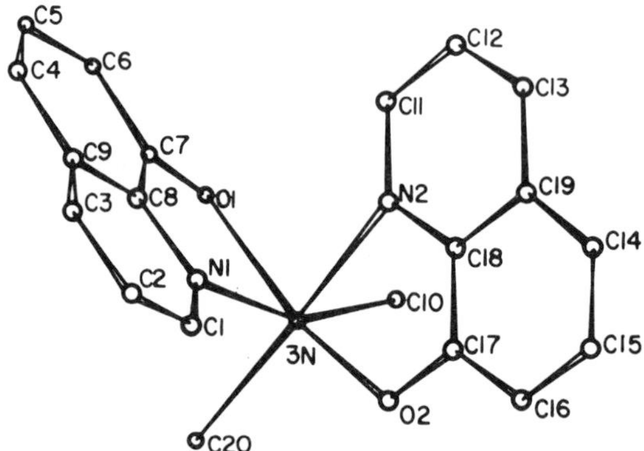

Fig. 15. Perspective view of the molecular structure of $(CH_3)_2Sn(C_9H_6NO)_2$.

close to the tetrahedral angle and lends support to the suggested relationship between tin-proton coupling constants and the tin orbital hybridization (*39*).

3. *Diorganotin Halide Oxinates and Other Salt Oxinates*, $R_2SnX(OX)$

Dimethyltin chloride oxinate can be prepared from the reaction of dimethyltin dichloride with oxine in absolute ethanol (*47*). This compound can also be obtained through disproportionation of $(CH_3)_2SnCl_2$ and $(CH_3)_2Sn(OX)_2$ in refluxing benzene (*71*). Dialkyltin isocyanate oxinate and acetate oxinate are similarly prepared (*41*). Dialkyltin nitrate oxinate and sulfate oxinate are prepared by the following methods (*41*):

$$R_2Sn(OX)_2 + AgNO_3 \xrightarrow{\text{in } CH_3OH} R_2Sn(NO_3)(OX) + Ag(OX)$$

$$2\ R_2SnCl(OX) + Ag_2SO_4 \xrightarrow{\text{in } CH_3OH} [R_2Sn(OX)]_2SO_4 + 2\ AgCl$$

Dialkyltin halide oxinates are monomeric in nonpolar solvents and are considered to have a pentacoordinated tin atom. In the presence of donor molecules, such as pyridine and water, these compounds disproportionate to the more stable hexacoordinated compounds, $R_2Sn(OX)_2$ (*47*). Disproportionation also occurs by heating these compounds with pyridine in a sealed tube (*47*).

4. *Monoorganotin Chloride Bisoxinates*, $RSnCl(OX)_2$ (*29, 47*)

Methyltin trichloride, prepared by treating methyltin oxide with dilute hydrochloric acid, is heated with oxine in ethanol followed by neutralization with aqueous ammonia to give methyltin chloride bisoxinate. An ethanol solution of oxine and *n*-butyltin trichloride react to give *n*-butyltin chloride bisoxinate. A novel compound, $[n\text{-}C_4H_9Sn(OX)_2]_2S$, is prepared from *n*-butyltin sesquisulfide and oxine in boiling toluene (*56*).

5. *Monoorganotin Trisoxinates*, $RSn(OX)_3$

Methyltin sesquisulfide is treated with oxine in a 1:2 mole ratio in boiling toluene for 20–30 h to give methyltin trisoxinate (*56*). *n*-Butyltin oxide and oxine react to give water and *n*-butyltin trisoxinate (*46*). *n*-Butyltin trichloride and sodium oxinate in methanol also give yellow crystals of $n\text{-}C_4H_9Sn(OX)_3$. This is a stable compound showing a constant melting point and showing no OH stretching band in the solid state, after remaining in air for several months.

It has been concluded from spectroscopic techniques that *n*-butyltin trisoxinate may have a heptacoordinated tin atom without the influence of water but is easily hydrolyzed in dilute solution to hexacoordinated *n*-butyltin bisoxinate hydroxide which is stable to further hydrolysis:

$$C_4H_9Sn(OX)_3 + H_2O \longrightarrow C_4H_9Sn(OX)_2(OH) + OX\cdot H$$

B. Organotin Acetylacetonates and Other β-Diketonates

Organotin acetylacetonates have been prepared, since this ligand is simple and it is useful to have structural information through their infrared, Raman, and nmr spectra.

Diorganotin bisacetylacetonates, $R_2Sn(acac)_2$ (acac $= CH_3COCHCOCH_3$) (*117*), are obtained by the reaction of the diorganotin dichloride and sodium methoxide in methanol, followed by the addition of acetylacetone:

$$R_2SnCl_2 + 2\ CH_3ONa \xrightarrow{\text{in } CH_3OH} R_2Sn(OCH_3)_2 + 2\ NaCl$$

$$R_2Sn(OCH_3)_2 + 2\ \text{acacH} \xrightarrow{\text{in } CH_3OH} R_2Sn(\text{acac})_2 + 2\ CH_3OH$$

Dimethyltin bisacetylacetonate is also prepared by the reflux of dimethyltin oxide in acetylacetone for several hours (*71*).

Organohalogenotin bisacetylacetonates, $RXSn(\text{acac})_2$ ($X = Cl$, Br), are prepared by the reaction of acetylacetone with organotin trihalides in water (*117*):

$$RSnX_3 + 2\ \text{acacH} \xrightarrow{\text{in } H_2O} RXSn(\text{acac})_2 + 2\ HX$$

The iodides, $(CH_3)ISn(\text{acac})_2$, can be prepared by the reaction of methyltin triiodide with sodium acetylacetonate in chloroform:

$$CH_3SnI_3 + 2\ \text{acacNa} \xrightarrow{\text{in } CHCl_3} (CH_3)ISn(\text{acac})_2 + 2\ NaI$$

Some other β-diketonates,

$$R_2Sn(CH_3COCHCOC_6H_5)_2 \quad \text{and} \quad R_2Sn(C_6H_5COCHCOC_6H_5)_2$$

are prepared by the reflux of 1-phenyl-1,3-butanedione or 1,3-diphenyl-1,3-propanedione in benzene in the presence of the organotin oxide (*71*).

The C—Sn—C group in $(CH_3)_2Sn(\text{acac})_2$ in the solid is thought to be linear by its infrared and Raman spectra, and the following configuration has been proposed (*51*):

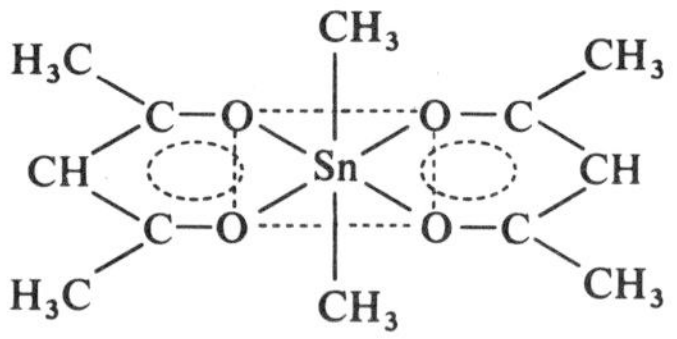

```
                    CH3
H3C                  |                 CH3
   \C—O·······|·······O—C/
  //   ⋮    \  |  /    ⋮   \\
CH     ⋮      Sn        ⋮    CH
  \\   ⋮    /  |  \    ⋮   //
   /C—O·······|·······O—C\
H3C                  |                 CH3
                    CH3
```

This compound is insoluble in common organic solvents, but the other acetylacetonates shown in Table 18 are soluble.

In Table 16 are shown the nmr data and ν(C═O) and ν(Sn—O) in the infrared spectra of organotin acetylacetonates, from which the following important relations have been deduced (*52*, *53*).

(i) There is a linear relationship between the chemical shifts of the γ proton and those of the methyl protons of acetylacetonate ligands.

(ii) Both τ values decrease with the increasing electron-attracting power of the substituents on the tin atom and they can be correlated with the C═O and Sn—O stretching frequencies. These relationships can be explained in terms of the inductive effect of the substituents on the tin atom.

TABLE 16

PROTON CHEMICAL SHIFTS AND RELEVANT INFRARED BANDS OF ORGANOTIN AND TIN BISACETYLACETONATES

$XYSn(acac)_2$ X	Y	(>C—H) (ppm)[a]	(>C—CH_3) (ppm)[a]	ν(C=O) (cm^{-1})[b]	ν(Sn—O) (cm^{-1})[b]
CH_3	CH_3	—	—	1566	406
C_2H_5	C_2H_5	4.73	8.04	1572	404
C_2H_3	C_2H_3	4.61	8.03	1563	408
C_6H_5	C_6H_5	4.64	8.06	1564	408
C_2H_5	Br	4.47	7.95	1560	433
CH_3	I	4.44	7.95	1562	432
CH_3	Br	4.46	7.95	1558	436
CH_3	Cl	4.46	7.95	1588	435
C_6H_5	Br	4.43[c]	7.93[d]	1552	438
C_6H_5	Cl	4.42[c]	7.94[d]	1555	438
I	I	4.27	7.90[c]	1546	445
Br	Br	4.28	7.86[c]	1543	453
Cl	Cl	4.26	7.84[c]	1543	461

[a] In chloroform and TMS as an internal standard.
[b] In Nujol mulls.
[c] Mean value of doublet.
[d] Mean value of quartet.

The nmr spectra of $C_6H_5XSn(acac)_2$ show an interesting behavior of acetylacetone protons due to the internal rotation of the phenyl group around the Sn—C bonds (*50*).

In some alkylhalogenotin and dihalogenotin bisacetylacetonates, long range spin-spin couplings between the tin and the γ and methyl protons of the acetylacetonate ring are observed (*48, 50*).

C. ORGANOTIN TROPOLONATES

Dimethyltin bistropolonate (*58*), $(CH_3)_2Sn(C_7H_5O_2)_2$, is obtained by the reaction of dimethyltin dichloride and sodium tropolonate (mole ratio, 1:2) in methanol. Phenyltin trichloride in benzene added to a solution of tropolone in ether, followed by addition of dimethoxyethane gives phenyltin chloride bistropolonate (*74*). Reflux of a mixture of $C_6H_5SnCl(C_7H_5O_2)_2$ sodium tropolonate and acetonitrile followed by the addition of water and methanol yields phenyltin tristropolonate, $C_6H_5Sn(C_7H_5O_2)_3$ (*74*). The molecular weight is that of a monomer in dichloroethane. This is indicative of a seven-coordinate structure for this compound.

D. Organotin Kojates (*85*)

Chelated kojate complexes of tin(IV) have a five-membered ring as shown below, and the ν(Sn—O) is close to those of the tin(IV) chelates of tropolonate and oxinate (cf. Table 17).

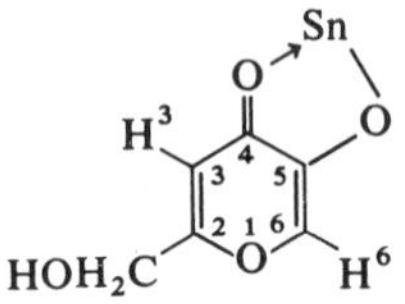

TABLE 17

Range of ν(Sn—O) in Tin(IV) Compounds

Compound	ν(Sn—O) (cm^{-1})	References
$(R_3Sn)_2O$	740–780	(*59, 122*)
$R_2Sn(OR')_2$	600–680	(*15*)
	530–620	
$XYSn(T)_2$	550–590	(*58*)
$XYSn(Kj)_2$	550–580	(*85*)
$XYSn(OX)_2$	510–540	(*113*)
$R_2SnX_2 \cdot 2$ DMSO	405–430	(*110*)
$XYSn(acac)_2$	400–460	(*53*)
$R_2SnX_2 \cdot 2$ DMSeO	385–410	(*112*)
Pyridine *N*-oxide adducts	310–390	(*24, 49, 127*)
$(C_6H_5)_3PO$ and $(C_6H_5)_3AsO$ adducts	300–345	(*23*)
$R_3SnOOCH$	300	(*80*)
Carbonyl donor adducts	180–200	(*69*)

Treatment of an aqueous solution of dimethyltin dichloride and kojic acid in a 1:2 mole ratio by aqueous ammonia gives dimethyltin biskojate.

A distorted octahedral *trans*-configuration has been suggested for this compound. The presence of the two Sn—C bands at 584 cm^{-1} (strong $\nu Sn—C_{asym}$) and 504 cm^{-1} (weak, $\nu Sn—C_{sym}$) indicates a nonlinear C—Sn—C moiety, and the coupling constant ($J_{^{119}Sn—CH_3}$), 83.3 cps, is an intermediate value between that of tetrahedral tetramethyltin (54.0 cps) and that of dimethyltin bisacetylacetonate (99.3 cps) with a linear C—Sn—C moiety.

Methylchlorotin and methylbromotin biskojates are formed upon addition of kojic acid to solutions obtained by reacting methyltin oxide with either aqueous hydrochloric or hydrobromic acid.

The long range spin-spin coupling of the 3-proton of the kojate ring with tin (J_{H^3-Sn}) is observed in these compounds and in dihalogenotin biskojates. However, the coupling of the 6-proton with tin has not been detected. Neither of the long-range spin-spin couplings is found in dimethyltin biskojate.

E. Mixed Organotin Chelates

Although many organotin chelates have been well studied, there are few reports on mixed chelates, in which two kinds of ligands coordinate to one tin atom. Westlake and Martin (*125*) tried to prepare various diorganotin mixed chelates using thallium β-diketonates and they could assume only $(C_6H_5)_2Sn(OX)(C_6H_5COCHCOC_6H_5)$ as a pure compound. The method of preparation of one such mixed chelate, dimethyltin oxinate tropolonate (*57*), is outlined as follows: Dimethyltin bisoxinate (**6**) (1.1 g; mp 231–3°C) and dimethyltin bistropolonate (**7**) (1.2 g; mp 181–3°C) (*58*) are dissolved in 100 ml of ethanol etc. and the solution is refluxed for 3 h. A yellow solid obtained by vacuum evaporation (1.8 g, 80% yield) is recrystallized from ethanol to give yellow crystals of $(CH_3)_2Sn(OX)(T)$ (**8**) (mp 168–170.5°C) which are air-stable and soluble in common organic solvents. The infrared spectrum in Nujol, the x-ray powder pattern, and the sharp melting point of (**8**) are quite different from those of (**6**) and (**7**). Thus it is shown that (**8**) is not a mixture of (**6**) and (**7**) in the solid state.

The existence of the mixed chelate in solution is also supported by the nmr studies. The nmr spectra of (**8**) in *sym*-tetrachloroethane and in chloroform showed three methyl proton signals at 20°C (Fig. 16), of which the signals at the highest and the lowest field have been independently assigned to those of (**6**) and (**7**), respectively. Then the middle field is assigned to that of the mixed chelate (**8**). Thus in solution, the following equilibrium is suggested:

$$\underset{(\mathbf{8})}{2\,(CH_3)_2Sn(OX)(T)} \rightleftharpoons \underset{(\mathbf{6})}{(CH_3)_2Sn(OX)_2} + \underset{(\mathbf{7})}{(CH_3)_2Sn(T)_2}$$

As shown in Fig. 16, at 30–55°C, the signals of (**7**) and (**8**) coalesce to an intermediate signal in both solvents. This fact suggests a rapid exchange of (T) and (OX) between (**7**) and (**8**). At higher temperatures, the signal of (**6**) begins to merge with the intermediate signal. These results can also be explained on the basis of a similar rapid ligand-exchange. A single methyl proton signal observed at 87°C is caused by a sufficiently rapid ligand-exchange, since each methyl group in the three species is in the same average chemical environment.

Dimethyltin kojate tropolonate (*111*) has been obtained by reaction of methanol solutions of dimethyltin dichloride and dimethyltin biskojate with sodium tropolonate in methanol. The compound obtained differs from the original reactants as shown by its infrared spectrum, x-ray powder pattern, and sharp melting point.

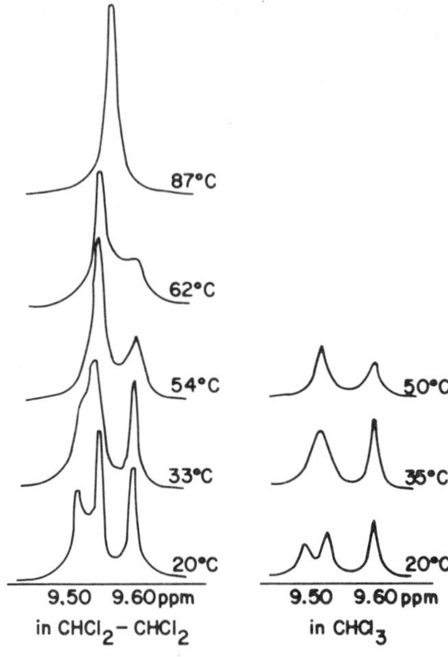

Fig. 16. The nmr spectra of $(CH_3)_2Sn(Ox)T$ (internal standard, TMS).

TABLE 18

ORGANOTIN CHELATES

Compound		mp or (bp) (°C)	References
Oxinates			
$R_3Sn(OX)$			
R = CH_3		(108.5–109.5/0.2)	(*47*)
C_2H_5		(132–134/0.05)	(*47*)
n-C_4H_9		(149–151/0.007)	(*47*)
C_6H_5		145–146.5	(*47, 92*)
$R_2Sn(OX)_2$			
R = CH_3		231–233, 236.5–238	(*34, 113*)
C_2H_5		167–168, 177–178	(*34, 113*)
n-C_3H_7		151–153, 162–163	(*34, 113*)
n-C_4H_9		150–152, 154.5–155.5	(*12, 34, 113*)
i-C_4H_9		188–189	(*34*)
n-C_8H_{17}		78–79	(*34*)
C_6H_5		231–233, 251–252	(*12, 113, 125*)
$C_6H_5CH_2$		118–120	(*34*)
C_6F_5		277 decomp.	(*40*)
$R_2SnX(OX)$			
R = CH_3	X = Cl	137	(*41, 47*)
CH_3	I	134–135	(*125*)
C_2H_5	Cl	119–120.5	(*47*)

TABLE 18—*continued*

Compound		mp or (bp) (°C)	References
n-C_3H_7	Cl	90–91	(*47*)
n-C_4H_9	Cl	48	(*41*)
n-C_4H_9	$Cl \cdot H_2O$	130	(*41*)
CH_3	NCS	123–124	(*47, 123*)
n-C_3H_7	NCS	134, 144	(*47, 123*)
n-C_4H_9	O_2CCH_3	>142 decomp.	(*41*)
C_6H_5	Cl	156	(*41, 125*)
C_6H_5	NO_3	>175 decomp.	(*41*)
C_6H_5	$\frac{1}{2}$ SO_4	>310 decomp.	(*41*)
$RSnX(OX)_2$			
R = CH_3	X = Cl	246–247	(*47*)
CH_3	Br	268–260	(*29*)
n-C_4H_9	Cl	182–183	(*47*)
C_6H_5	Cl	150, 218–219	(*29, 75*)
n-C_4H_9	$\frac{1}{2}$ S	225–226	(*56*)
$RSn(OX)_3$			
R = CH_3		280	(*56*)
n-C_4H_9		223, 229–231	(*31, 46*)
Acetylacetonates and other β-diketonates			
$XYSn(acac)_2$			
X = CH_3	Y = CH_3	177–178, 300 decomp.	(*71, 117*)
C_2H_5	C_2H_5	86.5–87	(*117*)
C_2H_3	C_2H_3	87–88	(*117*)
C_6H_5	C_6H_5	125 decomp.	(*71, 75*)
CH_3	Cl	135–136	(*117*)
CH_3	Br	129	(*117*)
CH_3	I	115–116	(*117*)
C_2H_5	Br	94	(*117*)
C_6H_5	Cl	149–152	(*117*)
$(CH_3)_2Sn(CH_3COCHCOC_6H_5)_2$		134–135	(*71*)
$(C_6H_5)_2Sn(CH_3COCHCOC_6H_5)_2$		181–182	(*75*)
$(CH_3)_2Sn(C_6H_5COCHCOC_6H_5)_2$		189–191	(*71*)
$(C_6H_5)_2Sn(C_6H_5COCHCOC_6H_5)_2$		238–241	(*75*)
Tropolonates			
$(CH_3)_2Sn(T)_2$		181–183	(*111*)
$C_6H_5SnCl(T)_2$		248–252	(*74*)
$C_6H_5Sn(T)_3$		298–302	(*74*)
Kojates			
$(CH_3)_2Sn(kj)_2$		194–195.5 decomp.	(*85*)
$CH_3SnCl(kj)_2$		205–208 decomp.	(*85*)
$CH_3SnBr(kj)_2$		113–114 decomp.	(*85*)
Mixed chelates			
$(C_6H_5)_2Sn(OX)(C_6H_5COCHCOC_6H_5)_2$		182–186	(*125*)
$(CH_3)_2Sn(OX)(T)$		168–170.5	(*57*)
$(CH_3)_2Sn(kj)(T)$		230–231	(*111*)

F. Adducts of Organotin Halides

1. *Trialkylphosphine Oxide and Trialkylarsine Oxide Adducts*

Dimethyltin dihalides and triphenylphosphine oxide in equimolar proportions are mixed in dry ethanol. The adducts separate gradually, are collected on a filter crucible, washed successively with ethanol and light petroleum, and dried under high vacuum. Thus, $(CH_3)_2SnX_2 \cdot 2\,(C_6H_5)_3PO$ (X = Cl, Br, I) are obtained (*24*). $CH_3SnI_3 \cdot 2\,(C_6H_5)_2PO$ and $(CH_3)_2SnX_2 \cdot 2\,(C_6H_5)_3AsO$ (X = Cl, Br, I) can be obtained in a similar way (*23*).

Triethylphosphine oxide-trialkyltin halide adducts, $R_3SnX \cdot (C_2H_5)_3PO$ (R = CH_3, C_2H_5, n-C_3H_7, n-C_4H_9, X = Cl, I) are obtained as liquids. These adducts are also synthesized by the reaction of R_3SnX with $(C_2H_5)_2POC_2H_5$ heated to 100–110°C for 1–2 h (*86*).

The melting points of these and the following adducts are summarized in Table 19.

2. *Pyridine N-oxide Adducts* (*24, 49*)

The adducts are prepared by mixing pyridine *N*-oxide in a benzene solution of the corresponding organotin(IV) halides. The adducts of trimethyltin and dimethyltin are obtained after the mixture has stood overnight. They are recrystallized from benzene and chloroform, respectively.

The Sn—O bond in the pyridine *N*-oxide adducts has been assumed to have a rather ionic character, from the fact that the position of the ν(Sn—O) is close to that of the trialkyltin carboxylates (cf. Table 17), and that the dependence of the stability constant of the complexes on the substituent of the pyridine ring is similar to that of the pK_a of the ligand (*49*).

3. *Dimethylsulfoxide and Diphenylselenoxide Adducts*

The complexes, $R_2SnX_2 \cdot 2\,(CH_3)_2SO$ and $R_2SnX_2 \cdot 2\,(CH_3)_2SeO$, are prepared by the reaction of 1 mole of organotin halide with 2 moles of the ligand in carbon tetrachloride or ether (*63, 110, 112*).

A different method is used to obtain the dimethylsulfoxide complex of methyltin trichloride (*62*). Dimethyltin dichloride (0.05 mole) is suspended in 50 ml of DMSO and cooled with ice water. To this is added dropwise a solution of tin tetrachloride (0.05 mole) in approximately 20 ml of DMSO. The mixture is refluxed for 24 h and after addition of benzene to the clear solution the complex $CH_3SnCl_3 \cdot 2\,(CH_3)_2SO$ crystallizes. The crude material (yield 90%) is recrystallized from hot ethanol and chloroform, or ether.

From the results of their far-infrared spectra, the following structures have been suggested (*110, 112*).

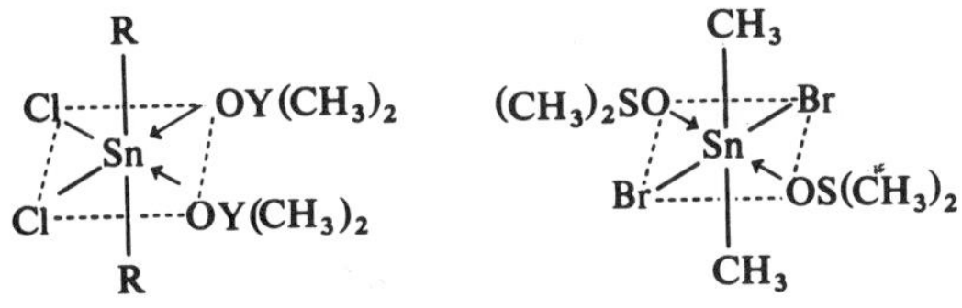

(R = CH_3, C_2H_5, C_6H_5)
(Y = S, Se)

4. *Weak Carbonyl Donor Adducts* (*69*)

Dimethyltin dichloride, or dibromide, is mixed with *N*,*N*-dimethylformamide in carbon tetrachloride in less than a 1:2 mole ratio. After standing at about 4°C, white crystals of 1:2 adducts are formed, which are filtered off and washed with the solvent. The dimethyltin dichloride adduct can be dried at reduced pressure, but the dimethyltin dibromide adduct loses 1 mole of the ligand at a reduced pressure of 5 mm for about 90 min to give the 1:1 adduct. The 1:1 adduct is also obtained from a 1:1 mixture of dimethyltin dichloride, or dibromide and the ligand in carbon tetrachloride. Similarly, *p*-substituted aromatic carbonyl donors give 1:1 adducts with dimethyltin halides as shown in Table 19. However, acetophenone or benzaldehyde adducts cannot be isolated by a similar method.

Information obtained from the infrared and nmr spectra shows that the oxygen of the C=O interacts with the tin atom. Thus the C=O stretching frequencies of these ligands shift to about 35 cm^{-1} lower frequencies upon coordination. The distorted octahedral configuration as shown in (**9**) has been assumed for the *N*,*N*-dimethylformamide 1:2 complexes since the geometry of the C—Sn—C group is non-linear. The weak coordination of the oxygen to tin atom can be understood from the ν(Sn—O) shown in Table 17. As to the characteristic pentacoordinated 1:1 adduct, the geometry of the C—Sn—C and Cl—Sn—Cl are assumed from the infrared information and the probable configuration, as shown in (**10**) has been suggested.

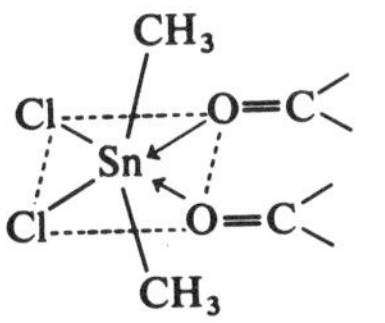

$(CH_3)_2SnCl_2 \cdot 2$ DMF
(**9**)

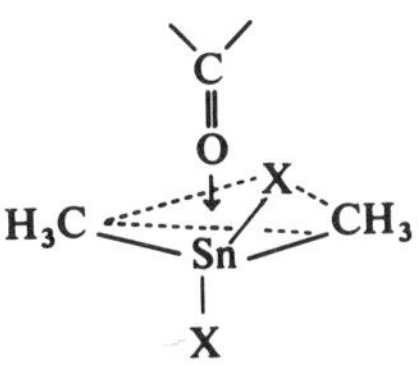

$(CH_3)_2SnX_2 \cdot$ DMF
(**10**)

Mixing of a concentrated (0.5 *M*) solution of phenalen-1-one and phenyltin trichloride in ether, under anhydrous conditions, leads to the rapid precipitation of a colored phenyltin trichloride-phenalen-1-one 1:1 solid adduct (*72*).

G. Adducts of Organotin Cations (*61*)

$[(CH_3)_2SnL_4]^{2+} \cdot 2\,(C_6H_5)_4B^-$: To a warm solution of dimethyltin dichloride and sodium tetraphenyl borate in a 2:1 mole ratio in DMSO is added an excess of water while stirring. A white crystalline material is precipitated which is filtered and washed repeatedly with warm water and finally dried. This compound, $[(CH_3)_2Sn(DMSO)_4]^{2+} \cdot 2\,(C_6H_5)_4B^-$, has no serious survival problem and may be kept for considerable periods under vacuum. The corresponding DMF and DMA complexes can also be obtained similarly. The $(CH_3)_2Sn(IV)$ complexes are insoluble in alcohol, ether, chloroform, benzene, and dioxane but dissolve in acetone, dichloromethane, DMF, DMA, DMSO, and sulfone.

$[(CH_3)_3SnL_2]^+ \cdot (C_6H_5)_4B^-$: Sodium tetraphenylborate (0.005 mole) and trimethyltin bromide (0.005 mole) are dissolved in a slight excess of warm dimethylacetamide. Ice is added and the mixture is refrigerated. White needles of the complex appear which are filtered, dried and stored under vacuum in the dark. In a similar way, DMSO and DMF adducts can be also obtained. The $(CH_3)_3Sn(IV)$ complexes are more soluble than their $(CH_3)_2Sn(IV)$ relatives, and chloroform, benzene, alcohol, and acetone are suitable solvents.

TABLE 19

Adducts of Organotin Halides and Organotin Salts

Compound			mp or (bp) (°C)	References
$RSnX_3 \cdot 2\,L$				
R	X	L		
CH_3	I	$(C_6H_5)_3PO$		(*23*)
CH_3	I	$(C_6H_5)_3AsO$		(*23*)
CH_3	Br	C_5H_5NO	138–142	(*49*)
CH_3	Cl	$(CH_3)_2SO$	175, 188–190	(*62, 63*)
C_6H_5	Cl	$(CH_3)_2SO$	152	(*63*)
$RSnX_3 \cdot L$				
R	X	L		
C_6H_5	Cl	$C_{13}H_8O$ (phenalen-1-one)		(*72*)
$R_2SnX_2 \cdot 2\,L$				
R	X	L		
CH_3	Cl	$(C_6H_5)_3PO$		(*24*)
CH_3	Br	$(C_6H_5)_3PO$		(*24*)
CH_3	I	$(C_6H_5)_3PO$		(*24*)
CH_3	Cl	$(C_6H_5)_3AsO$		(*23*)

TABLE 19—*continued*

Compound			mp or (bp) (°C)	References
CH_3	Br	$(C_6H_5)_3A_5O$		(*23*)
CH_3	I	$(C_6H_5)_3A_5O$		(*23*)
CH_3	Cl	C_5H_5NO	135–136.5	(*24, 49*)
CH_3	Br	C_5H_5NO		(*24*)
CH_3	I	C_5H_5NO		(*24*)
C_6H_5	Cl	C_5H_5NO	166–168	(*49*)
CH_3	Cl	$(CH_3)_2SO$	110–110.5, 113	(*63, 110*)
CH_3	Br	$(CH_3)_2SO$	119–120	(*110*)
C_2H_5	Cl	$(CH_3)_2SO$	64	(*110*)
C_6H_5	Cl	$(CH_3)_2SO$	134–135	(*63, 110*)
CH_3	Cl	$(CH_3)_2SeO$	188–189	(*112*)
CH_3	Br	$(CH_3)_2SeO$	119–120	(*112*)
C_2H_5	Cl	$(CH_3)_2SeO$	159–160	(*112*)
C_6H_5	Cl	$(CH_3)_2SeO$	158–160	(*112*)
C_6H_5	Cl	$(CH_2)_4SO$	136	(*63*)
CH_3	Cl	$(CH_3)_2NCHO$	84–85.5	(*69*)
CH_3	Br	$(CH_3)_2NCHO$	68–70.5	(*69*)
$R_2SnX_2 \cdot L$				
R	X	L		
CH_3	Cl	$(CH_3)_2NCHO$	56–58.5	(*69*)
CH_3	Br	$(CH_3)_2NCHO$	72.5–74	(*69*)
CH_3	Cl	$CH_3OC_6H_4COCH_3$	49–51	(*69*)
CH_3	Cl	$(CH_3)_2NC_6H_4CHO$	86.5–88.5	(*69*)
CH_3	Br	$(CH_3)_2NC_6H_4CHO$	86–88.5	(*69*)
CH_3	Cl	$(CH_3)_2NC_6H_4COCH_3$	>60 decomp.	(*69*)
CH_3	Cl	$[(CH_3)_2NC_6H_4]_2CO$	140–142.5	(*69*)
$R_3SnX \cdot L$				
R	X	L		
CH_3	I	$(C_2H_5)_3PO$	(95/0.035)	(*86*)
C_2H_5	Cl	$(C_2H_5)_3PO$	(82/0.035)	(*86*)
C_2H_5	I	$(C_2H_5)_3PO$	(92/0.031)	(*86*)
n-C_3H_7	I	$(C_2H_5)_3PO$	(92/0.032)	(*86*)
n-C_4H_9	I	$(C_2H_5)_3PO$	(104/0.038)	(*86*)
CH_3	Cl	C_5H_5NO	86–87	(*49*)
C_6H_5	Cl	C_5H_5NO	133–135	(*49*)
CH_3	Cl	$(CH_3)_2SO$	49	(*63*)
C_6H_5	Cl	$(CH_3)_2SO$	112	(*63*)
$(CH_3)_2SnSO_4 \cdot (CH_3)_2SO$			136	(*20*)
$[(CH_3)_2SnL_4]^{2+} \cdot 2\ (C_6H_5)_4B^-$				
L = $(CH_3)_2SO$			92–93	(*61*)
$(CH_3)_2NCHO$			116–118	(*61*)
$(CH_3)_2NCOCH_3$			80–85	(*61*)
$[(CH_3)_3SnL_2]^+ \cdot (C_6H_5)_4B^-$				
L = $(CH_3)_2SO$			107	(*61*)
$(CH_3)_2NCHO$			144–145	(*61*)
$(CH_3)_2NCOCH_3$			80	(*61*)

REFERENCES

1. C. C. Addison, W. B. Simpson, and A. Walker, *J. Chem. Soc.*, 2360 (1964).
2. A. Y. Aleksandrov, N. N. Delyagin, K. P. Mitrofanov, L. S. Polak, and V. S. Shpinel, *Dokl. Akad. Nauk S.S.S.R.*, **148**, 126 (1963).
3. D. L. Alleston and A. G. Davies, *J. Chem. Soc.*, 2050 (1962).
4. D. L. Alleston, A. G. Davies, M. Hancock, and R. F. M. White, *J. Chem. Soc.*, 5469 (1963).
5. D. L. Alleston, A. G. Davies, and M. Hancock, *J. Chem. Soc.*, 5744 (1964).
6. H. H. Anderson, *Inorg. Chem.*, **3**, 108 (1963).
7. H. H. Anderson, *Inorg. Chem.*, **1**, 647 (1962).
8. H. H. Anderson, *Inorg. Chem.*, **3**, 912 (1964).

8a. H. H. Anderson and J. A. Vasta, *J. Org. Chem.* **19**, 1300 (1954).

9. T. M. Andrews, F. A. Bower, B. R. Laliberte, and J. C. Montermoso, *J. Am. Chem. Soc.*, **80**, 4102 (1958).
10. R. Barbieri, G. Faraglia, M. Guistiniani, and L. Roncucci, *J. Inorg. Nucl. Chem.*, **26**, 203 (1964).
11. I. R. Beattie and T. Gilson, *J. Chem. Soc.*, 2585 (1961).
12. D. Blake, G. E. Coates, and J. M. Tate, *J. Chem. Soc.*, 756 (1961).
13. A. G. Borisov, A. N. Abranova, and Z. N. Parnev, *Izvest. Akad. Nauk S.S.S.R.*, *Ser. Khim.*, 941 (1964).
14. V. A. Bryukhanov, V. I. Gol'danskii, N. N. Delyagin, L. A. Korytko, E. F. Makarov, I. P. Suzdalev, and V. S. Shpinel, *Zhur. Eksp. i. Thoret. Fiz.*, **43**, 448 (1962).
15. F. K. Buchev, W. Greeard, E. F. Mooney, R. G. Rees, and H. A. Willis, *Spectrochim. Acta*, **20**, 51 (1964).
16. I. G. M. Campbell, G. W. A. Fowles, and L. V. Nixon, *J. Chem. Soc.*, 1389 (1964).
17. B. L. Chamberland and A. G. MacDiarmid, *J. Chem. Soc.*, 445 (1961).
18. H. C. Clark and R. J. O'Brien, *Inorg. Chem.*, **2**, 1020 (1963).
19. H. C. Clark and R. G. Goel, *Inorg. Chem.*, **4**, 1428 (1965).
20. H. C. Clark and R. G. Goel, *J. Organometal. Chem.*, **7**, 263 (1967).
21. H. C. Clark and R. J. O'Brien, *Inorg. Chem.*, **2**, 740 (1963).
22. H. C. Clark, R. J. O'Brien, and A. L. Pickard, *J. Organometal. Chem.*, **4**, 43 (1965).
23. J. P. Clark, V. M. Langford, and C. J. Wilkins, *J. Chem. Soc. A*, 792 (1967).
24. J. P. Clark and C. J. Wilkins, *J. Chem. Soc. A*, 871 (1966).
25. R. A. Cummins and P. Dunn, *Australian J. Chem.*, **17**, 185 (1964).
26. R. A. Cummins, *Australian J. Chem.*, **17**, 594 (1964).
27. A. G. Davies and P. G. Harrison, *J. Chem. Soc. C*, 298 (1967).
28. P. Dunn and T. Norris, *Rep. Defence Std. Lab.*, Dept. Supply, Commonwealth, Australia, **269**, 21 (1964); through *CA*, **61**, 3135*b* (1964).
29. G. Faraglia, L. Roncucci, and R. Barbieri, *Ric. Sci. Rend. Sez. A*, **8** (2), 205 (1965); through *CA*, **63**, 12654 (1965).
30. A. N. Fenster and E. I. Becker, *J. Organometal. Chem.*, **11**, 549 (1968).
31. I. Foldesi and G. Straner, *Acta Chim. Acad. Sci. Hung.*, **45**, 313 (1965); through *CA*, **64**, 3591 (1966).
32. M. Frankel, D. Gertner, D. Wagner, and A. Zilkha, *J. Org. Chem.*, **30**, 1596 (1965).
33. J. P. Freeman, *J. Am. Chem. Soc.*, **80**, 5954 (1958).
34. W. Gerrard, E. F. Mooney, and R. G. Rees, *J. Chem. Soc.*, 740 (1964).
35. J. J. Gormley and R. G. Rees, *J. Organometal. Chem.*, **5**, 291 (1966).
36. T. Harada, *Sci. Papers Inst. Phys. Chem. Research*, **57**, 25 (1963).
37. A. Henderson and A. K. Holliday, *J. Organometal. Chem.*, **4**, 377 (1965).
38. R. H. Herber, H. A. Stocklee, and W. T. Reichle, *J. Chem. Phys.*, **42**, 2447 (1965).

39. J. R. Holmes and H. D. Kaesz, *J. Am. Chem. Soc.*, **83**, 3903 (1961).
40. J. M. Holmes, R. D. Peacock, and J. C. Tatlow, *J. Chem. Soc. A*, 150 (1966).
41. F. Huber and R. Kaiser, *J. Organometal. Chem.*, **6**, 126 (1966).
42. R. K. Ingham, S. D. Rosenberg, and H. Gilman, *Chem. Rev.*, 459–539 (1960).
43. K. Issleib and B. Walther, *J. Organometal. Chem.*, **10**, 177 (1967).
44. M. J. Janssen, J. G. A. Luijten, and G. J. M. van der Kerk, *Rec. trav. chim.*, **82**, 90 (1963).
45. L. V. Kaabak and A. P. Tomilov, *Zhur. Obshchei. Khim.*, **23**, 2808 (1963).
46. K. Kawakami, Y. Kawasaki, and R. Okawara, *Bull. Chem. Soc. Japan*, **40**, 2963 (1967).
47. K. Kawakami and R. Okawara, *J. Organometal. Chem.*, **6**, 249 (1966).
48. Y. Kawasaki, *J. Inorg. Nucl. Chem.*, **29**, 840 (1967).
49. Y. Kawasaki, M. Hori, and K. Uenaka, *Bull. Chem. Soc. Japan*, **40**, 2463 (1967).
50. Y. Kawasaki and T. Tanaka, *J. Chem. Phys.*, **43**, 3396 (1965).
51. Y. Kawasaki, T. Tanaka, and R. Okawara, *Bull. Chem. Soc. Japan*, **37**, 903 (1964).
52. Y. Kawasaki, T. Tanaka, and R. Okawara, *Bull. Chem. Soc. Japan*, **40**, 1562 (1967).
53. Y. Kawasaki, T. Tanaka, and R. Okawara, *Spectrochim. Acta*, **22**, 1571 (1966).
54. W. Kitching, *J. Organometal. Chem.*, **6**, 586 (1966).
55. K. A. Kocheshkov, E. M. Panov, and N. N. Zemlyanskii, *Izvest. Akad. Nauk S.S.S.R. Otdel. Khim. Nauk*, 2255 (1961).
56. M. Komura and R. Okawara, *Inorg. Nucl. Chem. Letters*, **2**, 93 (1966).
57. M. Komura, T. Tanaka, T. Mukai, and R. Okawara, *Inorg. Nucl. Chem. Letters*, **3**, 17 (1967).
58. M. Komura, T. Tanaka, and R. Okawara, *Inorg. Chim. Acta*, **2**, 321 (1968).
59. H. Kriegsman, H. Hoffman, and S. Pitschtschan, *Z. Anorg. Allgem. Chem.*, **315**, 283 (1962).
60. H. Kubo, *Agr. Biol. Chem.*, **29**, 43 (1960); through *CA*, **63**, 7032 (1965).
61. V. G. Kumar Das and W. Kitching, *J. Organometal. Chem.*, **10**, 59 (1967).
62. H. G. Langer, *Tetrahedron Letters*, 43 (1967).
63. H. G. Langer and A. H. Blut, *J. Organometal. Chem.*, **5**, 288 (1966).
64. D. H. Lohmann, *J. Organometal. Chem.*, **4**, 382 (1965).
65. J. G. A. Luijten and G. J. M. van der Kerk, *Rec. trav. chim.* **83**, 295 (1964).
66. Y. Maeda, C. R. Dillard, and R. Okawara, *Inorg. Nucl. Chem. Letters*, **2**, 197 (1966).
67. Y. Maeda and R. Okawara, *J. Organometal. Chem.*, **10**, 247 (1967).
68. Y. Maeda, C. R. Dillard, and R. Okawara, unpublished data.
69. G. Matsubayashi, T. Tanaka, and R. Okawara, *J. Inorg. Nucl. Chem.*, **30**, 1831 (1968).
70. M. M. McGrady and R. S. Tobias, *Inorg. Chem.*, **3**, 1157 (1964).
71. M. M. McGrady and R. S. Tobias, *J. Am. Chem. Soc.*, **87**, 1909 (1965).
72. A. Mohammad, D. P. N. Satchell, and R. S. Satchell, *J. Chem. Soc. B*, 723 (1967).
73. A. S. Mufti and R. C. Poller, *J. Chem. Soc. C*, 1362 (1967).
74. E. L. Muetterties and C. M. Wright, *J. Am. Chem. Soc.*, **86**, 5132 (1964).
75. W. H. Nelson and D. F. Martin, *J. Inorg. Nucl. Chem.*, **27**, 89 (1965).
76. M. Ohara and R. Okawara, *J. Organometal. Chem.*, **3**, 484 (1965).
77. M. Ohara, R. Okawara, and Y. Nakamura, *Bull. Chem. Soc. Japan*, **38**, 1379 (1965).
78. R. Okawara, B. J. Hathaway, and D. E. Webster, *Proc. Chem. Soc.*, 13 (1963).
79. R. Okawara and M. Ohara, *Bull. Chem. Soc. Japan*, **36**, 624 (1963).
80. R. Okawara and M. Ohara, *J. Organometal. Chem.*, **1**, 360 (1964).
81. R. Okawara and E. G. Rochow, *J. Am. Chem. Soc.*, **82**, 3285 (1960).
82. R. Okawara and H. Sato, *Int. Symp. Mol. Struct. Spectry.*, Tokyo, A308, 1962.
83. R. Okawara and M. Wada, *Advances in Organometallic Chemistry*, Vol. 5, Academic, New York, 1967, pp. 137–167.

> 84. R. Okawara, D. E. Webster and E. G. Rochow, *J. Am. Chem. Soc.*, **82**, 3287 (1960).
85. J. Otera, Y. Kawasaki, and T. Tanaka, *Inorg. Chim. Acta*, **1**, 294 (1967).
86. A. N. Pudovik, A. A. Muratova, and E. P. Semkina, *Zhur. Obshchei. Khim.*, **33**, (10) 3350 (1963); through *CA*, **60**, 4175 (1964).
87. G. A. Razuvaev, N. S. Vyazankin, and O. A. Shchepetkova, *Tetrahedron*, **18**, 667 (1962).
88. G. A. Razuvaev, O. A. Shchepetkova, and N. S. Vyazankin, *Zhur. Obshchei. Khim.* **32**, 2152 (1962).
89. W. T. Reichle, *Inorg. Chem.*, **5**, 87 (1966).
90. G. H. Reifenberg and W. J. Considine, *J. Organometal. Chem.*, **9**, 495 (1967).
91. E. G. Rochow, D. Seyferth, and A. C. Smith Jr., *J. Am. Chem. Soc.*, **75**, 3099 (1953).
92. L. Roncucci, G. Faraglia, and R. Barbieri, *J. Organometal. Chem.* **1**, 427 (1964).
93. L. Roncucci, G. Faraglia, and R. Barbieri, *J. Organometal. Chem.*, **6**, 278 (1966).
94. H. Sato, *Bull. Chem. Soc. Japan*, **40**, 410 (1967).
95. H. Sato and R. Okawara, *Int. Symp. Mol. Struct. Spectry.*, Tokyo, Japan, September, 1962.
96. A. K. Sawyer and H. G. Kuivila, *J. Am. Chem. Soc.*, **82**, 5958 (1960).
97. A. K. Sawyer and H. G. Kuivila, *Chem. Ind.*, 260 (1961).
98. A. K. Sawyer and H. G. Kuivila, *J. Org. Chem.*, **27**, 610 (1962).
99. A. K. Sawyer and H. G. Kuivila, *J. Org. Chem.*, **27**, 837 (1962).
100. E. D. Schlemper, *Inorg. Chem.*, **6**, 2012 (1967).
101. H. Schmidt, C. Blohm, and G. Jander, *Angew. Chem.*, **59**, 233 (1947).
102. D. Seyferth and F. G. A. Stone, *J. Am. Chem. Soc.*, **79**, 515 (1957).
103. D. Seyferth, F. M. Armbrecht Jr., B. Prokai, and R. J. Cross, *J. Organometal. Chem.*, **6**, 573 (1966).
104. P. Shapiro and E. I. Becker, *J. Org. Chem.*, **27**, 4688 (1962).
105. P. B. Simons and W. A. G. Graham, *J. Organometal. Chem.*, **8**, 479 (1967).
106. P. B. Simons and W. A. G. Graham, *J. Organometal. Chem.* **10**, 457 (1967).
107. W. B. Simpson, *Chem. Ind.*, 854 (1966).
108. N. A. Slovokhotova, N. A. Faizi, N. N. Zemlyanskii, E. M. Panov, and K. A. Kocheshkov, *Zhur. Obshchei. Khim.*, **33**, 2610 (1963).
109. T. N. Srivastava and S. K. Tandon, *Indian J. Appl. Chem.*, **26** (5–6), 171 (1963); through *CA*, **60**, 15900 (1964); *Indian J. Chem.*, **3**, 535 (1965); *Indian J. Appl. Chem,.* **27**, 116 (1964); through *CA*, **62**, 4359 (1965).
110. T. Tanaka, *Inorg. Chim. Acta*, **1**, 217 (1967).
111. T. Tanaka et al., to be published.
112. T. Tanaka and T. Kamitani, *Inorg. Chim. Acta*, **2**, 175 (1968).
113. T. Tanaka, M. Komura, Y. Kawasaki, and R. Okawara, *J. Organometal. Chem.*, **1** 484 (1964).
114. G. Tanberger and O. R. Klimmer, *Arch. expt. Pathol. Pharmakol.*, **242**, 370 (1961); through *CA*, **56**, 16099 (1962).
115. R. S. Tobias, I. Orgins, and B. A. Nevett, *Inorg. Chem.*, **1**, 638 (1962).
116. T. T. Tsai, A. Culter, and W. L. Lehn, *J. Org. Chem.*, **30**, 3049 (1965).
117. R. Ueeda, Y. Kawasaki, T. Tanaka, and R. Okawara, *J. Organometal. Chem.*, **5**, 194 (1966).
118. G. J. M. van der Kerk and J. G. A. Luijten, *J. Appl. Chem.*, **6**, 93 (1956).
119. G. J. M. van der Kerk and J. G. Noltes, *J. Appl. Chem.*, **9**, 113 (1959).
120. M. Vilarem and J. C. Maire, *Compt. Rend. Ser. C*, **262**, 480 (1966).
121. N. S. Vyanzankin, G. A. Razuvaev, O. S. D'yachkovskaya, and O. A. Shchepetkova, *Dokl. Akad. Nauk S.S.S.R.*, **143**, 1348 (1962).
122. N. N. Vyshiniskii and N. K. Rudnevskii, *Opt. Spectra*, **10**, 421 (1961).

123. M. Wada, K. Kawakami, and R. Okawara, *J. Organometal. Chem.*, **4**, 159 (1965).
124. M. Wada, M. Shindo, and R. Okawara, *J. Organometal. Chem.*, **1**, 95 (1963).
125. A. H. Westlake and D. F. Martin, *J. Inorg. Nucl. Chem.*, **27**, 1579 (1965).
126. E. Wiberg and H. Behringer, *Z. Anorg. Allgem. Chem.*, **329**, 290 (1964).
127. C. J. Wilkins and H. M. Haendler, *J. Chem. Soc.*, 3174 (1965).
128. K. Yasuda and R. Okawara, *J. Organometal. Chem.*, **3**, 76 (1965).
129. K. Yasuda, H. Matsumoto, and R. Okawara, *J. Organometal. Chem.*, **6**, 528 (1966).
130. M. Yokoo, J. Ogura, and T. Kanzawa, *Polymer Letters*, **5**, 57 (1967).
131. N. N. Zemlyanskii, E. M. Panov, and K. A. Kocheshkov, *Zhur. Obshchei. Khim.*, **32**, 291 (1962).
132. N. N. Zemlyanskii, E. M. Panov, N. A. Slovokhotova, O. P. Shamagina, and K. A. Kocheshkov, *Dokl. Akad. Nauk S.S.S.R.*, **149**, 312 (1963).
133. N. N. Zemlyanskii, I. P. Gol'dstein, E. N. Gur'yanova, O. P. Syutkina, E. M. Panov, N. A. Slovokhotova, and K. A. Kocheshkov, *Izvest. Akad. Nauk S.S.S.R. Ser. Khim.*, 728 (1967).
134. S. M. Zhivukhin, E. D. Dudikova, and E. M. Ter-Sarkisyan, *Zhur. Obshchei. Khim.*, **32**, 3059 (1962).
135. H. Zimmer, O. A. Homberg, and M. Jayawant, *J. Org. Chem.*, **31**, 3857 (1966).

6. ORGANOTIN COMPOUNDS WITH Sn—S, Sn—Se, AND Sn—Te BONDS

HERBERT SCHUMANN, INGEBORG SCHUMANN-RUIDISCH,

Institute for Inorganic and Analytical Chemistry of the Technical University of Berlin Berlin, Germany

AND MAX SCHMIDT

Institute for Inorganic Chemistry of the University of Würzburg Würzburg, Germany

I. Introduction

In 1860 Kulmitz (*115*), in the course of investigations of "stannethyl," obtained "stannethyl sulfide" by the reaction of triethyltin bromide with hydrogen sulfide. This first organotin compound containing a covalent tin-sulfur bond was the basis for the class of organotin sulfide compounds, which are continually gaining significance at the present time. Organotin compounds with covalent tin-selenium and tin-tellurium bonds, on the other hand, were first mentioned by Backer (*27*) in 1942 and Weinberg (*P122*) in 1957, respectively. These organotin chalcogens like all other organotin compounds had a shadowy existence for a long time. Only when the specific technical significance of organotin sulfides became known around 1950 did this area of organometallic chemistry become a center of interest and lead to an enormous increase in the relevant technical and patent literature.

In response to a rapidly growing need for the systematic consideration of synthesis, and properties of these organotin compounds, several monographs and review articles appeared (*1, 13, 46, 57a, 64, 77, 81, 87a, 91a, 93, 95, 98, 100a, 107, 110a, 119, 126a, 127, 128a, 131, 137, 155, 161c, 172, 184a, 185, 186, 187, 189, 190, 198a, 203, 216a, 227e, 227f, 227h, 233*), some of which were specifically directed toward the preparation of compounds of technical interest (*90, 99a, 104a, 125, 144a, 169, 170, 175c, 222a,b, 223a, 224*). In this connection, one is referred to the monograph covering the organic chemistry of tin appearing in 1967 (*137*), as well as to the extensive article on "Organosulfur Derivatives of Silicon, Germanium, Tin and Lead" (*1*).

In this monograph, the accumulated past literature covering the whole area of specialization of organotin chalcogens is reviewed, and for the sake of coherency, and in view of the fact that the major part of this work was carried out after 1960, the article by Ingham et al. (*93*) was not selected as the starting point.

This article is limited to a review of covalent organometallic compounds of tetravalent tin with sulfur, selenium, and tellurium. Tin(II) compounds, thiocyanates, isothiocyanates, and organotin esters of oxyacids of sulfur and selenium are not reviewed. Likewise, publications and patents which only deal with analytical problems or with technical and biological applications but do not give detailed information concerning the synthesis and properties of compounds are omitted.

II. Organotin Sulfides

A. Preparation

As expected, organotin sulfides exhibit many similarities to their oxygen analogs. Thus there are certain parallels in preparative methods for these two classes of compounds. However, the great technical interest in organotin sulfides led to the development of a large number of new synthetic routes specifically directed toward the preparation of these compounds; those with relatively broad applicability will be mentioned briefly here. Individual examples are cited in the following sections.

Organotin halides are the most important starting materials in the preparation of many types of organotin sulfur compounds. They react with many compounds bearing an SH-function; for example, with hydrogen sulfide, mercaptans, dithiols, mercaptoesters, etc., in organic solvents or even in water, alone or in the presence of hydrogen halide acceptors, with the removal of hydrogen halide and formation of the corresponding organotin sulfides:

$$R_{(4-n)}SnX_n + n\,RSH \longrightarrow R_{(4-n)}Sn(SR)_n + n\,HX$$

With alkali thiolates or sodium sulfide, organotin halides are converted to organotin sulfides with elimination of alkali halides:

$$R_{(4-n)}SnX_n + (n/2)\,Na_2S \longrightarrow [R_{(4-n)}Sn]S_{n/2} + n\,NaX$$

In a similar manner, organotin hydroxides, oxides, or esters are able to react with mercaptans or thioacids. Organotin alkoxides or oxides form organotin thiocarbonates or bis(triorganotin) sulfides by reaction with carbon disulfide or thiourea derivatives:

$$R_3SnOR' + CS_2 \longrightarrow R_3SnSC(S)OR'$$

$$R_3SnOSnR_3 + (R_2N)_2C{=}S \longrightarrow (R_3Sn)_2S + (R_2N)_2C{=}O$$

Of comparable value are the reactions of organotin-nitrogen or -phosphorus compounds with hydrogen sulfide, thiols, or carbon disulfide, which proceed with cleavage of the tin-nitrogen or tin-phosphorus bond and formation of organotin mercaptides, sulfides, or thiocarbamates.

The tin-carbon bond in various organotin compounds can also be attacked under certain conditions by mercaptans, thioacids, hydrogen sulfide, or at higher temperatures by elemental sulfur, with the formation of monomeric or polymeric organotin sulfides.

$$R_4Sn + R'SH \longrightarrow R_3SnSR' + RH$$

$$4\,R_4Sn + S_8 \longrightarrow \tfrac{4}{3}\,(R_2SnS)_3 + 4\,R_2S$$

Hexaorganoditin compounds, as well as polymeric organotin compounds add sulfur and thus are converted to bis(triorganotin) sulfides, or trimeric, or polymeric diorganotin sulfides:

$$4\,R_6Sn_2 + \tfrac{1}{2}\,S_8 \longrightarrow 4\,R_3SnSSnR_3$$

$$(R_2Sn)_n + S_n \longrightarrow n/3\,(R_2SnS)_3$$

Alkali derivatives of organotin compounds and organotin hydrides are also important starting materials for the synthesis of organotin sulfides. The former cleave the sulfur-sulfur bond in disulfides or elemental sulfur with formation of organotin mercaptides and alkali organotin sulfides; the latter add to polar double bonds such as those in isothiocyantes, thiourea, or carbon disulfide.

$$8\,R_3SnLi + S_8 \longrightarrow 8\,R_3SnSLi$$

$$R_3SnH + C_6H_5NCS \longrightarrow R_3SnSCH{=}NC_6H_5$$

Alkali organotin sulfides can further be coupled with many halogen-containing compounds, with elimination of alkali halide, to give new organotin sulfides:

$$R_3SnSLi + ClMR'_n \longrightarrow R_3SnSMR'_n + LiCl$$

The majority of known organotin sulfides were prepared by one of the methods briefly mentioned above. Although the compounds known up until 1950 were obtained almost exclusively by the simplest method, namely by thiolysis of tin-halogen or tin-oxygen compounds, the technical interest in organotin sulfides in the subsequent years led to a wealth of investigations concerning their synthesis. Consequently, a large number of organic sulfur compounds were found useful as starting materials. These investigations, which were carried on, partly out of technical interest and partly out of purely scientific interest, are continuing at a rapid rate. It can be expected that, in the course of the next few years, a variety of new methods of synthesis for such compounds will be evolved. In subsequent sections, individual methods for the synthesis of specifically mentioned organotin sulfides will be described in detail.

B. Chemical Properties and Reactions

All organotin sulfides are, under normal conditions, liquid or crystalline compounds, which, with the exception of high polymers, usually dissolve easily or moderately in organic solvents without decomposition.

Their thermal stability varies widely and strongly depends on the nature of the organic substituents bound to tin. Generally, tetraalkyltins are light-sensitive and polymerize in sunlight.

The great majority of organotin sulfides are stable toward oxidation. Only bis(trimethyltin) sulfide and alkali organotin sulfides are oxidized by atmospheric oxygen.

No investigations concerning hydration or reduction reactions have appeared thus far.

Organotin sulfides are, in general, more stable than the corresponding oxygen derivatives, a fact which is explained by the decreased difference in size between tin and the group VI atom and thus by an increased overlapping of the bonding orbitals. Even at low temperatures, however, the tin-sulfur bond is cleaved by mineral acids.

Covalent halogen compounds effect cleavage of the tin-sulfur bond in organotin sulfides. Reactions of this type have been investigated exhaustively in the past few years because they allow the synthesis of elemental sulfur compounds which could not be obtained at all or, if so, only unsatisfactorily, by other means. Thus, for example, one can obtain boron sulfide from organotin sulfides and boron trichloride. The reaction with alkyl halides proceeds with the breaking of the tin-sulfur bond and formation of organotin halides and alkyl sulfonium halides. Organotin sulfonium halides cannot be isolated. Tetrahalides and organometallic halides of silicon, germanium, tin, and lead attack organotin sulfides with the formation of organotin halides and the corresponding silicon, germanium, tin, and lead-sulfur compounds. Phosphorus trichloride, arsenic trichloride, and antimony trichloride also react with organotin mercaptides or bis(triorganotin) sulfides to give phosphorus, arsenic, or antimony mercaptides which could not be prepared by other methods. At the same time organotin halides and disulfides are formed.

The tin-sulfur bond is also attacked by transition metals. Detailed investigations were carried out on displacement reactions between organotin sulfides and silver halides, silver cyanide, cyanate, and isothiocyanate. Mercuric chloride and organotin mercaptides undergo an exchange to organotin chlorides and mercuric mercaptides; *N*-mercuric carbamate and bis(trialkyltin) sulfides form *N*-stannylcarbamates and mercuric sulfide. Hexameric nickel dimercaptide can be isolated as the product of the reaction between diorganotin dimercaptides and nickel chloride.

Recently, it was discovered that organotin sulfides, especially diorganotin mercaptides, could serve as ligand sources in the synthesis of interesting polydentate-transition metal sulfur complexes. Thus, for example, chromium or manganese carbonyls could be converted into thiocarbonyl complexes.

The coordination number of four, for tin in covalent organotin sulfides, can be increased to five or six by complex formation with various bases, particularly those containing nitrogen. Such complexes, which were investigated with tin- and sulfur-containing heterocycles in particular, proved to be highly temperature-sensitive. Specific chemical properties and reactions of individual members of classes of compounds described in the following sections will be given in detail at the appropriate place.

C. Physical Properties

Although organotin sulfides have long been known, very little investigation has been carried out on their physical properties. Only the great interest in an understanding of the bonding theory of covalent metal bonds in organometallic compounds provided an impetus for such research. Because the question of bonding is by no means, as yet, answered, it is to be expected that investigations in this area will increase significantly in the coming years.

In the following section, only the major physical properties of those already investigated for organotin sulfur compounds will be mentioned. Detailed accounts are to be found in the individual sections.

The most reliable information concerning the structure and bonding properties of organotin sulfides would come from x-ray structure analysis. Apart from some information on the dimensions of the unit cells in tin tetramercaptides, there are only two complete x-ray structure analyses, those for trimeric diphenyltin sulfide and for triphenylstannyl triphenylplumbyl sulfide. It was found, in these investigations, that the distance of 2.42 Å between tin and sulfur corresponds exactly to the sum of the covalent radii of these two atoms, that the tin angle of 112° deviates only slightly from the tetrahedral angle, and that the bond angle of 104°, formed between ligands and sulfur, corresponds to the known value for organic sulfides.

Calculations carried out from the infrared and Raman spectra lead to the same result.

A large number of investigations of the vibrational spectra of organotin sulfides lead to the conclusion that the tin-sulfur stretching vibration is found in the region 300–400 cm^{-1}. From band shifts, it was established in some cases that intermolecular coordination through formation of five-coordinate tin was taking place.

Several ^{1}H-nmr, ^{119}Sn-nmr, Mössbauer, and ultraviolet spectra have been carried out. For example, the ^{1}H-nmr spectra of compounds of the type $(CH_3)_3M—S—M'(CH_3)_3$ (M and M′ = Si, Ge, Sn, Pb) have been interpreted to mean that the degree of π-bonding in the metal-sulfur bond falls off sharply from silicon to lead, and that there is scarcely any inclusion of the free sulfur electron pairs into the tin-sulfur bonds.

In contrast to this, surprisingly low values for the atomic susceptibility of tin are derived from magnetic susceptibility measurements of various organotin sulfides, which indicate a definite double bond character for the tin-sulfur bond.

D. Individual Classes of Compounds

1. *Compounds of the Type* $R_nSn(SR')_{4-n}$

a. Synthesis. The simplest method for preparing organotin mercaptides of the general formula $R_nSn(SR')_{4-n}$ is the reaction of organotin halides or tin tetrachloride with mercaptans. This was employed for the first time in 1877 by Claesson for the synthesis of tin tetra(ethylmercaptide) (*45*):

$$SnCl_4 + 4\,C_2H_5SH \longrightarrow Sn(SC_2H_5)_4 + 4\,HCl$$

As shown in the following equations, this reaction can be extended to organotin halides.

$$R_3SnX + R'SH \longrightarrow R_3SnSR' + HX$$

$$R_2SnX_2 + 2\,R'SH \longrightarrow R_2Sn(SR')_2 + 2\,HX$$

$$RSnX_3 + 3\,R'SH \longrightarrow RSn(SR')_3 + 3\,HX$$

This is usually carried out in the presence of amines, sodium hydroxide, or sodium carbonate as HX-acceptor, and leads to the synthesis of a great number of aromatic, aliphatic, and heterocyclic substituted triorganotin mercaptides (*7, 32, 56, 107a, 123a, 128b, 144b, 180, 191, 214a, 219, P13, P19, P31, P59, P82, P83, P85, P89e, P110, P136, P140, P152, P158k,*), diorganotin dimercaptides (*7, 32, 56, 128b, 134a, 191, 219, P12, P13, P15, P19, P59, P60, P70, P76g, P83, P110, P111, P121, P136, P137, P152*), organotin trimercaptides (*7, 128b, 219, P13, P19, P59, P110, P136, P152*), and tin tetramercaptides (*28, 128, 128b, 145b, 214e, P5, P19, P59, P110*). As particularly described in the patent literature (*128b, P53, P58, P65, P90*), the removal of liberated hydrogen halide can be simplified by prior synthesis of organotin halide-amine complexes:

$$(C_4H_9)_2SnCl_2\cdot 2\,NH_2C_6H_5 + 2\,C_{12}H_{25}SH \longrightarrow (C_4H_9)_2Sn(SC_{12}H_{25})_2 + 2\,C_6H_5NH_3Cl$$

If, in this reaction, one replaces the organotin halides with the corresponding oxygen compounds (*7, 47, 55, 117a, 123a, 128, 145a, 175–177, 215, P1, P6, P64, P66, P67, P68, P69, P76e, P92, P96, P100, P102, P108, P109, P112–P115, P121, P124, P125, P131, P137, P140, P142, P145, P152–P154, P158b, P159*), or uses mercaptides instead of mercaptans (*28, 29, 30, 30a, 99, 118, 161, 215, P6, P21, P95, P95b, P140*), one usually obtains better yields, just as in the direct reaction of organotin oxygen compounds with mercaptides (*24*). This latter procedure is also much more widely useful:

$$R_3SnOH + R'SH \longrightarrow R_3SnSR' + H_2O$$

$$(R_3Sn)_2O + 2\,R'SH \longrightarrow 2\,R_3SnSR' + H_2O$$

$$R_2SnO + 2\,R'SH \longrightarrow R_2Sn(SR')_2 + H_2O$$

$$RSnOOH + 3\,R'SH \longrightarrow RSn(SR')_3 + 2\,H_2O$$

$$R_2SnCl_2 + 2\,NaSR' \longrightarrow R_2Sn(SR')_2 + 2\,NaCl$$

$$SnCl_4 + 4\,NaSR' \longrightarrow Sn(SR')_4 + 4\,NaCl$$

$$(R_3Sn)_2O + 2\,AgSR' \longrightarrow 2\,R_3SnSR' + Ag_2O$$

All these reactions lead to the corresponding organotin mercaptides, usually in high yield, either at room temperature or with heating. Solvents employed are inert hydrocarbons, benzene, alcohols, ether, amines, and in many cases water or one of the reactants itself.

Procedures for the synthesis of organotin mercaptides have been discovered in the past few years which allow organotin amines to react smoothly with mercaptans (*3*), with elimination of the easily removed amine, and formation of organotin sulfur compounds:

$$(CH_3)_3SnN(C_2H_5)_2 + C_4H_9SH \longrightarrow (CH_3)_3SnSC_4H_9 + (C_2H_5)_2NH$$

Organotin thioalkali compounds react with alkyl-, aryl-, and alkoxy halides (*P34*) and cleave disulfides with the formation of organotin mercaptides and sodium mercaptides (*32*, *43*). This reaction can also be extended to organotin phosphines (*43*):

$$(C_6H_5)_3SnNa + C_6H_5SSC_6H_5 \longrightarrow (C_6H_5)_3SnSC_6H_5 + C_6H_5SNa$$

$$[(CH_3)_3Sn]_3P + 2\,C_2H_5SSC_2H_5 \longrightarrow (CH_3)_3SnP(SC_2H_5)_2 + 2\,(CH_3)_3SnSC_2H_5$$

The organotin hydrides, recently investigated in great detail, react with mercaptans (*48*, *118*, *145*, *151*), azothioethers (*138a*), phenylsulfenyl chloride (*145*) and diphenyl disulfide (*145*) to give organotin mercaptides:

$$(C_6H_5)_3SnH + C_{10}H_7SH \longrightarrow (C_6H_5)_3SnS\text{-}C_{10}H_7 + H_2$$

$$2\,(C_6H_5)_3SnH + C_6H_5SCl \longrightarrow (C_6H_5)_3SnSC_6H_5 + H_2 + (C_6H_5)_3SnCl$$

$$(C_6H_5)_3SnH + C_6H_5SSC_6H_5 \longrightarrow (C_6H_5)_3SnSC_6H_5 + C_6H_5SH$$

Tin-carbon bonds of various organotin compounds are cleaved under certain conditions by mercaptans (*56*, *178*, *179*, *182*) and even by elemental sulfur (*191*, *202*), with the formation of organotin mercaptides:

$$(C_2H_5)_4Sn + C_6H_5SH \longrightarrow (C_2H_5)_3SnSC_6H_5 + C_2H_6$$

$$8\,(C_4H_9)_3SnCl + S_8 \longrightarrow 8\,(C_4H_9)_2SnCl(SC_4H_9)$$

Dibutyltin chloride butylmercaptide is also formed by disproportionation of dibutyltin bis(butylmercaptide) and dibutyltin dichloride (*58*). The preparation of compounds, which are difficult to obtain in other ways, is effected, in many cases, by so-called "transmercaptation" of organotin mercaptides with other mercaptans (*3*, *56*, *128*, *177*, *215*) of organotin halides with organosilicon- or organolead mercaptides (*3*, *56*), or from the decomposition of dibutyltin sulfide with alcohols (*P103–P106*):

$$(CH_3)_3SnSC_4H_9 + C_6F_5SH \longrightarrow (CH_3)_3SnSC_6F_5 + C_4H_9SH$$

$$(CH_3)_2SnCl_2 + Pb(SCH_3)_2 \longrightarrow (CH_3)_2Sn(SCH_3)_2 + PbCl_2$$

As unusual reactions can be mentioned the preparation of diphenyltin bis-benzylmercaptide from diphenyltin and dibenzyl disulfide (*113*) or from $(CH_3)_2Sn(SNa)_2$ and $C_6H_5CH_2Cl$ (*P38f*), the insertion of thiophosgene or thiourea into the tin-phosphorus bond of triphenyltin diphenylphosphine (*199*, *201*), the reaction of tin(II) chloride with mercaptans (*158*, *233*), the fission of the Sn—As bond in $(CH_3)_3Sn{-}As(CH_3)_2$ by mercaptans (*176*), the reaction of $(CH_3)_3Sn{-}P(C_6H_5)_2$ with $(C_6H_5)_2BSC_4H_9$ (*79b*), and the cleavage of organic disulfides with tin(II) chloride (*233*), powdered tin (*105a*) and triethyl triphenyl ditin (*48*):

$$(C_6H_5)_2Sn + C_6H_5CH_2SSCH_2C_6H_5 \longrightarrow (C_6H_5)_2Sn(SCH_2C_6H_5)_2$$

$$(C_6H_5)_3SnP(C_6H_5)_2 + (NH_2)_2C{=}S \longrightarrow (C_6H_5)_3SnSC(NH_2)_2P(C_6H_5)_2$$

$$4\,C_{10}H_7SH + SnCl_2 + \tfrac{1}{2}\,O_2 \longrightarrow (C_{10}H_7S)_4Sn + 2\,HCl + H_2O$$

$$Sn + 2\,HCl \longrightarrow SnCl_2 + H_2$$

$$H_2 + RSSR \longrightarrow 2\,RSH$$

$$2\,RSH + RSSR + SnCl_2 \longrightarrow (RS)_4Sn + 2\,HCl$$

$$2\,C_6F_5S{-}SC_6F_5 + Sn \longrightarrow Sn(SC_6F_5)_4$$

$$(C_2H_5)_3Sn{-}Sn(C_6H_5)_3 + C_6H_5S{-}SC_6H_5 \longrightarrow (C_2H_5)_3SnSC_6H_5 + (C_6H_5)_3SnSC_6H_5$$

Organotin mercaptides could be found also by free-radical reactions between mercaptans and 1,3-enyne systems (*215b*).

The following interesting organotin dimercaptide complexes might be cited as particular cases in this class of compounds. They could be synthesized by the reaction of $[\pi C_5H_5Fe(CO)_2]_2SnCl_2$, $[\pi C_5H_5(CO)_3Mo]_2SnCl_2$, $[\pi C_5H_5(CO)_3W]_2SnCl_2$, and $\pi C_5H_5(CO)_3Mo-SnCl_3$ with ethylmercaptan (*24a, 135, 136a*), of $Mn(CO)_5Br$ and $Cr(CO)_6$ with $(CH_3)_2Sn(SCH_3)_2$ (*14*), or of $Cr(CO)_6$ with $(CH_3)_3SnSC_6H_5$ (*181b*):

```
            SC2H5                      CH3                          CH3
     OC       |      CO          OC Br  |                         CO |
       \      |     /              \ |  S      CH3            OC   \ S      CH3
 (C5H5)-Fe - Sn - Fe-(C5H5) ;        Mn     Sn          ;       Cr     Sn         ;
       /      |     \              / |  S      CH3            OC   / S      CH3
     OC       |      CO          OC CO  |                         CO |
            SC2H5                      CH3                          CH3

                               Sn(CH3)
                                  |
                    (CO)5Cr <---- S
                                  |
                                 C6H5
```

b. Chemical Properties and Reactions. Triorganotin mercaptides and diorganotin dimercaptides are liquid or crystalline under normal conditions and, in general, can be distilled without decomposition. Compounds of the type $RSn(SR')_3$ and $Sn(SR')_4$ are usually not distillable, are liquids or solids, and are monomeric in solution. Their thermal stability widely varies and depends strongly on the nature of the substituents R and R′. While, for example, diphenyltin bis-thiophenoxide (*191*) and dibutyltin chloride butylmercaptide (*202*) decompose readily above 110°C to give off diphenylsulfide and dibutylsulfide, respectively:

$$3\,(C_6H_5)_2Sn(SC_6H_5)_2 \longrightarrow 3\,(C_6H_5)_2S + [(C_6H_5)_2SnS]_3$$

$$2\,Cl{-}\overset{\displaystyle C_4H_9}{\underset{\displaystyle C_4H_9}{Sn}}{-}S{-}C_4H_9 \longrightarrow Cl{-}\overset{\displaystyle C_4H_9}{\underset{\displaystyle C_4H_9}{Sn}}-S-\overset{\displaystyle C_4H_9}{\underset{\displaystyle C_4H_9}{Sn}}{-}Cl + (C_4H_9)_2S$$

trimethyltin methylmercaptide decomposes only at 270°C with formation of tetramethyltin, dimethylsulfide, low-boiling hydrocarbons, and tin sulfide (*7*).

All organotin-sulfur compounds in the class being discussed here are stable toward oxygen. No reports of reduction reactions are available.

In general, one states that the tin-sulfur bond is more stable than the tin-oxygen bond, so that it is not usually possible to convert organotin mercaptides to their corresponding oxygen derivatives by hydrolysis or alcoholysis. In special cases, however, it is possible to convert organotin mercaptides into organotin alkoxides by the addition of a catalyst such as *p*-toluenesulfonic acid (*215*):

$$(C_4H_9)_2Sn(SC_3H_7)_2 + 2\,C_2H_5OH \longrightarrow (C_4H_9)_2Sn(OC_2H_5)_2 + 2\,C_3H_7SH$$

The substitution of the mercaptide group proceeds somewhat more easily with acids (*177*), lactones (*93b*), or salts of oxyacids (*22*), as is shown by the reaction of triethyltin isopropylmercaptide with acetic acid, which leads to triethyltin acetate (*177*). For reactions of organotin mercaptides with other mercaptans, which lead to various substituted organotin mercaptides (*3*, *177*, *215*), refer to Section II.D.la. The tin-sulfur bond is particularly easily cleaved by a variety of halogen compounds, a reaction which is attractive for the synthesis of new, simple sulfur compounds. Thus organotin mercaptides react with alkyl halides (*233*), $SnCl_2$ (*158*), organotin halides (*58*), BCl_3 (*9*), PCl_3 (*9*), $AsCl_3$ (*9*), $HgCl_2$ (*9*, *158*), $NiCl_2$ (*11*, *232*), $HgBr_2$ (*9*), HgI_2 (*9*), $CdCl_2$ (*9*), $CdBr_2$ (*9*), and K_2PtCl_4 (*9*), with formation of organotin halides and the corresponding mercaptides:

$$[(CH_3)_2NC_6H_4S]_4Sn + 8\,CH_3I \longrightarrow SnI_4 + 4\,[(CH_3)_3\overset{\oplus}{N}C_6H_4SCH_3]I^{\ominus}$$

$$3\,(CH_3)_3SnSC_4H_9 + BCl_3 \longrightarrow B(SC_4H_9)_3 + 3\,(CH_3)_3SnCl$$

Bromine and iodine cleave trimethyltin thiophenolate, with the formation of trimethyltin halide and diphenyldisulfide (*7*, *145a*), a reaction which can be used for the quantitative determination of sulfur in organotin mercaptides.

The reaction between certain organotin mercaptides and organotin hydrides results exclusively in an exchange of functional groups (*49*, *50*):

$$C_2H_5)_3SnSC_4H_9 + (C_6H_5)_3SnH \longrightarrow (C_2H_5)_3SnH + (C_6H_5)_3SnSC_4H_9$$

Aldehydes like Cl_3CCHO insert into the Sn—S— bond of organotin mercaptides with the formation of compounds, which split off organotin halides at low temperatures (*92e*, *94a*):

$$(CH_3)_3SnSCH_3 + Cl_3CCHO \rightleftharpoons (CH_3)_3SnOCH(CCl_3)SCH_3$$

$$(CH_3)_3SnOCH(CCl_3)SCH_3 \xrightarrow{+2\,(CH_3)_3SnSCH_3} 3\,(CH_3)_3SnCl + (CH_3S)_3CCHO$$

$(CH_3)_3SnSCH_3$ reacts with 1,1-bis(trihalogeno-methyl)-2,2-dicyanoethylenes under 1,4-insertion (*17a*):

$$(CH_3)_3SnSCH_3 + (CF_3)_2C = C(CN)_2 \longrightarrow (CF_3)_2C(SCH_3) - C(CN) = C = N - Sn(CH_3)_3$$

Organotin sulfur compounds of the type R_3SnSAr react under mild conditions with sulfenyl halides in aprotic solvents to give triorganotin halides and diphenyldisulfides in high yields (*227i*):

$$R_3SnSC_6H_5 + C_6H_5SBr \longrightarrow C_6H_5S - SC_6H_5 + R_3SnBr$$

The reaction between $(CH_3)_3SnSR$ and C_6H_5NCO invariably gave 80–90% 1,3,5-triphenyltriazine-2,4,6-trione (*92c*).

Particularly remarkable reactions are that of triphenyltin thiophenolate with sulfur, which leads to trimeric diphenyltin sulfide by cleavage of a tin-carbon bond (*191*) and that of trimethyltin methylmercaptide with sulfur which forms bis(trimethyltin) sulfide (*7*):

$$8\,(C_6H_5)_3SnSC_6H_5 + S_8 \longrightarrow \tfrac{8}{3}\,[(C_6H_5)_2SnS]_3 + 8\,(C_6H_5)_2S$$

$$2\,(CH_3)_3SnSCH_3 + S_x \longrightarrow (CH_3)_3Sn—S—Sn(CH_3)_3 + (CH_3)_2S_{x+1}$$

The interesting synthesis of sulfur-bridged transition metal complexes must also be mentioned (*6*, *11*, *12*, *14*, *17*, *96a*, *170a*, *232*):

$$3\,(CO)_5MBr + 3\,(CH_3)_2Sn(SCH_3)_2 \longrightarrow [(CO)_3M(SCH_3)]_3 + 6\,CO + 3\,(CH_3)_2SnBr(SCH_3)$$

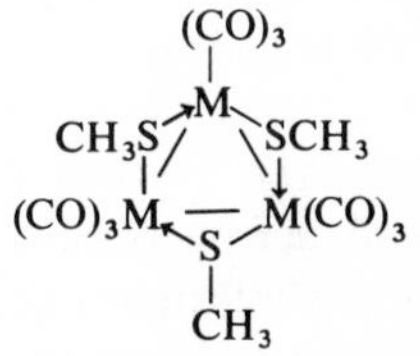

(M = Mn, Re)

$$6\,NiCl_2 + 6\,(CH_3)_2Sn(SC_2H_5)_2 \longrightarrow [Ni(SC_2H_5)_2]_6 + 6\,(CH_3)_2SnCl_2$$

Of special significance is the capture reaction between organotin mercaptides and hydrogen chloride; it is the basis for the use of such organotin mercaptides as heat stabilizers for PVC (*83e*, *90b*, *99b*, *133*, *227k*). The analytical determination of sulfur in such stabilizers or in other organotin sulfides can be determined by thiomercurimetrical titration (*44a*).

c. Physical Properties. All organotin mercaptides prepared to date are listed in Appendix 1. As far as there are available values in the literature, boiling points, melting points, refractive index, and density are included. Information on such values is lacking for most compounds described in the patent literature. For the sake of completeness those compounds are also listed in Appendix 1.

Although the crystal type and unit cell parameters of various tin tetramercaptides were determined as early as 1933 (*28–30*, *99*), there have been no detailed x-ray structure analyses of organotin mercaptides to date.

The infrared spectra of some organotin mercaptides have been described and interpreted (*7*, *52*, *83a*, *83b*, *144b*, *145a*, *b*, *152*, *197*, *221a*). The Sn$-$S stretching vibration in $(C_6H_5)_3SnSC_6H_5$ was found at 348 cm^{-1} (*197*). The observed shifts in the C$-$D stretching vibration in $CDCl_3$ solutions of trimethyltin mercaptides allow conclusions to be drawn concerning the basicity of these compounds relative to their silicon germanium and lead analogs and the corresponding alcoholates and amines (*2*) respectively. Basicity increases steadily in the order: Si, Ge, Sn, Pb. A comparison of the C$-$H and C$-$D shifts in the nmr spectra of solutions of the same compounds in chloroform and deuterochloroform leads to similar conclusions (*5*). The chemical shifts as well as the constants of ^1H coupling with the isotopes ^{119}Sn, ^{117}Sn, and ^{13}C of all four methyltin methylmercaptides, by comparison with the corresponding values for tetramethyltin (Appendix 2), show that progressive substitution of methyl groups by more electronegative CH_3S-groups goes along with a decrease in shielding of the remaining CH_3—Sn protons. The changes in coupling constants are interpreted as an increase in $(p \rightarrow d)\pi$ double bond character of the tin-sulfur bond in going from trimethyltin methyl-mercaptide to tetra(methylthio)tin (*8*, *219*). Other ^1H- and ^{19}F-nmr measurements have been made on some organotin pentafluorothiophenolates (*90a*, *144b*, *145a*).

Molar refractivities have been determined for a great number of organotin mercaptides (*181*). Refractivity for the tin-sulfur bond is 7.631.

For Mössbauer measurements see (*83c*) and (*135a*). The relative values of the quadrupole splittings obtained for dibutyl- and diphenyltin complexes of 2-pyridinethiol-1-oxide with 3.20 and 1.45 mm/sec, respectively, indicate a trans arrangement of the butyl groups and a cis arrangement of the phenyl groups. The dipole moment, 5.0 *D*, of the dibutyl compound in benzene is indicative of cis arrangement of the oxygen and sulfur atoms (*145c*).

Diamagnetic susceptibility values for dimethyltin bis(*n*-alkylmercaptides) and trimethyltin *n*-alkylmercaptides were determined by Abel et al. (*10*). The K_M-values assembled in Table 1 increase with increasing length of the alkyl group attached to sulfur.

TABLE 1

MAGNETIC SUSCEPTIBILITIES FOR METHYLTIN ALKYLMERCAPTIDES

Compound	K_M
$(CH_3)_3SnSC_2H_5$	129.6
$(CH_3)_3SnS$-*n*-C_3H_7	140.1
$(CH_3)_3SnS$-*n*-C_4H_9	151.6
$(CH_3)_3SnS$-*n*-C_8H_{17}	195.6
$(CH_3)_2Sn(SCH_3)_2$	136.1
$(CH_3)_2Sn(SC_2H_5)_2$	158.4
$(CH_3)_2Sn(S$-*n*-$C_3H_7)_2$	180.3
$(CH_3)_2Sn(S$-*n*-$C_4H_9)_2$	203.3
$(CH_3)_2Sn(S$-*n*-$C_8H_{17})_2$	295.9

Gas chromatographic investigations have been reported for trimethyltin ethylmercaptide (*151*).

From uv-measurements on some organotin mercaptides, $(p \rightarrow d)\pi$-interactions between Sn and S could be established (*83a, 83b*).

2. *Organotin Mercaptoesters*

A brief section about organotin mercaptoesters is justified by the large number of compounds of this class, described mainly in the patent literature. It should be noted that practically no mention is made in the patents concerning the physical properties of these compounds but the possibility of their uses is pointed out. This subject will be discussed in greater detail in a special chapter (Applications and Biological Effects of Organotin Compounds, Vol. 3).

Organotin mercaptoesters can generally be prepared by the treatment of organotin halides, oxides, or sulfides with mercaptoesters in solvents such as benzene, alcohols, amines, or water (*130, 166, 183, 222, P2, P4, P9, P10, P15–P17, P19, P20, P27, P28, P33, P35, P37, P38, P38a, P38d,f, P39, P40–P43, P45, P46, P50, P58a–d,f, P59a, P65, P67, P69, P76a, P76c,d,f, P78, P89, P89a, P89b, P91–P93, P95a, P95b, P98, P100–P107, P116, P121, P123, P127–P129,*

P132–P135, P138–P140, P142, P152, P160). This is illustrated with the following examples:

$$(C_4H_9)_3SnCl + HSCH_2C(O)OC_4H_9 \longrightarrow (C_4H_9)_3SnSCH_2C(O)OC_4H_9 + HCl$$

$$(C_8H_{17})_3SnOC_4H_9 + HSCH_2C(O)OC_4H_9 \longrightarrow (C_8H_{17})_3SnSCH_2C(O)OC_4H_9 + C_4H_9OH$$

$$(C_4H_9)_2SnCl_2 + 2\ HSCH\begin{matrix} \diagup C(O)OC_4H_9 \\ \diagdown CH_2C(O)OC_4H_9 \end{matrix} \longrightarrow (C_4H_9)_2Sn\left[-SCH\begin{matrix} \diagup C(O)OC_4H_9 \\ \diagdown CH_2C(O)OC_4H_9 \end{matrix}\right]_2 + 2\ HCl$$

$$[(NCCH_2CH_2)_2SnOH]_2O + 4\ HSCH_2C(O)OC_8H_{17} \longrightarrow 3\ H_2O + 2\ (NCCH_2CH_2)_2Sn[-SCH_2C(O)OC_8H_{17}]_2$$

$$C_4H_9SnOOH + 3\ HSCH_2CH_2C(O)OC_6H_{13} \longrightarrow 2\ H_2O + C_4H_9Sn[-SCH_2CH_2C(O)OC_6H_{13}]_3$$

A particularly advantageous method is the treatment of amine complexes of organotin halides with mercaptoesters in ether solution (*P53, P58, P65, P90*):

$$(C_4H_9)_2SnCl_2 \cdot 2\ NH_2C_6H_5 + 2\ HSCH_2C(O)OC_8H_{17} \longrightarrow 2\ C_6H_5NH_2 \cdot HCl + (C_4H_9)_2Sn[SCH_2C(O)OC_8H_{17}]_2$$

The amine hydrochloride by-product which precipitates can be easily removed by filtration or centrifugation.

Appendix 3 contains a large number of such organotin mercaptoesters with their physical constants, wherever these are available.

Radioactively labeled compounds of this type were used to study the mechanism of PVC-stabilization (*72–76*). Dibutyltin dithioglycolate and dioctyltin bis(thioglycolic acid octyl ester) were studied polarographically (*80*), and various other organotin mercaptoesters were examined by thin layer chromatography for the same purpose (*92a, 220, 221*) and in the viewpoint of their extractability (*40b, 44c, 211a, 232a*).

3. *Organotin Thiocarboxylates, Thiocarbonate Esters, and Related Compounds*

In this section all compounds containing the groups —Sn—S—C(O)R, —Sn—S—C(O)O—, —Sn—S—C(S)O—, —Sn—S—C(S)S—, —Sn—S—C(S)N, —Sn—S—C(S)P, —Sn—S—C(N)R, —Sn—S—C(N)O— and —Sn—S—C(N)S— will be discussed. Cyclic compounds of this type will be taken up in Section II.D.6.

a. Preparation. These types of compounds are most easily obtained by treatment of organotin halides or oxides with thioacids, thioacid esters, thiolates, xanthates, and similar reagents (*24b, 39, 39a, 56, 83, 117, 117b, 196a, P8, P11, P23–P26, P33, P38e, P44, P58e, P60, P64, P76, P76b, P95b, P97, P109, P112, P114, P115, P117, P119, P120, P148, P158, P158h*:

$$(C_4H_9)_3SnCl + C_6H_5C(O)SH + (C_2H_5)_3N \longrightarrow (C_4H_9)_3SnSC(O)C_6H_5 + (C_2H_5)_3N \cdot HCl$$

$$(C_4H_9)_2SnO + C_6H_5C(O)SH + C_{12}H_{25}SH \longrightarrow H_2O + (C_4H_9)_2Sn\begin{matrix} \diagup SC_{12}H_{25} \\ \diagdown SC(O)C_6H_5 \end{matrix}$$

$$(C_6H_5)_3SnCl + C_2H_5OC(O)SK \longrightarrow (C_6H_5)_3SnSC(O)OC_2H_5 + KCl$$

$$(C_4H_9)_2SnCl_2 + 2\ C_4H_9OC(S)SK \longrightarrow (C_4H_9)_2Sn[\text{—}SC(S)OC_4H_9]_2 + 2\ KCl$$

$$(C_6H_5)_3SnI + C_{12}H_{25}SC(S)SK \longrightarrow (C_6H_5)_3SnSC(S)SC_{12}H_{25} + KI$$

$$(C_2H_5)_3SnCl + (C_2H_5)_2NC(S)SNa \longrightarrow (C_2H_5)_3SnSC(S)N(C_2H_5)_2 + NaCl$$

$$SnCl_4 + 2\ (C_2H_5)_2NC(S)SNa \longrightarrow Cl_2Sn[\text{—}SC(S)N(C_2H_5)_2]_2 + 2\ NaCl$$

$$2\ (C_4H_9)_3SnCl + CaS_2C_2N_2 \longrightarrow (C_4H_9)_3SnS\underset{\underset{N\text{—}C\equiv N}{\|}}{C}SSn(C_4H_9)_3 + CaCl_2$$

Organotin thiocarboxylates are also obtainable by the treatment of lithium or sodium organotin sulfides with acid chlorides (*205, 206, 207, P38f*):

$$(C_6H_5)_3SnSLi + C_6H_5C(O)Cl \longrightarrow (C_6H_5)_3SnSC(O)C_6H_5 + LiCl$$

$$(C_6H_5)_2Sn(SLi)_2 + 2\ C_6H_5C(O)Cl \longrightarrow (C_6H_5)_2Sn[\text{—}SC(O)C_6H_5]_2 + 2\ LiCl$$

A very elegant preparative method is the insertion reaction of carbon disulfide, isothiocyanates, and related compounds at various tin-element bonds. Thus organotin alkoxides and oxides form xanthates with carbon disulfide (*34, 35, 36, 36a, 57, 59, 175a, P48*) or, in the presence of amines, dithiocarbamates (*39, 40a, 79a, 94c, 116, 117*). The reaction with phenylisothiocyanate or thiobenzoylisocyanate proceeds with the formation of thermally quite unstable products (*36, 59, 94c, 140*):

$$(C_4H_9)_3SnOCH_3 + CS_2 \longrightarrow (C_4H_9)_3SnSC(S)OCH_3$$

$$(C_2H_5)_3SnOH + CS_2 + (C_2H_5)_2NH \longrightarrow (C_2H_5)_3SnSC(S)N(C_2H_5)_2 + H_2O$$

$$[(C_4H_9)_3Sn]_2O + C_6H_5NCS \longrightarrow (C_4H_9)_3SnOC(NC_6H_5)SSn(C_4H_9)_3$$

$$(C_2H_5)_3SnOC_2H_5 + C_6H_5NCS \longrightarrow (C_2H_5)_3SnSC(NC_6H_5)OC_2H_5$$

The insertion of carbon disulfide, $CH_3\overline{CH - CH_2 - S}$ and C_2H_5NCS into the tin-nitrogen bond (*48, 57, 60b, 79, 79a, 92c, 96b, 97*), or into the tin-phosphorus bond (*199, 201*), leads to formation of organotin dithiocarbamates or their phosphorus analogs:

$$(CH_3)_3SnN(CH_3)_2 + CS_2 \longrightarrow (CH_3)_3SnSC(S)N(CH_3)_2$$

$$(C_6H_5)_3SnP(C_6H_5)_2 + CS_2 \longrightarrow (C_6H_5)_3SnSC(S)P(C_6H_5)_2$$

In contrast, the reaction of triphenyltin diphenylphosphine with carbon oxysulfide leads to the formation of a tin-oxygen bond (*199*). Lappert et al. (*79*) propose for this insertion reaction an ionic, four-center mechanism with a cyclic transition state:

$$\begin{array}{ccc} R_3Sn & - & \underline{N}R_2 \\ \uparrow & & \downarrow \\ |\underline{S} & = C = & \bar{\underline{S}} \end{array}$$

Noltes et al. (*48*, *120*, *121*, *141*, *142*) were able to show that the tin-hydrogen bond in organotin hydrides is also cleaved by isothiocyanates. One obtains the interesting organotin-sulfido-azomethines, the mechanism of whose formation has been studied kinetically (*120*, *142*):

$$(C_2H_5)_3SnH + C_6H_5NCS \longrightarrow (C_2H_5)_3SnSCH{=}NC_6H_5$$

Studies of displacement reactions between mono- or dithiocarbonic acid derivatives of IV elements revealed a preferred tendency of formation for compounds with Sn – S bonds (*191a*):

$$(C_6H_5)_3Pb - S - C(X)OC_2H_5 + (C_6H_5)_3SnCl \longrightarrow$$
$$(C_6H_5)_3Sn - S - C(X)OC_2H_5 + (C_6H_5)_3PbCl \quad (X = O, S)$$

Mention should also be made of the cleavage of diorganotin sulfides with carboxylic acids, which leads to various polymeric forms of diorganotin hydroxothioacid esters (*P103*, *P104*, *P105*, *P106*):

$$[(C_4H_9)_2SnS]_3 + 3\,HSCH_2C(O)OH \longrightarrow 3\,(C_4H_9)_2Sn\begin{matrix} \diagup OH \\ \diagdown SC(O)CH_2SH \end{matrix}$$

b. Chemical Properties and Reactions. As a rule, organotin thiocarboxylates and thio- or dithiocarbonic acid derivatives are liquid or solid substances at room temperature. Although they are stable toward oxygen, they are sensitive to heat. Thus organotin-sulfido azomethines decompose rapidly to bis(triorganotin) sulfides (*140*, *142*). The carbamates decompose with heat to organotin sulfides and thiourea derivatives (*116*, *117*), or to organotin sulfides and isocyanates with liberation of hydrogen sulfide (*117*):

$$2\,(C_6H_5)_3SnSC(S)N\begin{matrix} \diagup H \\ \diagdown CH_2C_6H_5 \end{matrix} \xrightarrow{130°C} [(C_6H_5)_3Sn]_2S + [C_6H_5CH_2NHC(S)]_2S$$

$$3\,(C_6H_5)_2Sn[-SC(S)NHCH_2C_6H_5]_2 \xrightarrow{50°C} [(C_6H_5)_2SnS]_3 + 3\,H_2S + 6\,C_6H_5CH_2NCS$$

The same kind of decomposition is observed with compounds of the type

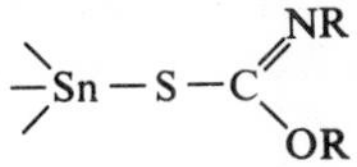

$$-\overset{|}{\underset{|}{Sn}}-S-C\begin{matrix}\nearrow NR \\ \searrow OR\end{matrix} \quad (35, 36, 57, 59)$$

N,N-Dimethyldithiocarbamate complexes of tin(IV) react with 1,2-dihalogenoethanes to give 2-dimethylamino-1,3-dithiolanyliumdialkyltin tetrahalides (*216b*).

c. Physical Properties. All the known organotin compounds of the types described in this section, together with the available physical constants, are assembled in Appendix 4. A considerable number of investigations of infrared (*36, 39, 40a, 52, 59, 79, 83, 90e, 152, 196a, 197, 199, 201, 206*) and ultraviolet (*39, 40a, 117*) spectra have been reported. From these come the following frequencies for the tin-sulfur stretching vibration:

$(C_6H_5)_3SnSC(O)C_6H_5$: νSnS 369 cm^{-1} (*197*)

$(C_6H_5)_3SnSC(O)OC_2H_5$: νSnS 378 cm^{-1} (*197*)

$(C_6H_5)_3SnSC(S)OC_2H_5$: νSnS 411,347 cm^{-1} (*197*)

$(C_6H_5)_3SnSC(S)P(C_6H_5)_2$: νSnS 349 cm^{-1} (*199, 201*)

$(CH_3)_3SnSC(S)N(CH_3)_2$: νSnS 507,448 cm^{-1} (*79*)

In the case of $(CH_3)_3SnSC(S)N(CH_3)_2$, the spectra suggest a coordination of the C = S sulfur atom to tin (*79, 90c*), which can be represented by two resonance structures:

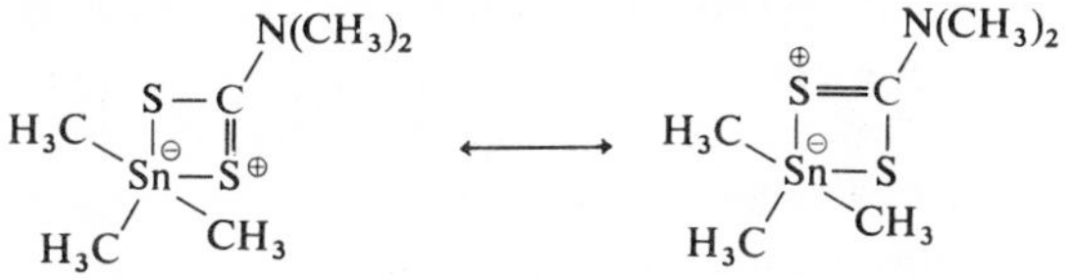

$$\begin{matrix} & & N(CH_3)_2 \\ & S-C & \\ H_3C\searrow & |^{\ominus}\ \ \| & \\ & Sn-S^{\oplus} & \\ H_3C\swarrow & \searrow CH_3 & \end{matrix} \longleftrightarrow \begin{matrix} & & N(CH_3)_2 \\ & \overset{\oplus}{S}=C & \\ H_3C\searrow & |^{\ominus}\ \ | & \\ & Sn-S & \\ H_3C\swarrow & \searrow CH_3 & \end{matrix}$$

In support of this hypothesis is the fact that this compound is monomeric in solution.

By comparing the ^{1}H-nmr spectra of organotin dithiocarbamates in benzene and chloroform solvents, Bonati et al. (*39*) and Honda et al. (*91*) showed that an interaction occurs between organotin dialkyldithiocarbamates and benzene (*91*). From ir, uv, and ^{1}H-nmr spectra, as well as the dipole moments (Table 2) of compounds of the type $R_2Sn[SC(S)NR_2]_2$, it is concluded that the ligands are arranged about the tin atom in a *cis*-octahedral structure. For compounds of the type $R_3SnSC(S)NR_2$ a pentacoordinated tin is proposed (*39, 90c*). ^{119}Sn chemical shifts in some triorganotin dithiocarbamates have been obtained by $^1H-(^{119}Sn)$ double resonance experiments (*60a*) and by direct measurements for $(C_6H_5)_2Sn[SC(O)C_6H_5]_2 = +101.2$

TABLE 2

DIPOLE MOMENTS FOR COMPOUNDS $R_nSn(SSCNR_2)_{4-n}$

Compounds	α_0	$-\beta$	∞^{P2} (cm^3)	E^P (cm^3)	∞^{P+A^P} (cm^3)	μ (D)	$\mu_{20\%}$ (D)
$Et_2Sn(SSCNEt_2)_2$	3.00	0.45	363.8	118.9	244.9	3.46	3.29
$Ph_2Sn(SSCNEt_2)_2$	3.78	0.47	519.2	149.0	370.2	4.26	4.08
$Ph_2Sn(SSCNPh_2)_2$	4.75	0.49	828.4	209.2	619.2	5.50	5.31
$Ph_3SnSSCNEt_2$	2.49	0.47	332.9	130.2	202.6	3.15	2.94
$Ph_3SnSSCNPh_2$	1.57	0.44	299.2	160.3	138.9	2.61	2.28
$Et_3SnSSCNEt_2$	2.66	0.37	259.0	85.0	174.0	2.92	2.77

ppm and for $(C_6H_5)_3Sn-S-C(S)N(CH_3)_2 = +94.5$ ppm (*92b*). Mössbauer spectra of some organotin bis(dithiocarbamates) have been recorded and discussed (*70a, 83c*).

4. *Bis(triorganotin) Sulfides*

a. Synthesis. The first organotin sulfur compound belongs to this class. Prepared in 1860 by Kulmitz, bis(triethyltin) sulfide was written at that time as $Sn_2(C_4H_5)_3S$ (*114, 115*).

Bis(triorganotin) sulfides are formed by the reaction of bis(triorganotin) oxides, or triorganotin hydroxides, or -halides with hydrogen sulfide or sodium sulfide (*7, 55, 64b, 84, 86, 105, 108, 109, 126, 129, 134, 162, 165, 176, 213b, P1, P58g, P77, P94a, P156*). As solvents, aromatic hydrocarbons, alcohols, ether, or, in special cases, water are employed. Bis(triorganotin) sulfides are obtained in particularly high yield by the reaction of triorganotin iodides with freshly prepared silver sulfide (*214*), and by the reaction of bis(triorganotin) oxides and triorganotin halides with polysulfides (*P22, P29, P47, P61, P66*) or thiosulfates (*P88*) as the source of sulfur:

$$[(C_2H_5)_3Sn]_2O + H_2S \longrightarrow [(C_2H_5)_3Sn]_2 + H_2O$$

$$2[C_6H_5C(CH_3)_2CH_2-]_3SnOH + H_2S \longrightarrow \{[C_6H_5C(CH_3)_2CH_2-]_3Sn\}_2S + 2\,H_2O$$

$$[(C_4H_9)_3Sn]_2O + Na_2S + H_2O \longrightarrow [(C_4H_9)_3Sn]_2S + 2\,NaOH$$

$$2\,[(CH_3)_2CH-C_6H_4-CH_2]_3SnCl + Na_2S \longrightarrow 2\,NaCl + \{[(CH_3)_2CH-C_6H_4-CH_2]_3Sn\}_2S$$

$$2\,(C_6H_5)_3SnI + Ag_2S \longrightarrow [(C_6H_5)_3Sn]_2S + 2\,AgI$$
$$[(C_4H_9)_3Sn]_2O + Na_2S_x + H_2O \longrightarrow [(C_4H_9)_3Sn]_2S + 2\,NaOH + S_{x-1}$$
$$2\,(C_3H_7)_3SnCl + Na_2S_2O_3 \cdot 5\,H_2O \longrightarrow [(C_3H_7)_3Sn)_2S + 2\,NaCl + H_2SO_4 + 4\,H_2O$$

Bis(triorganotin) polysulfides can also be prepared (*P29*). Several of these compounds were synthesized by Schwartz and Post (*211*) from sulfanes and allyltriorganotins:

$$2\,R_3SnCH_2CH{=}CH_2 + H_2S_x \longrightarrow (R_3Sn)_2S_{x-1} + 2\,CH_2{=}CHCH_3 + S$$

The reaction of alkali triorganotin sulfides with organotin- or organometal halides of other types, is a widely useful method, and is also applicable to the synthesis of unsymmetrical triorganotin triorganometal sulfides (*88*, *171*, *173*, *205*, *206*, *209*, *210*, *227c*, *227d*):

$$(C_6H_5)_3SnSLi + R_3SnCl \longrightarrow (C_6H_5)_3Sn{-}S{-}SnR_3 + LiCl$$
$$(CH_3)_3SnSLi + (CH_3)_3GeCl \longrightarrow (CH_3)_3Sn{-}S{-}Ge(CH_3)_3 + LiCl$$

An equally good method for the synthesis of such mixed organotin organometal sulfides or unsymmetrical bis(triorganotin) sulfides consists of the reaction of organotin hydrides with organometal sulfides or hydrosulfides (*139*, *225*, *226*, *227*):

$$(C_2H_5)_3SnH + [(C_6H_5)_3Sn]_2S \longrightarrow (C_2H_5)_3Sn{-}S{-}Sn(C_6H_5)_3 + (C_6H_5)_3SnH$$
$$(C_2H_5)_3SnH + (C_2H_5)_3GeSH \longrightarrow (C_2H_5)_3Sn{-}S{-}Ge(C_2H_5)_3 + H_2$$

Organotin halides, oxides, alkoxides, amines, and hydrides react with carbon disulfide with the formation of organotin sulfides. In a few cases it was possible to isolate those thermally labile addition products, which had been proposed as intermediates (*36*, *36a*, *48*, *59*, *60b*, *79*, *93a*, *94*, *94b*, *116*, *117*, *142*, *164*, *175a*). Analogous reactions are also observed between organotin halides or oxides and derivatives of carbon disulfide, such as dithiocarbamates (*39*, *48*, *140*), isocyanates (*33*, *35*, *36*, *57*, *59*, *94*, *117*, *123*), or carbon oxysulfide (*164*), as well as between organotin halides or $R_3Sn-SnR_3$ and triorganoantimony sulfide (*144c*, *212*).

$$(R_3Sn)_2O + CS_2 \longrightarrow (R_3Sn)_2S + COS$$
$$2\,(C_6H_5)_3SnBr + 4\,NH_3 + CS_2 \longrightarrow [(C_6H_5)_3Sn]_2S + NH_4SCN + 2\,NH_4Br$$
$$[(CH_3)_3Sn]_2NCH_3 + CS_2 \longrightarrow [(CH_3)_3Sn]_2S + CH_3NCS$$
$$2\,(C_2H_5)_3SnH + CH_3NCS \longrightarrow [(C_2H_5)_3Sn]_2S + ?$$

In a similar manner, the reactions of organotin hydrides with mercaptans (*143*, *145*), S_2Cl_2 (*23*), organic disulfides (*145*), and elemental sulfur (*225*, *227*) proceed with the formation of bis(triorganotin) sulfides:

$$2\,(C_6H_5)_3SnH + (C_6H_5CH_2)_2S_2 \longrightarrow [(C_6H_5)_3Sn]_2S + C_6H_5CH_3 + C_6H_5CH_2SH$$
$$8\,(C_2H_5)_3SnH + S_8 \longrightarrow 4\,[(C_2H_5)_3Sn]_2S + 4\,H_2S$$

Elemental sulfur cleaves the metal-metal bond in hexaorganoditins (*48*, *225*, *227*) as well as the tin-carbon (*160*, *202*), tin-sulfur (*7*), tin-phosphorus (*200*), and tin-sodium (*106*) bond in a great variety of organotin compounds, with the formation of bis(triorganotin) sulfides:

$$8\,(C_6H_5)_3Sn{-}Sn(C_6H_5)_3 + S_8 \longrightarrow 8\,[(C_6H_5)_3Sn]_2S$$
$$8\,(C_2H_5)_3SnCH_2C(O)CH_3 + S_8 \longrightarrow 4\,[(C_2H_5)_3Sn]_2S + 4\,S[CH_2C(O)CH_3]_2$$
$$2\,[(C_6H_5)_3Sn]_2PC_6H_5 + S_8 \longrightarrow 2\,[(C_6H_5)_3Sn]_2S + (C_6H_5PS_3)_2$$
$$8\,(CH_3)_3SnNa + S_8 \longrightarrow 4\,[(CH_3)_3Sn]_2S + 4\,Na_2S$$

Bis(triethyltin) selenides and tellurides react with sulfur forming bis-(triethyltin) sulfide and selenium or tellurium, respectively (*36b*, *227b*):

$$8\,[(C_2H_5)_3Sn]_2X + S_8 \longrightarrow [(C_2H_5)_3Sn]_2S + 8\,X \quad (X = Se, Te)$$

As the first example of a tin compound of the type $(XR_2Sn)_2S$, Schumann and Schmidt (*202*) obtained bis(dibutylchlorotin) sulfide from tributyltin chloride and sulfur at 190°C, according to the equation:

$$8\,(C_4H_9)_3SnCl + S_8 \longrightarrow 4\,[(C_4H_9)_2ClSn]_2S + 4\,(C_4H_9)_2S$$

In repeating the reaction, however, Poller and Spillman (*159*) obtained only the decomposition products of bis(dibutylchlorotin) sulfide, namely, dibutyltin dichloride and dibutyltin sulfide:

$$3\,[(C_4H_9)_2ClSn]_2S \longrightarrow 3\,(C_4H_9)_2SnCl_2 + [(C_4H_9)_2SnS]_3$$

Confirmation of the existence and the properties of bis(dibutylchlorotin) sulfide was accomplished by Harrison (*60*), who was able to synthesize this compound by another route:

$$3\,R_2SnX_2 + (R_2SnS)_3 \longrightarrow 3\,(XR_2Sn)_2S$$
$$(R = CH_3, C_4H_9;\; X = F, Cl, Br, SCN, C_7H_{15}COO{-})$$

Eventually Migdal et al. (*129*) succeeded in preparing a whole series of such compounds as follows:

$$2\,R_2SnX_2 + Na_2S \longrightarrow (XR_2Sn)_2S + 2\,NaX$$

They showed conclusively that these compounds, when distilled around 190°C, decompose by the path proposed by Poller et al.

b. Chemical Properties and Reactions. Bis(trialkyltin) sulfides or bis(triaryltin) sulfides are generally crystalline solids or thermally stable liquids which can be distilled. By contrast, the unsymmetrically substituted members of this class are unstable. These are compounds in which either the two tin atoms bound to sulfur are differently substituted, or compounds in which another group IV element as well as tin is bonded to sulfur.

They disproportionate more or less easily, with the formation of the corresponding symmetrical compounds (*129, 171, 173, 197, 204a*):

$$2\,(CH_3)_3Sn\text{—}S\text{—}Ge(CH_3)_3 \longrightarrow [(CH_3)_3Sn]_2S + [(CH_3)_3Ge]_2S$$

All uniformly substituted bis(triorganotin) sulfides are inert toward water and, with the exception of bis(trimethyltin) sulfide, stable toward atmospheric oxygen. They are usually easily soluble in all organic solvents. However, the tin-sulfur bond is cleaved by acids or strong bases; in the case of hydrochloric acid, this leads to hydrogen sulfide and organotin chlorides (*84*).

$R_3Sn-S-SnR_3$ compounds do not react with $HC \equiv CR'$ in contrast to the $R_3Sn-O-SnR_3$ compounds (*213a*). Bis(trialkyltin) sulfides react readily with hydrogenperoxide to yield bis(trialkyltin) sulfates (*214c,d*).

The tin-sulfur bond is also cleaved by AgF, Ag_2O, $AgOOCH_3$, and AgCNO (*24*), by Grignard reagents (*P89c, 54a*), by $Hg(CH_2COR')_2$ (*139a*), by various esters as well as by halides and azides of silicon, germanium, tin, phosphorus, arsenic, antimony, and mercury (*9, 22, 61, 62, 181a, 227a*). Many of these reactions are assuming importance as new and excellent methods for the preparation of simple sulfur compounds which are difficult to obtain in other ways.

$$R_3Sn\text{—}S\text{—}SnR_3 + 2\,HCl \longrightarrow 2\,R_3SnCl + H_2S$$

$$R_3Sn\text{—}S\text{—}SnR_3 + 2\,AgF \longrightarrow 2\,R_3SnF + Ag_2S$$

$$[(C_4H_9)_3Sn]_2S + \left(H\text{—}\overset{O}{\overset{\|}{C}}\text{—}\overset{C_6H_5}{\overset{|}{N}}\right)_2 Hg \longrightarrow 2(C_4H_9)_3Sn\text{—}\overset{C_6H_5}{\overset{|}{N}}\text{—}C\begin{smallmatrix}\nearrow O \\ \searrow H\end{smallmatrix} + HgS$$

Because of the weaker nucleophilicity of the —S—SnR_3 group, as compared to various organotin oxygen, nitrogen, and phosphorus compounds, the bis(triorganotin) sulfides do not react with organotin hydrides (*139*) while phenyllithium reacts with the formation of tin tetraorganyls (*168a*). Thus no detectable change occurs below 160°C between bis(tributyltin) sulfide and tributyltin hydride.

Bis(triphenyltin) sulfide reacts with elemental sulfur with elimination of diphenylsulfide and formation of polymeric organotin sulfides (*194*).

Bis(triorganotin) sulfides react with copper with desulfurization, forming hexaorganoditins, or, in the presence of oxygen, bis(triorganotin) oxides and copper(II) sulfide (*161b*).

$$(R_3Sn)_2S + Cu \longrightarrow R_3Sn\text{—}SnR_3 + CuS$$

$$(R_3Sn)_2S + Cu \xrightarrow{\frac{1}{2}O_2} (R_3Sn)_2O + CuS$$

Bis(trimethyltin) sulfide-pentacarbonyl-chromium (O) complexes are obtained from chromium hexacarbonyl and bis(trimethyltin) sulfide by uv-irradiation (*181b*):

$$Cr(CO)_6 + \left[(CH_3)_3Sn\right]_2S \xrightarrow{h\nu} (CO)_5Cr-S\begin{matrix} \diagup Sn(CH_3)_3 \\ \diagdown Sn(CH_3)_3 \end{matrix} + CO$$

c. Physical Properties. The symmetrical and unsymmetrical compounds of the general formula RR′R″Sn—S—SnR″R′R and $R_3Sn—S—MR_3$ known to date are assembled in Appendix 5.

An x-ray crystal structure determination has only been carried out for $(C_6H_5)_3Sn—S—Pb(C_6H_5)_3$ (*197*) (Fig. 1, Table 3). The rhombic crystals

TABLE 3

Bond Lengths and Bond Angles in $(C_6H_5)_3Sn—S—Pb(C_6H_5)_3$

	Length (Å)		Angle (deg)
Sn—S	2.442	Sn—S—Pb	107.18
Pb—S	2.476	S—Sn—C	107.0 ± 6
Sn—C	2.224 ± 0.09	S—Pb—C	107.7 ± 8
Pb—C	2.21 ± 0.12	C—Sn—C	111.8 ± 4
		C—Pb—C	111.1 ± 4

have lattice constants $a = 17.81$ Å, $b = 18.60$ Å, and $c = 9.94$ Å. Four molecules make up a unit cell. The bond lengths and bond angles suggest that the sulfur in this molecule is bonded through a simple covalent linkage with the sp^3-hybridized tin and lead atoms.

The infrared and Raman spectra of a number of bis(triorganotin) sulfides were examined (*51*, *52*, *64c*, *78*, *108*, *109*, *122*, *152*, *173*, *197*, *204*, *206*, *211*). The bands assigned to the tin-sulfur stretching vibration are listed in Table 4. From these vibrational spectra Kriegsmann (*109*) calculated the force constant of the tin-sulfur bond to be 1.7–1.8 mdyn/Å, which corresponds to a single bond. Calculation of the Sn—S—Sn bond angle furnished a value of 103° for bis(trialkyltin) sulfides, and 104° for bis(triphenyltin) sulfide. These values agree well with those found from the x-ray structure analysis of $(C_6H_5)_3Sn—S—Pb(C_6H_5)_3$.

Abel et al. (*2*) were able to measure the basicities of these compounds in a manner analogous to that used for mercaptides (see page 308), by comparison of the position of the C—D stretching vibration in deuteriated and non-

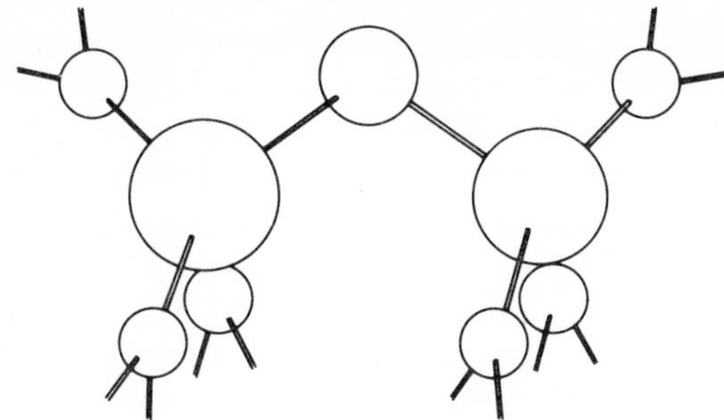

Fig. 1. Structure of triphenyltin-triphenyllead sulfide. (○ = C, ○ = S, ◯ = Sn, ◯ = Pb)

deuteriated solvents of bis(trimethylmetal) sulfides, $[(CH_3)_3M]_2S$, where M = C, Si, Ge, Sn, and Pb.

Cummins (*51*) used the position of the Sn—C stretching vibration in bis(tributyltin) sulfide as evidence for *trans*-gauche isomerism in the alkyl chain.

TABLE 4

INFRARED AND RAMAN FREQUENCIES OF THE Sn_2S GROUP IN COMPOUNDS OF THE TYPE $R_3Sn—S—SnR_3$, IN WAVE NUMBERS

Compound	$\nu_{as}Sn_2S$		ν_sSn_2S		δ Sn_2S		References
	ir	Raman	ir	Raman	ir	Raman	
$(CH_3)_3Sn—S—Sn(CH_3)_3$	—	366	—	320	—	65	(*109*)
$(C_2H_5)_3Sn—S—Sn(C_2H_5)_3$	—	368	—	318	—	65	(*109*)
	371	—	317	—	—	—	(*122*)
$(C_4H_9)_3Sn—S—Sn(C_4H_9)_3$	—	370	—	314	—	63	(*109*)
$(C_6H_5)_3Sn—S—Sn(C_6H_5)_3$	—	378	—	328	—	—	(*109*)
	376	—	330	—	—	—	(*204*)
$(C_6H_5)_3Sn—S—Ge(C_6H_5)_3$	404	—	355	—	—	—	(*204*)
$(C_6H_5)_3Sn—S—Pb(C_6H_5)_3$	365	—	305	—	—	—	(*204*)

The uv spectrum of bis(tributyltin) sulfide exhibits two maxima at 208.0 and 246.5 mμ (*53*).

The nmr spectra of bis(trimethyltin) sulfide (*5, 8, 171, 173, 188*), trimethylstannyl trimethylgermyl sulfide (*171, 173*), bis(triethyltin) sulfide (*65*), triethylstannyl triethylgermyl sulfide (*65*) and $[(o\text{—}CH_3C_6H_4)_3Sn]_3S$ (*63c*) have been measured and the chemical shifts and coupling constants determined. Conclusions were also made about the basicity of bis(trimethyltin) sulfide (*5*). Tin-119 chemical shifts have been obtained in a number of such compounds by $^1H\text{—}(^{119}Sn)$ double-resonance experiments (*60a*) and by direct measurements for $[(C_6H_5)_3Sn]_2S = +48.7$ ppm (*92b*).

Molecular refractivities for bis(triethyltin) sulfide were determined by Sayre (*181*). From the dipole moment of 2.45 D for bis(tributyltin) sulfide, a group moment of 1.91 D was assigned for the $(C_4H_9)_3SnS$ group (*54*).

From the Mössbauer spectrum of bis(triphenyltin) sulfide, an isomer shift of 1.222 mm/sec, a quadrupole splitting of 1.167 mm/sec, and a ρ-value of 0.955 were obtained, all of which confirm a pure sp^3-hybridized tin atom (*83c, 83d, 89, 144d*). In the mass spectrum of bis(dibutylchlorotin) sulfide, the highest peak was found at m/e 569, which corresponds to a monomer (*129*). Polarographic measurements have been made on $(R_3Sn)_2S$ compounds (*39b*).

5. *Diorganotin Sulfides*

a. Synthesis. Diorganotin sulfides of the general formula R_2SnS have been known for a long time (*148*) and were originally formulated as monomers. It was soon recognized, however, that these compounds contain no tin-sulfur double bond but instead, form polymers, principally cyclic trimers (*85*).

By analogy with the synthesis of bis(triorganotin) sulfides, they are obtained from diorganotin dihalides or oxides and hydrogen sulfide (*68, 82, 100, 101, 104, 136, 148, 149, 154, 184, 216, P122, P130, P151*), alkali sulfides (*55, 85, 86, 87, 159, 162, 163, 166, 216, P76f,h, P95b, P158g*), alkali polysulfides (*136, P22, P47, P61, P150*), or sodium thiosulfate (*P87*):

$$3\,(CH_3)_2SnCl_2 + 3H_2S \longrightarrow \text{cyclo-}[(CH_3)_2SnS]_3 \text{ (six-membered Sn}_3\text{S}_3\text{ ring, each Sn bearing }(CH_3)_2) + 6HCl$$

$$[(C_6H_5)_2SnO]_n + n\,H_2S \longrightarrow n/3\,[(C_6H_5)_2SnS]_3 + n\,H_2O$$

$$3\,(N{\equiv}C\text{—}CH_2\text{—}CH_2)_2SnBr_2 + 3\,Na_2S \longrightarrow [(N{\equiv}C\text{—}CH_2-CH_2)_2SnS]_3 + 6\,NaBr$$

$$3\,(C_8H_{17})_2SnCl_2 + 3\,Na_2S_2O_3 \longrightarrow [(C_8H_{17})_2SnS]_3 + 6\,NaCl + 3\,SO_3$$

Carbon disulfide and carbon oxysulfide react with polymeric organotin oxides in a sealed tube above 140°C to give trimeric diorganotin sulfides and carbon oxysulfide or carbon dioxide, respectively (*164*). When heated at 160°C for three hours, diphenyltin oxide and carbon disulfide form an amorphous product, mp 56–64°C, which corresponds in composition to diphenyltin sulfide, mp 183–184°C, but is a heretofore unknown isomer.

An analogous reaction mechanism is assumed for the reactions of diorganotin oxides with carbon disulfide in the presence of amines. They lead to diorganotin sulfides (*60b*, *117*) by way of thermally unstable intermediates which are rarely isolable:

$$[(C_6H_5)_2SnO]_n + n\,CS_2 \longrightarrow n/3\,[(C_6H_5)_2SnS]_3 + n\,COS$$

$$3\,[(C_6H_5)_2SnO]_n + 6n\,C_6H_5CH_2NH_2 + 6n\,CS_2 \longrightarrow n\,[(C_6H_5)_2SnS]_3 + 6n\,C_6H_5CH_2NCS + 3n\,H_2S + 3n\,H_2O$$

Elemental sulfur reacts rapidly with diphenyltin in toluene at reflux temperature, but only after several days at room temperature with cleavage of the tin-tin bond and formation of diphenyltin sulfide (*113*):

$$24\,[(C_6H_5)_2Sn]_n + 3n\,S_8 \longrightarrow 8n\,[(C_6H_5)_2SnS]_3$$

Tin-carbon bonds are likewise cleaved by sulfur. Bost and Borgstrom (*40*) found in 1929 that diphenylsulfide was formed from tetraphenyltin and sulfur at temperatures above 170°C. This reaction was investigated further by Schumann and Schmidt (*191*) and they found that triphenyltin thiophenolate and diphenyltin bis(thiophenolate) are formed first by the nucleophilic attack of sulfur. These compounds, however, are unstable at the reaction temperature and redistribute to the stable end products diphenyltin sulfide and diphenyl sulfide:

$$8\,(C_6H_5)_4Sn + S_8 \longrightarrow 8\,(C_6H_5)_3SnSC_6H_5$$

$$8\,(C_6H_5)_3SnSC_6H_5 + S_8 \longrightarrow 8\,(C_6H_5)_2Sn(SC_6H_5)_2$$

$$3\,(C_6H_5)_2Sn(SC_6H_5)_2 \longrightarrow [(C_6H_5)_2SnS]_3 + 3\,(C_6H_5)_2S$$

From triphenyltin chloride or diphenyltin dichloride and sulfur, phenyl-substituted tin-sulfur six-membered ring compounds are obtained, with some chlorine atoms still bonded to sulfur. In addition polymeric organotin sulfides, thiantrene, and tin sulfide are formed (*202*).

Diphenyltin sulfide is also formed by the reaction of bis(triphenyltin) sulfide with diphenyltin dichloride (*159*); the analogous butyl compound arises from the thermal decomposition of bis(dibutylchlorotin) sulfide (*129*).

Further possible synthetic methods are; the reaction of diorganotin dihalides with trimethylantimony sulfide (*212*, *213*) or the addition of polysulfanes to diorganotin diallyl compounds (*211*), the latter leading to compounds of the type $(R_2SnS_x)_3$:

$$3\,(C_6H_5)_2SnCl_2 + 3\,(CH_3)_3SbS \longrightarrow [(C_6H_5)_2SnS]_3 + 3\,(CH_3)_3SbCl_2$$

$$3\,(C_2H_5)_2Sn(CH_2CH{=}CH_2)_2 + 3\,H_2S_{5.2} \longrightarrow [(C_2H_5)_2SnS_{1.2}]_3 + 6\,CH_3CH{=}CH_2 + S_x$$

Particularly interesting is the reaction between bis(π-cyclopentadienyl iron dicarbonyl) tin dichloride and hydrogen sulfide (*135*), which leads to the corresponding sulfur derivative by substitution of chlorine:

$$[(C_5H_5)Fe(CO)_2]_2SnCl_2 + H_2S \longrightarrow [(C_5H_5)Fe(CO)_2]_2SnS + 2\,HCl$$

b. Chemical Properties and Reactions. Dialkyl and diaryltin sulfides are crystalline, colorless compounds at room temperature, which are stable toward oxygen and hydrolysis. As a rule they are trimeric and exist as six-membered sulfur-tin rings. Only in the case of $[(CH_3)_3SiCH_2]_2SnS$ is there evidence for the occurrence of a tetramer (*184*). The compounds exhibit rather high thermal stability. For instance, dialkyltin sulfides decompose only above 250°C (*31*). In the case of dipropyltin sulfide, hexapropylditin, sulfur, and tin(II) sulfide were shown to be the decomposition products (*87*). Diorganotin sulfides are attacked by hydrogen chloride with the formation of diorganotin dichlorides and hydrogen sulfide (*159*). RMgX compounds react with diorganotin sulfides under formation of tin tetraorganyls (*54a*). They are converted to 1,1,3,3-tetraalkyl-1,3-dihaloditin sulfides by dialkyltin dihalides; to 1,1,3-trialkyl-1,3,3-trihaloditin sulfides by alkyltin trihalides; and to dialkyltin dihalides by tin tetrahalides. By contrast, no reaction occurs with trialkyltin halides (*60*):

$$(R_2SnS)_3 + 3\,R_2SnX_2 \longrightarrow 3\,XR_2Sn{-}S{-}SnR_2X$$

$$(R_2SnS)_3 + 3\,R'SnX_3 \longrightarrow 3\,XR_2Sn{-}S{-}SnR'X_2$$

When $SnCl_4$ is used, a more complex reaction is observed. The resulting $Sn_2Cl_2S_3$ is a white, reactive solid, which is amorphous to x-rays (*214b*):

$$[(C_4H_9)_2SnS]_3 + 2\,SnCl_4 \longrightarrow 3\,(C_4H_9)_2SnCl_2 + Sn_2Cl_2S_3$$

Of considerable technical importance are the reactions between dibutyltin sulfide and mercaptoacids, mercaptoesters, and alcohols, which lead to diorganotin mercaptoesters or mercaptides (*P103*, *P104*, *P105*, *P106*):

$$[(C_4H_9)_2SnS]_3 + 6\,HSCH_2COOC_4H_9 \longrightarrow 3\,(C_4H_9)_2Sn(SCH_2COOC_4H_9)_2 + 3\,H_2S$$

$$[(C_4H_9)_2SnS]_3 + 3\,HO\overset{\overset{\displaystyle O}{\|}}{C}-CH_2-\overset{\overset{\displaystyle SH}{|}}{C}H-\overset{\overset{\displaystyle O}{\|}}{C}-OH \longrightarrow 3\,H_2O + 3\,(C_4H_9)_2Sn\begin{cases} O-\overset{\overset{\displaystyle O}{\|}}{C}-CH_2 \\ S-\underset{\underset{\displaystyle O}{\|}}{C}-CH-SH \end{cases}$$

$$[(C_4H_9)_2SnS]_3 + 6\,C_2H_5OH \longrightarrow (C_4H_9)_2Sn\begin{cases} OC_2H_5 \\ SC_2H_5 \end{cases}$$

One observes a rapid exchange of halogen atoms and bridged sulfur atoms in solutions of dimethyltin sulfide in molten dimethyltin dihalides. As can be shown by ^1H-nmr measurements, the rate constant of this exchange reaction decreases with the increasing atomic weight of the halogen; i.e., in the order Cl, Br, and I (*132*):

$$(CH_3)_2SnX_2 + n/3\,[(CH_3)_2SnS]_3 \longrightarrow X-\underset{\underset{\displaystyle CH_3}{|}}{\overset{\overset{\displaystyle CH_3}{|}}{Sn}}\left[-S-\underset{\underset{\displaystyle CH_3}{|}}{\overset{\overset{\displaystyle CH_3}{|}}{Sn}}-\right]_{n-1}S-\underset{\underset{\displaystyle CH_3}{|}}{\overset{\overset{\displaystyle CH_3}{|}}{Sn}}-X$$

Just as tetraphenyltin, diphenyltin sulfide reacts above 200°C with elemental sulfur with the formation of diphenylsulfide, polymeric phenyltin sulfides, and tin(II) sulfide (*194*). The reaction of trimeric dimethyltin sulfide with triethylindium etherate leads to ethylindium sulfide (*234*):

$$3\,(C_2H_5)_3In + [(CH_3)_2SnS]_3 \longrightarrow 3\,C_2H_5InS + 3\,(CH_3)_2Sn(C_2H_5)_2$$

Dibutyltin sulfide reacts with Na_2S in water/isopropanol under formation of $(C_4H_9)_2Sn(SNa)_2$ (*P38f*).

Dibutyltin sulfide is split by C_6H_5Li, $CH_3C\equiv CLi$, $C_6H_5CH_2CH_2C\equiv CLi$ forming tetraorganotin compounds (*168a*).

c. *Physical Properties.* Appendix 6 contains a list of the known diorganotin sulfides with an indication of their physical properties.

The x-ray crystal structure determination (*198*) of trimeric diphenyltin sulfide shows that three tin and three sulfur atoms form a six-membered ring in the boat conformation, but that the boat is somewhat distorted (Fig. 2). The bond angles and distances in this molecule correspond to those expected on the basis of covalent single bonds between tin and sulfur (Table 5). The Sn—S—Sn bond angle of 104° is exactly that calculated by Kriegsmann (*109*) from spectroscopic measurements of compounds $R_3Sn—S—SnR_3$; the

S—Sn—S bond angle of 112° deviates only slightly from that expected for a tetrahedral sp^3-hybridized tin. The tin-sulfur and tin-carbon bond distances coincide within experimental error with the sum of the covalent radii of these atoms.

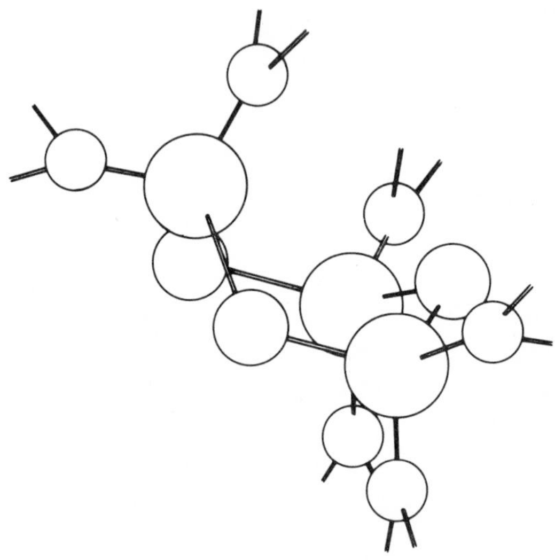

Fig. 2. Structure of trimeric diphenyltin sulfide.
(○ = C, ○ = S, ◯ = Sn)

The infrared spectra of dimethyltin sulfide (*42*, *52*, *83a*, *83b*, *108*, *109a*, *198*, *212*, *213*), diphenyltin sulfide (*52*, *152*, *196*, *198*), dibutyltin sulfide (*51*, *52*, *78*, *108*, *109a*, *196*, *198*, *211*), and diethyltin sulfide (*108*, *109a*, *122*, *211*) were measured, and some of the bands were assigned. Table 6 contains a list of the stretching bands assigned to the tin-sulfur group. Assuming C_s symmetry for these cyclic molecules, there should be two $\nu_s Sn_2S$, two $\nu_{as} Sn_2S$, and one δ_s-ring of the A'' type. Of the four bands found down to 250 cm^{-1}, two each are assigned to $\nu_{as} Sn_2S$ and $\nu_s Sn_2S$ of the tin-sulfur six-membered ring (*198*).

TABLE 5

BOND LENGTHS AND BOND ANGLES IN $[(C_6H_5)_2SnS]_3$

	Length (Å)		Angle (deg)
Sn—S	2.42 ± 0.02	S—Sn—S	111.9 ± 0.3
Sn—C	2.24 ± 0.06	Sn—S—Sn	104.0 ± 1.6
		S—Sn—C	109.0 ± 6.0

The remaining five frequencies are believed to be either at longer wave length or unobservable because of degeneracy. The slight frequency differences found for alkyl- and phenyl-substituted compounds can be explained by a greater distortion of the boat structure for the latter. The results of the Raman spectra of $[(CH_3)_2SnS]_3$, $[(C_2H_5)_2SnS]_3$ and $[(C_4H_9)_2SnS]_3$ (*109a*) are in agreement with this explanation.

TABLE 6

INFRARED FREQUENCIES OF THE Sn—S GROUP IN COMPOUNDS OF THE TYPE $(R_2SnS)_3$, IN WAVE NUMBERS

Compound	ν_{as}SnS	ν_sSnS	References
$[(CH_3)_2SnS]_3$	363	326	(*212*)
	357, 344	324, 290	(*198*)
	364, 343, 322 (Raman)		(*109a*)
$[(C_2H_5)_2SnS]_3$	368	330	(*122*)
	367, 327, 314 (Raman)		(*109a*)
$[(C_4H_9)_2SnS]_3$	357, 345	331, 294	(*198*)
	368, 318 (Raman)		(*109a*)
$[(C_6H_5)_2SnS]_3$	371, 343	321, 303	(*198*)

The ^{1}H-nmr spectrum has only been measured (*132*) for dimethyltin sulfide. Burke and Lauterbur (*44*) determined the ^{119}Sn-chemical shift in dibutyltin sulfide to be 124 ppm, relative to tetramethyltin. Davies et al. (*60a*) found a chemical shift of -128 ppm for $[(CH_3)_2SnS]_3$ in C_6H_6, and Hunter et al. (*92b*) found a shift of -19.5 ppm for $[(C_6H_5)_2SnS]_3$ in $CHCl_3$ and a shift of -125.6 ppm for $[(CH_3)_2SnS]_3$ in CS_2.

The Mössbauer spectra of dibutyltin sulfide (*20*, *21*, *83c*, *105b*, *144d*, *236*) show an isomer shift of 0.9 ± 0.2 mm/sec and a quadrupole splitting of 1.9 ± 0.2 mm/sec.

6. *Compounds of the Type* $(RSnS)_2S$

In 1903 Pfeiffer and Lehnhardt (*147*) obtained a product from the reaction of methyltin triiodide and hydrogen sulfide, which was insoluble in all solvents except aqueous ammonium sulfide, and to which they assigned the following structure containing tin-sulfur double bonds:

$$2\,CH_3SnI_3 + 3\,H_2S \longrightarrow CH_3\overset{\overset{\displaystyle S}{\|}}{Sn}{-}S{-}\overset{\overset{\displaystyle S}{\|}}{Sn}CH_3 + 6\,HI$$

Besides this methyl derivative, which was subsequently described by other authors (*148*, *P49*), analogous ethyl (*63b*), butyl (*P38c*, *P49*, *P55*, *P76f*), *p*-tolyl (*102*, *103*) chloro, bromo, and iodo (*104*, *136*) derivatives were prepared somewhat later. But only recently, it was recognized that these compounds exist in polymeric form.

The phenyl derivative is obtained by the reaction of tetraphenyltin, bis-(triphenyltin) sulfide, or diphenyltin sulfide with sulfur in a sealed tube above 200°C (*194*):

$$2n\,(C_6H_5)_4Sn + \tfrac{6}{8}n\,S_8 \longrightarrow [(C_6H_5SnS)_2S]_n + 3n\,(C_6H_5)_2S$$

or by the reaction of organotin trihalides and triorganoantimony sulfides (*212*):

$$2\,RSnCl_3 + 3\,R'_3SbS \longrightarrow (RSnS)_2S + 3\,R'_3SbCl_2$$

In 1966 Forstner and Muetterties (*71*) were able to show that analogous organosilicon sulfides exist in a low-molecular weight form of adamantane-like structure $(RSi)_4S_6$. Methyl-, butyl, and phenyltin trichloride react with aqueous sodium sulfide to form analogous compounds, which, however, cannot be distilled and for which molecular weight determinations gave no clear evidence for the existence of "tin-sulfur adamantanes." It is very likely that the reaction products consist of a mixture of oligomeric and polymeric compounds.

Komura and Okawara (*105*) investigated the reaction of methyl, *n*-propyl-, and *n*-butyltin oxides with sodium sulfide in aqueous hydrochloric acid. Only in the case of the *n*-butyl compound was an oligomeric substance formed whose molecular weight could be determined because of its solubility in benzene.

$$2\,[(C_4H_9SnO)_2O]_n + 6n\,Na_2S + 12n\,HCl \longrightarrow n\ [\text{adamantane-type } (C_4H_9Sn)_4S_6 \text{ structure: four } C_4H_9\text{–Sn units bridged by six S}] + 12n\,NaCl + 6n\,H_2O$$

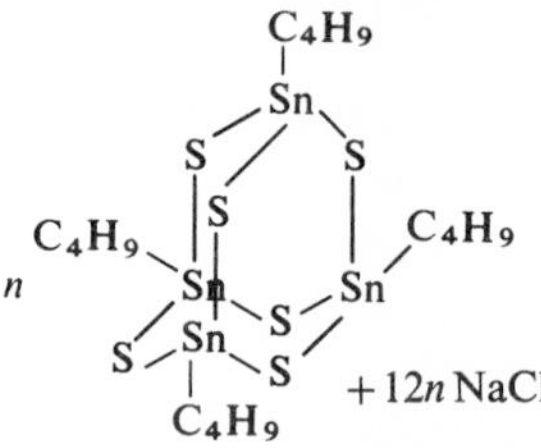

This tetrameric butyltin sesquisulfide, like the polymeric methyl and *n*-propyl derivatives, is cleaved by 8-hydroxyquinoline, with formation of hydrogen sulfide and compounds of the formula $[RSn(C_9H_6NO)_2]_2S$ and $RSn(C_9H_6NO)_3$. Grignard reagents split Sn—S bonds in such compounds under formation of tetraorganotin compounds (*54a*, *P89d*).

The infrared bands at 527 and 536 cm^{-1}, found for methyltin sesquisulfide by Kriegsmann and Hoffman (*108*), were not assigned.

The compounds of this class reported to date are listed in Appendix 7.

A relatively new paper describes the x-ray structural determination of $[(CH_3)Sn]_4S_6$ and $[(C_4H_9)Sn]_4S_6$ (*63a*). Butylthiostannonic acid is triclinic with $a = 12.26$ Å, $b = 9.76$ Å, $c = 12.74$ Å, $\alpha = 70°\ 40'$, $\beta = 110°\ 20'$, and $\gamma = 89°$. Methylthiostannonic acid is monoclinic with the space group C_{2c}: $a = 9.768$ Å, $b = 17.39$ Å, $c = 10.96$ Å, and $\beta = 109.0°$. The adamantane structure for both compounds could be established. The Sn—Sn distances are 3.8 Å, the Sn—S— distances 2.35 Å, and the Sn—C— distances 2.1 Å.

7. *Heterocycles Containing Ring Tin-Sulfur Bonds*

a. Synthesis. The only known four-membered rings containing tin-sulfur bonds namely, dialkyltin-dithioimidocarbonates, are obtained by the reaction of calcium cyanamide with carbon disulfide and subsequent treatment with dialkyltin dichlorides (*P158*) or from dipotassium cyanodithioimidocarbonate and dialkyltin dichlorides (*211b, P158c*), or from diorganotin dichlorides and disodium dimercaptomethylenemalonitrile (*P158i*):

$$(C_4H_9)_2SnCl_2 + (NaS)_2C{=}C(CN)_2 \longrightarrow (C_4H_9)_2Sn\langle S_2 \rangle C{=}C(CN)_2 + 2\ NaCl$$

$$CaCN_2 + CS_2 \longrightarrow Ca\langle S_2 \rangle C{=}N{-}C{\equiv}N$$

$$Ca\langle S_2 \rangle C{=}N{-}C{\equiv}N + R_2SnCl_2 \longrightarrow R_2Sn\langle S_2 \rangle C{=}N{-}C{\equiv}N$$

Organotin dihalides or oxides, when treated with ethanedithiol, ethylenedithiol, mercaptoethanol, or thioglycolic acid, form stannacyclopentane derivatives containing —S—Sn—S— or —S—Sn—O— units (*4, 7, 18, 56, 144, 215a, 229, P2, P14, P99, P123, P136, P158a, P158d*):

$$(CH_3)_2SnCl_2 + HS(CH_2)_2SH + 2(C_2H_5)_3N \longrightarrow \begin{matrix} H_2C{-}S \\ | \\ H_2C{-}S \end{matrix} \rangle Sn(CH_3)_2 + 2\ (C_2H_5)_3N{\cdot}HCl$$

$$(CH_3)_2SnCl_2 + NaSCH{=}CHSNa \longrightarrow \begin{matrix} HC{-}S \\ \| \\ HC{-}S \end{matrix} \rangle Sn(CH_3)_2 + 2\ NaCl$$

$$(C_4H_9)_2SnO + HS(CH_2)_2OH \longrightarrow \begin{matrix} H_2C{-}O \\ | \\ H_2C{-}S \end{matrix} \rangle Sn(C_4H_9)_2 + H_2O$$

$$(C_4H_9)_2SnO + HSCH_2COOH \longrightarrow \begin{matrix} H_2C{-}S \\ | \\ (O{=})C{-}O \end{matrix} \rangle Sn(C_4H_9)_2 + H_2O$$

But the dibutyltin derivative of 2-mercaptoethanol was dimeric, forming a ten-membered ring system (*215a*).

In an analogous manner, heterocycles condensed to aromatic systems are formed from the appropriate organotin halides or oxides and toluene-3,4-dithiol (*153, 156, 231*) or 2,3-dimercaptoquinoline (*180, P31*); interesting rings with tin-manganese (*217*), tin-rhenium (*217*) or tin-iron bonds (*161a*) can also be prepared:

$$H_3C\text{-}C_6H_3(SH)_2 + (CH_3)_2SnCl_2 + 2(C_2H_5)_3N \longrightarrow H_3C\text{-}C_6H_3S_2Sn(CH_3)_2 + 2\,(C_2H_5)_3N\cdot HCl$$

$$C_8H_4N_2(SH)_2 + (C_2H_5)_2SnO \longrightarrow C_8H_4N_2S_2Sn(C_2H_5)_2 + H_2O$$

$$[(CO)_5Mn]_2SnCl_2 + HS(CH_2)_2SH + 2\,(C_2H_5)_3N \longrightarrow (H_2C\text{-}S)_2Sn[Mn(CO)_5]_2 + 2\,(C_2H_5)_3N\cdot HCl$$

It was found by Abel et al. (*4*), that 2,2-dimethyl-1,3-dithia-2-stannacyclopentane was formed by the reaction of the corresponding silicon compound with dimethyltin dichloride, the by-product being dimethyldichlorosilane.

The analogous six-membered rings were obtained by Wieber (*230*) and Poller (*153, 156*) by the use of propanedithiol in the appropriate reactions. The synthesis of numerous derivatives of this type are described in the patent literature (*P2, P39, P59a, P73, P74, P92, P99, P140, P147, P154*):

$$(CH_3)_2SnCl_2 + HS(CH_2)_3SH + 2\,(C_2H_5)_3N \longrightarrow H_2C(CH_2S)_2Sn(CH_3)_2 + 2\,(C_2H_5)_3N\cdot HCl$$

$$(C_8H_{17})_2SnO + HS(CH_2)_2COOH \longrightarrow H_2C\text{-}CH_2\text{-}S\text{-}Sn(C_8H_{17})_2\text{-}O\text{-}C(=O) + H_2O$$

$$(C_4H_9)_2SnO + C_6H_4(SH)(COOH) \longrightarrow C_6H_4\text{-}S\text{-}Sn(C_4H_9)_2\text{-}O\text{-}C(=O) + H_2O$$

Until now, higher membered ring compounds have been reported mainly in the patent literature (*170, P13, P16, P30, P35, P43, P46, P62, P75, P76i, P103, P104, P105, P106, P116, P134, P135*). Thus the synthesis of tin- and sulfur-containing heterocycles with 7, 10, 11, 12, 14, 15 and 20 atoms in the ring have been described.

The first spirane with a tin-sulfur bond was obtained in 1934 by Backer et al. (*25, 29*) by the reaction of tin tetrachloride with the disodium salt of dimercaptodiethyl ether. Shortly thereafter, Brown et al. (*41*) found that the "Farblack," which was obtained by Pollak (*150*) in 1913 as a side product in the synthesis of dithiocatechol from *o*-benzenedisulfonyl chloride, tin, and hydrogen chloride, possesses the following spirane structure:

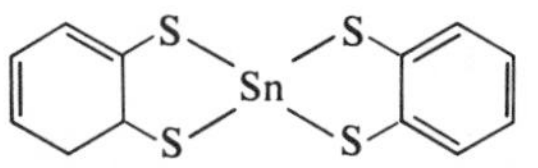

$$C_6H_4(S)_2Sn(S)_2C_6H_4$$

This compound as well as some of its various derivatives (*70, 92, 153, 156*) polymerizes readily and can scarcely be isolated as a monomer. Its formation is the basis of a microanalytical method for tin determination.

A method for the synthesis of spiranes, discovered by Abel and Brady (*7*), is the reaction of tin tetrachloride with ethandithiol in aqueous alkali:

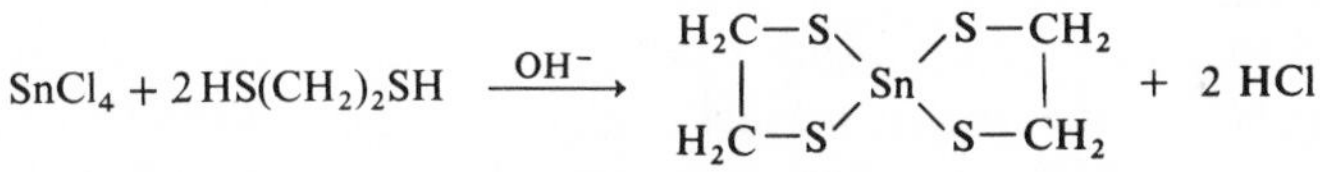

$$SnCl_4 + 2\,HS(CH_2)_2SH \xrightarrow{OH^-} \begin{matrix} H_2C-S \\ | \\ H_2C-S \end{matrix}\!\!>Sn<\!\!\begin{matrix} S-CH_2 \\ | \\ S-CH_2 \end{matrix} + 2\,HCl$$

b. Chemical Properties and Reactions. Heterocyclic compounds with tin and sulfur bonded covalently in the ring display the same chemical properties as simple organotin mercaptides. They are cleaved by $CdCl_2$, $HgCl_2$, $AgNO_3$, $CoBr_2$, $NiBr_2$, or $Hg(OAc)_2$ (*9, 18, 158*) with the formation of diorganotin dihalides, acetates, or nitrates and monomeric nickel, cobalt, and silver or polymeric cadmium or mercury mercaptides. With iodine (*158*), they form diorganotin diiodides and polymeric mercaptans. With transition metal carbonyl halides the analogous metalcarbonylmercaptides (*19*) are formed. Nitrogen- and oxygen-containing bases add, with the formation of complexes with pentacoordinated and hexacoordinated tin (*18, 157*):

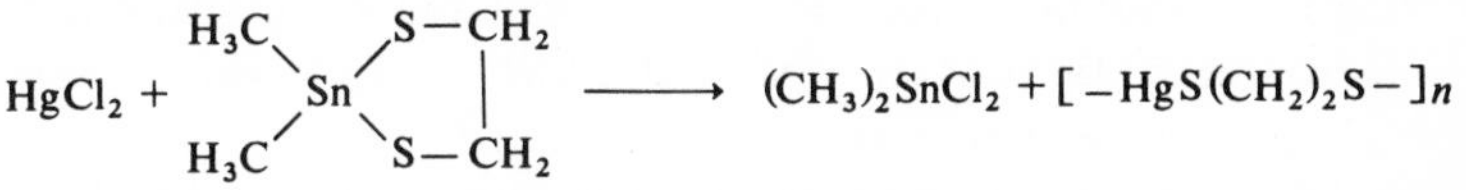

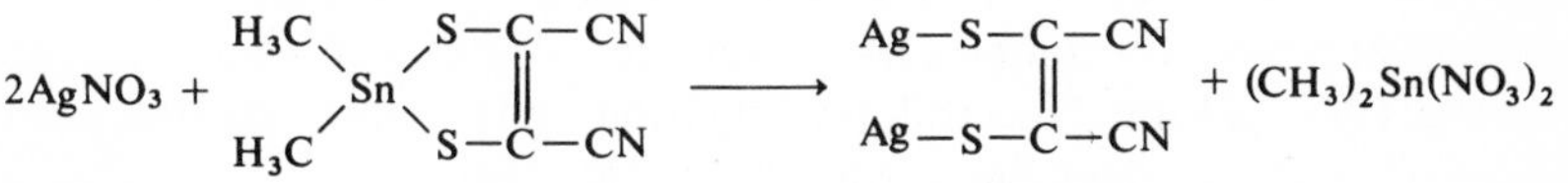

$$HgCl_2 + \begin{matrix} H_3C \\ \\ H_3C \end{matrix}\!\!>Sn<\!\!\begin{matrix} S-CH_2 \\ | \\ S-CH_2 \end{matrix} \longrightarrow (CH_3)_2SnCl_2 + [-HgS(CH_2)_2S-]_n$$

$$2AgNO_3 + \begin{matrix} H_3C \\ \\ H_3C \end{matrix}\!\!>Sn<\!\!\begin{matrix} S-C-CN \\ \| \\ S-C-CN \end{matrix} \longrightarrow \begin{matrix} Ag-S-C-CN \\ \| \\ Ag-S-C-CN \end{matrix} + (CH_3)_2Sn(NO_3)_2$$

$(H_2C{-}S)_2Sn(S{-}CH_2)_2 + 2\,(CH_3)_2SO \longrightarrow (H_2C{-}S)_2Sn[O{=}S(CH_3)_2]_2(S{-}CH_2)_2$

$(H_2C{-}S)_2Sn(S{-}CH_2)_2$ + 2,2′-bipyridine $\longrightarrow$ $(H_2C{-}S)_2Sn(S{-}CH_2)_2 \cdot$ 2,2′-bipyridine

Various ring systems such as $(C_8H_{17})_2\overline{SnSCH_2C(O)O}$ (*P73*) or $(CH_3)_2\overline{SnSCH_2CH_2CH_2S}$ (*228*) undergo reversible polymerization.

c. Physical Properties. In Appendix 8 the heterocycles of the types mentioned in this chapter are listed, together with their physical properties.

Backer and Drenth (*25*) determined the lattice constants for orthorhombic 1,4,6,9-tetrathia-5-stanna-5-spirononane, crystallized from benzene. Infrared spectra of various organotin-sulfur heterocycles have been determined by Poller et al. (*157*), Abel et al. (*7*), Finck et al. (*70*), Wieber et al. (*229*, *230*, *231*), Thompson et al. (*217*), and Finch et al. (*69*). In the region below 400 cm^{-1} Poller et al. (*157*) assigned the tin-sulfur stretching vibration in various complexes of bis-ethane-1,2-dithiolatotin to 293–345 cm^{-1}. These authors found three bands in the case of a *cis*-configuration, and two bands for a *trans*-configuration. Table 7 contains the bands assigned by Finch et al. (*69*) in the region 2000–110 cm^{-1} for some heterocycles. They assume a nearly planar structure with V_d symmetry for bis(ethane-1,2-dithiolato)-stannane and C_{2v} symmetry for unsubstituted dithiastannacyclopentanes. For calculation of the force constants (Table 8), the following bond distances and bond angles were used: $r_{CS} = 1.82$ Å, $r_{SnS} = 2.0$ Å, r_{CC}(Ring) $= 1.52$ Å, $r_{C—CH_3} = 1.54$ Å, ∡CCS = 113°, ∡SSnS = 94° 46′ and ∡CSSn = 109° 37′. The theoretical vibrational frequencies were calculated from these values and found to agree well with the observed values, a fact which supports an essentially planar ring structure.

Ultraviolet spectra have been recorded for tin bis(3,4-toluenedithiol) (*70*, *153*, *156*) and some tin complexes of *cis*-ethylene-bisthiol and 4,5-dimercapto-*o*-xylene (*92*); ^{1}H-nmr spectral investigations have been carried out on various five- and six-membered rings (*229*, *230*, *231*).

TABLE 7

INFRARED SPECTRA OF ORGANOTIN-SULFUR HETEROCYCLES (cm^{-1})

Assignment	$H_2C{-}S$, $H_2C{-}S$ >Sn< $S{-}CH_2$, $S{-}CH_2$	$H_2C{-}S$, $H_2C(CH_3){-}S$ >Sn< $S{-}CH_2$, $S{-}CH(CH_3)$	$H_2C{-}S$, $H_2C{-}S$ >Sn< C_4H_9, C_4H_9	$H_2C{-}S$, H_2C, $H_2C{-}S$ >Sn< C_4H_9, C_4H_9	$H_2C{-}S$, $H_2C{-}S$ >Sn< C_6H_5, C_6H_5	$H_2C{-}S$, H_2C, $H_2C{-}S$ >Sn< C_6H_5, C_6H_5
νCC(phenyl)					1480s	1482s
Butyl CH_3			1462s	1463s		
δCH_3		1445s				
νCC(phenyl)					1433s	1433s
δCH_2	1418w	1412s	1419s	1422s	1418m	1423s
	1410s					
Phenyl					1382w	1388w
δCH_3		1372s	1376m	1377m		
			1361w			
νCC(phenyl)					1334m	1338m
Phenyl					1306w	
νC—C		1302s		1306s		1308s
CH_2wag.	1289s		1280s		1287s	
	1283s					
				1273w	1264w	1266w
CH_2wag.		1264s				
νC—C	1248s		1246s	1258m	1251w	1259m
νC—CH_3			1238m			
						1242w
CH_2wag.		1222s	1193m	1197m		1180s
βCH(phenyl)					1167m	

TABLE 7 (*continued*)

Assignment	$[H_2C{-}S]_2Sn[S{-}CH_2]_2$ (two 5-membered rings)	$[H_2C{-}S{-}CH(CH_3){-}S]Sn[S{-}CH_2{-}CH(CH_3){-}S]$	$(H_2CS)_2Sn(C_4H_9)_2$ (5-membered ring)	$(H_2C)_3S_2Sn(C_4H_9)_2$ (6-membered ring)	$(H_2CS)_2Sn(C_6H_5)_2$ (5-membered ring)	$(H_2C)_3S_2Sn(C_6H_5)_2$ (6-membered ring)
CH_2twist	1156m	1176m	1161w			
	1122m		1114w	1118m		
νC—C		1099s				
νCC(butyl)			1074s	1076s		
βCH(phenyl)					1072s	1075s
			1050w	1050w		
CH_2rock		1048s				
			1035w			
δCH_3		1028m				
βCH(phenyl)					1022m	1023m
Butyl			1016m	1016s		
			1002w	1004w		
Phenyl ring					998m	997s
					976s	
CH_2rock	997m			981w	988w	986w
					972w	
ν_{as}S—C	930m	892s	920s	980m	924s	924m
	922s					
					915m	
			887m			
Butyl			866s	867s		
ν_sC—S	840s	812s	848s	854s	847s	857s

CH(phenyl)					732s	739s
					727s	731s
					696s	700s
		791w				
δCH_3(ring)		694s				
Butyl			674s	671s		
	660w	665w			657m	665m
	652m		638m		634m	636w
	625m					
			591m	591m		
δCH_3(ring)		597s				
		540s				
		537sh				
νSn—C(butyl)			508m	501w		
		485w	462w	479w		
		474w		453w		
δPhenyl ring					451vs	451vs
δRing	442s	441m	438s		440s	446s
						441m
			399m	398m		
ν_sSn—S	388s	392vs	357s	333s	368s	393w
					327w	348s
						330w
ν_{as}Sn—S	327s	363s	313m	305w	317w	270s
		341s	267s	270s	270s	
		310m				
	258m	270m				
δ Ring	233s	228s	201m	230m	247s	237m
		183w			203m	206w
Out-of-plane ring			158m	169s	179ms	177m

TABLE 8

FORCE CONSTANTS FOR DITHIACYCLOPENTANES

	mdyn/Å		mdyn/Å
f_{CS}	3.65	$r^2_{SSn}f_{SSnS}$	0.06
$f_{CS/CS}$	1.04	$r^2_{CC}f_{CCS}$	0.28
f_{CC}	4.58	$r^2_{CS}f_{CSSn}$	0.2
f_{SnS}	2.25		

Epstein et al. (*67*) have described the Mössbauer spectra of tin-1,2-ethanedithiolate and tin-3,4-toluenedithiolate (Table 9). Evidence was presented against a tetrahedral configuration in $Sn(EDT)_2$ and in favor of a polymeric structure for $Sn(TDT)_2$. Dimethyl and diphenyltin dithiolates show a large quadrupole coupling. Complexation of these compounds with pyridine or

TABLE 9

ISOMER SHIFTS AND QUADRUPOLE COUPLING CONSTANTS RELATIVE TO Mg_2Sn (LIQUID NITROGEN TEMPERATURE)

Compound	δ (mm/sec)	Δ (mm/sec)
$Sn(EDT)_2$	−0.48	0.98
$Sn(EDT)_2 \cdot 2py$	−0.77	1.84
$Sn(EDT)_2 \cdot o\text{-phen}$	−0.86	1.00
$Sn(TDT)_2$	−0.61	1.52
$Sn(TDT)_2 \cdot 2py$	−0.86	1.69
$Sn(TDT)_2 \cdot 2DMSO$	−0.87	1.62
$Sn(TDT)_2 \cdot 2(C_2H_5)_3N$	−1.32	0.84
$Sn(TDT)_2 \cdot o\text{-phen}$	−0.81	1.26
$Sn(TDT)_2 \cdot bipy$	−1.00	0.92
$(CH_3)_2Sn(EDT)$	−0.47	2.33
$(CH_3)_2Sn(EDT) \cdot py$	−0.57	2.25
$(CH_3)_2Sn(TDT)$	−0.46	2.62
$(CH_3)_2Sn(TDT) \cdot py$	−0.56	2.28
$(CH_3)_2Sn(TDT) \cdot o\text{-phen}$	−0.82	2.05
$(C_6H_5)_2Sn(EDT)$	−0.46	1.69
$(C_6H_5)_2Sn(EDT) \cdot py$	−0.60	1.76
$(C_6H_5)_2Sn(EDT) \cdot o\text{-phen}$	−0.82	1.76
$(C_6H_5)_2Sn(TDT)$	−0.49	1.93
$(C_6H_5)_2Sn(TDT) \cdot py$	−0.58	1.76

EDT = 1,2-ethane dithiolate, TDT = 3,4-toluene dithiolate, py = pyridine, *o*-phen = 1,10-phenanthroline, DMSO = dimethylsulfoxide, bipy = 2,2′-bipyridyl.

1,10-phenanthroline, which increases the coordination number of tin, causes a negative shift, suggesting a diminished electron density at tin.

Frye et al. (*72, 73, 74, 75, 76*) studied the mechanism of stabilization of polyvinyl chloride with various radioactively labeled dibutyltin-β-mercapto-propanoates.

8. *Polymers*

Within the large class of polymeric organotin compounds, various types are known, in which functional organic groups are attached to sulfur, which itself is bound to tin. Thus polymers of the following structure are formed by the reaction of polymeric alkyl or alkynyltin dihalides or oxides with mercaptides (*P36*):

$$\left[-\overset{\displaystyle SR}{\underset{\displaystyle SR}{\overset{|}{\underset{|}{Sn}}}}-(CH_2)_n- \right]_n$$

The reaction of diorganotin oxides with various dimercaptobenzene derivatives leads to high molecular weight compounds with bridging aromatic dimercapto groups (*P1, P70, P72, P94, P155*):

$$n\,R_2SnO + n\,HSCH_2-C_6H_4-CH_2SH \longrightarrow \left[-\overset{\displaystyle R}{\underset{\displaystyle R}{\overset{|}{\underset{|}{Sn}}}}-SCH_2-C_6H_4-CH_2S- \right]_n + n\,H_2O$$

In a similar way, polymers with Sn—O—Sn chains, which contain functional groups bound to tin through sulfur, (*P50, P93, P140*) are formed, as well as high molecular weight compounds with dithiols, mercapto acids, thioglycolic acid, or polysulfide groups as binding units between the diorganotin groups (*56, 129, 170, P20, P38b, P70, P73, P78, P94, P103, P104, P105, P106, P157, P158c*):

$$n(C_6H_5)_2SnCl_2 + n\,HSCH_2COONa \longrightarrow \left[-\overset{\displaystyle C_6H_5}{\underset{\displaystyle C_6H_5}{\overset{|}{\underset{|}{Sn}}}}-SCH_2C(O)O- \right]_n + n/2\,NaCl + n/2\,HCl$$

$$(n+1)\,R_2SnCl_2 + n\,Na_2S_x \longrightarrow Cl\left[-\overset{\displaystyle R}{\underset{\displaystyle R}{\overset{|}{\underset{|}{Sn}}}}-S_x- \right]_n \overset{\displaystyle R}{\underset{\displaystyle R}{\overset{|}{\underset{|}{Sn}}}}Cl + 2n\,NaCl$$

$$n\ \mathrm{Cl{-}\underset{C_4H_9}{\overset{C_4H_9}{Sn}}OC(O)(CH_2)_4C(O)O\underset{C_4H_9}{\overset{C_4H_9}{Sn}}Cl} + n\ \mathrm{Na_2S} \longrightarrow \left[\mathrm{-\underset{C_4H_9}{\overset{C_4H_9}{Sn}}OC(O)(CH_2)_4C(O)O\underset{C_4H_9}{\overset{C_4H_9}{Sn}}S-}\right]_n + n\ \mathrm{NaCl}$$

Various polymers with tin-sulfur fragments are also obtained by the reaction of allyl organotin compounds with polysulfanes (*211*), of tin tetrachloride with *cis*-ethylene bis-thiol (*92*), or by the redistribution reactions of dimethyltin dihalides and dimethyltin sulfide (*132*):

$$\mathrm{(CH_3)_2SnI_2} + n/3\,[\mathrm{(CH_3)_2SnS}]_3 \longrightarrow \mathrm{I}\left[\mathrm{\underset{CH_3}{\overset{CH_3}{Sn}}-S-\underset{CH_3}{\overset{CH_3}{Sn}}}\right]_n\mathrm{I}$$

Polymers containing six-membered tin-sulfur rings can be obtained by the reaction of tetraphenyltin (*194*) or organotin halides (*202*) with sulfur; alternatively they are formed in the reaction of bis(trimethylsilylmethyl)tin diiodide with hydrogen sulfide in ethanol (*184*), in which case low molecular weight compounds are also formed.

The reactions of various organotin sulfides with polyvinyl chloride, which give rise to commercially important polymers containing tin-sulfur bonds, are discussed in several review articles (*96*, *167*, *168*, *169*, *170*). The monomeric units in polymeric organotin sulfides are listed in Appendix 9.

9. *Miscellaneous Compounds*

Besides the compounds already mentioned in Sections 1–8, investigations have also been carried out on other classes of compounds containing the tin-sulfur bond as a structural feature. Although organotin thiols $R_nSn(SH)_{4-n}$ have not been, as yet, isolated in pure form, triethyltin thiol has been proposed by Vyazankin et al. (*227*) as an intermediate from triethyltinhydride and sulfur. The compound eliminates hydrogen sulfide instantly and is converted to the stable bis(triethyltin) sulfide. The simultaneous evolution of hydrogen is attributed to the reaction of the thiol with excess triethyltinhydride:

$$8\,\mathrm{(C_2H_5)_3SnH} + \mathrm{S_8} \longrightarrow 8\,\mathrm{(C_2H_5)_3SnSH}$$

$$2\,\mathrm{(C_2H_5)_3SnSH} \longrightarrow \mathrm{H_2S} + \mathrm{(C_2H_5)_3Sn{-}S{-}Sn(C_2H_5)_3}$$

$$\mathrm{(C_2H_5)_3SnSH} + \mathrm{(C_2H_5)_3SnH} \longrightarrow \mathrm{H_2} + \mathrm{(C_2H_5)_3Sn{-}S{-}Sn(C_2H_5)_3}$$

Pfeiffer et al. (*146*) presumed that the reaction of methyltin tribromide with hydrogen sulfide had led to methyldithiostannonic acid. In reality, however, they had probably prepared a polymeric compound with the structure $(CH_3SnS_{1.5})_n$.

A few analogous alkali thiolates have proved isolable. Thus lithium triphenyltin sulfide, dimeric in benzene, arises from the cleavage of S_8 with triphenyltin lithium in tetrahydrofuran (*205, 206*). The compound is extremely sensitive to air and decomposes on standing into bis(triphenyltin) sulfide and lithium sulfide. Like its sodium analog (*88*), diphenyltin bis (lithium sulfide) (*207*), dimethylstannyl-bis(sodium sulfide) (*P38f*), diethylstannyl-bis(sodium sulfide) (*P38f*), or dibutyltin bis(sodium sulfide) (*P34, P38f*), this compound is useful for the synthesis of unsymmetrical hexaphenyldimetallosulfides or even for new organotin sulfides:

$$8\ (C_6H_5)_3SnLi + S_8 \longrightarrow 4\ (C_6H_5)_3Sn\underset{\underset{\displaystyle Li}{\overset{|}{S}}}{\overset{\overset{\displaystyle Li}{|}}{\overset{S}{\rightleftarrows}}}Sn(C_6H_5)_3$$

$$[(C_6H_5)_3SnSLi]_2 + 2\ R_3MCl \longrightarrow 2\ (C_6H_5)_3Sn{-}S{-}MR_3 + 2\ LiCl$$

$$4\ (C_6H_5)_2SnLi_2 + S_8 \longrightarrow 4\ (C_6H_5)_2Sn(SLi)_2$$

$$(C_4H_9)_2Sn(SNa)_2 + 2\ C_4H_9Cl \longrightarrow (C_4H_9)_2Sn(SC_4H_9)_2 + 2\ NaCl$$

Organotin thiosulfonates are formed by the reaction of tetraalkyltins, organotin halides or oxides with alkylthiosulfonates (*P149*).

$$(C_4H_9)_4Sn + 2\ HSSO_2CH_3 \longrightarrow (C_4H_9)_2Sn(SSO_2CH_3)_2 + 2\ C_4H_{10}$$

Compounds with the group —Sn—S—P=S have been intensively investigated in the last few years. This class of compound is available from the reaction of organotin halides or oxides with alkali dithiophosphinates (*37, 38, 110, 111, 118a, P8, P24, P38e, P71, P86, P118, P158f*) or by cleavage of the tin-phosphorus bond in organotin phosphines with sulfur (*200*):

$$(C_6H_5)_3SnCl + NaS\overset{\overset{\displaystyle S}{\|}}{P}(C_2H_5)_2 \longrightarrow (C_6H_5)_3Sn{-}S\overset{\overset{\displaystyle S}{\|}}{P}(C_2H_5)_2 + NaCl$$

$$(C_4H_9)_2SnO + 2\ (C_4H_9O)_2\overset{\overset{\displaystyle S}{\|}}{P}SH \longrightarrow (C_4H_9)_2Sn\left[S\overset{\overset{\displaystyle S}{\|}}{P}(OC_4H_9)_2\right]_2 + H_2O$$

$$4\ (C_6H_5)_3SnP(C_6H_5)_2 + S_8 \longrightarrow 4\ (C_6H_5)_3SnS\overset{\overset{\displaystyle S}{\|}}{P}(C_6H_5)_2$$

Kuchen et al. (*111*) and Bonati et al. (*38*) found these compounds to be monomeric, presumably as chelates with the following structure:

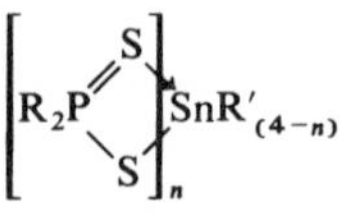

$$\left[R_2P\begin{matrix}\nearrow S \searrow \\ \searrow S \nearrow\end{matrix}\right]_n SnR'_{(4-n)}$$

From infrared (*38*), and [1]H-nmr (*38*) spectra, and dipole moment measurements (*38, 111*), it follows that the chelates $R_2Sn[SP(S)R'_2]_2$ contain two *cis*-oriented R groups on tin in the most likely octahedral complexes with hexacoordinated tin. Dichloro-bis(diethyldithiophosphinato) tin(IV) has a large dipole moment of 7.3 D, while the corresponding diphenyltin and dibutyltin compounds possess dipole moments of only 3.32 and 3.22 D, respectively (*111*). The infrared spectra have been reported for $Cl_2Sn[SP(S)(C_2H_5)_2]_2$ (*38*), $Br_2Sn[SP(S)(C_2H_5)_2]_2$ (*38*), and $(C_6H_5)_3SnSP(S)(C_6H_5)_2$ (*200*). The tin-sulfur stretching frequency for the last compound was found to be at 340 cm^{-1} (*200*).

It was discovered by Shindo et al. (*212, 213*) that dimethyl- and diethyltin dichloride and dibromide form complexes with trimethylantimony sulfide. These complexes partially decompose to dimethyltin sulfide and trimethylantimony dihalide by sulfur-halogen exchange:

$$R_2SnX_2 + 2\,(CH_3)_3SbS \longrightarrow R_2SnX_2 \cdot 2\,(CH_3)_3SbS$$

$$R_2SnX_2 \cdot 2\,(CH_3)_3SbS \longrightarrow (CH_3)_3SbS + (CH_3)_3SbX_2 + \tfrac{1}{3}\,(R_2SnS)_3$$

The structure of these complexes could be proved by x-ray analysis, infrared, and [1]H-nmr spectra.

Organotin sulfide silicates, which are used as stabilizers, have been prepared from organotin chlorides, organotin oxides, mercapto alcohols and organo-substituted silicates (*P158e*). SO_2 reacts with R_6Sn_2 and $R_3SnMn(CO)_5$ under fission of the metal-metal bonds. A structure with Sn—S bonds was discussed for one of the fission products (*44b*).

Finally, reference should be made to the synthesis of stanna-dibenzocycloheptadiene sulfide from stanna-dibenzocycloheptadiene and sulfur (*112*), as well as to the preparation of organotin sulfides with hexacoordinated tin (*P17, P126, P131*):

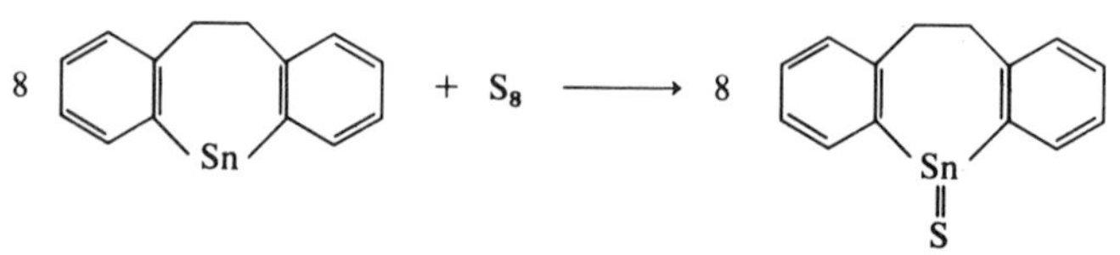

8 [stanna-dibenzocycloheptadiene] + S_8 ⟶ 8 [stanna-dibenzocycloheptadiene with Sn=S]

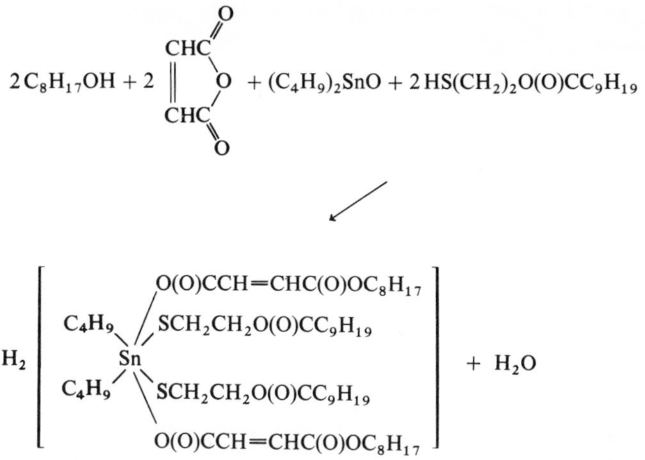

$$2\,C_8H_{17}OH + 2\ \begin{matrix} & & O \\ & & // \\ CHC & & \\ \| & & O \\ CHC & & \\ & & \backslash\backslash \\ & & O \end{matrix} + (C_4H_9)_2SnO + 2\,HS(CH_2)_2O(O)CC_9H_{19}$$

$$\swarrow$$

$$H_2\left[\begin{matrix} & & O(O)CCH{=}CHC(O)OC_8H_{17} \\ C_4H_9 & & SCH_2CH_2O(O)CC_9H_{19} \\ & Sn & \\ C_4H_9 & & SCH_2CH_2O(O)CC_9H_{19} \\ & & O(O)CCH{=}CHC(O)OC_8H_{17} \end{matrix}\right] + H_2O$$

A compilation of known compounds belonging in this section can be found in Appendix 10.

III. Organotin Selenides

A. Preparation

In 1942, Backer et al. (*26*, *27*) reported the first organometallic compounds with covalent tin-selenium bonds:

$$Sn(Se{-}t\text{-}C_4H_9)_4,\ Sn(SeC_6H_5)_4,\ Sn(Se{-}p\text{-}C_6H_4CH_3)_4,$$

$$Sn(Se{-}p\text{-}C_6H_4{-}t\text{-}C_4H_9)_4 \quad \text{and} \quad Sn(Se{-}p\text{-}C_6H_4Cl)_4.$$

They were prepared by the reaction of tin tetrachloride with the appropriate seleno-Grignard reagents, selenophenol, or sodium selenophenolates:

$$SnCl_4 + 4\ t\text{-}C_4H_9SeMgCl \longrightarrow Sn(Se\text{-}t\text{-}C_4H_9)_4 + 4\ MgCl_2$$

$$SnCl_4 + 4\ C_6H_5SeNa \longrightarrow Sn(SeC_6H_5)_4 + 4\ NaCl$$

In 1950, Tchakirian et al. (*218*) discovered that methyl- and ethyl-stannonic acid, in the presence of hydrogen selenide in aqueous solution under nitrogen, formed compounds of the type $(RSn)_2Se_3$, to which they assigned a structure with tin-selenium double bonds. They also isolated analogs of the corresponding polymeric sulfur compounds containing tin-selenium single bonds.

$$2\ RSnOOH + 3\ H_2Se \longrightarrow \begin{matrix} Se & & Se \\ \| & & \| \\ RSn & {-}Se{-} & SnR \end{matrix} + 4\ H_2O$$

Seven years later, the synthesis of dimethyltin selenide was described by Weinberg (*P122, P130*). Finally, by 1961, the systematic investigation of preparing organotin selenides got under way.

Bis(triorganotin) selenides and trimeric diorganotin selenides are formed by the reaction of the appropriate organotin halides with sodium selenide, in boiling benzene, in an atmosphere of nitrogen (*139*, *192*, *193*). Exclusion of air and moisture was not a necessary condition for synthesis, as was shown a year later by Abel et al. (*3*), who were able to carry out the synthesis of trimethyltin selenophenoxide by the reaction of trimethyltin chloride with selenophenol in aqueous solution. Analogous compounds were prepared in cyclohexane or ethanol (*16*).

$$2\,(C_6H_5)_3SnCl + Na_2Se \longrightarrow (C_6H_5)_3Sn{-}Se{-}Sn(C_6H_5)_3 + 2\,NaCl$$

$$3\,(CH_3)_2SnCl_2 + 3\,Na_2Se \longrightarrow [(CH_3)_2SnSe]_3 + 6\,NaCl$$

(cyclic trimer: six-membered ring of alternating Sn and Se, each Sn bearing two CH_3 groups)

$$(CH_3)_3SnCl + C_6H_5SeH \xrightarrow[-HCl]{[OH^-]} (CH_3)_3SnSeC_6H_5$$

Selenium is taken up gradually by triphenyltin lithium in tetrahydrofuran at room temperature, with the formation of lithium triphenyltin selenide (*205*, *208*), which, however, cannot be isolated in pure form. The solution, which is very sensitive to oxygen and heat, decomposes to lithium selenide and bis(triphenyltin) selenide. It reacts smoothly with organometallic halides with formation of triphenyltin-triphenylmetal selenides (*205*, *208*):

$$8\,(C_6H_5)_3SnLi + Se_8 \longrightarrow 8\,(C_6H_5)_3SnSeLi$$

$$(C_6H_5)_3SnSeLi + (C_6H_5)_3MCl \longrightarrow (C_6H_5)_3Sn{-}Se{-}M(C_6H_5)_3 + LiCl$$

$$(M = Ge, Sn, Pb)$$

In the reaction with diphenyltin dichloride, bis(triphenyltin) selenide and trimeric diphenyltin selenide were isolated (*208*):

$$6\,(C_6H_5)_3SnSeLi + 3\,(C_6H_5)_2SnCl_2 \longrightarrow 3\,(C_6H_5)_3Sn{-}Se{-}Sn(C_6H_5)_2{-}Se{-}Sn(C_6H_5)_3$$

$$\downarrow$$

$$3\,(C_6H_5)_3Sn{-}Se{-}Sn(C_6H_5)_3 + [(C_6H_5)_2SnSe]_3$$

A variety of unsymmetrical organometallic substituted selenoethers were isolated from the reaction of organotin halides with lithium-triphenylgermanium-, -triphenyllead-, or -trimethylgermanium selenide (*174*, *209*, *210*).

Vyazankin et al. (*225*, *226*, *227*, *227g*) discovered several novel, synthetic routes to organotin selenides. They obtained bis(triethyltin) selenide from the reaction of triethyltin hydride or hexaethylditin with elemental selenium or from triethyltin hydride and diethylselenide, with the elimination of ethane:

$$4\,(C_2H_5)_3SnH + Se_8 \longrightarrow 2\,(C_2H_5)_3Sn{-}Se{-}Sn(C_2H_5)_3 + 4\,H_2Se$$

$$8\,(C_2H_5)_3SnSn(C_2H_5)_3 + Se_8 \longrightarrow 8\,(C_2H_5)_3Sn{-}Se{-}Sn(C_2H_5)_3$$

$$2\,(C_2H_5)_3SnH + (C_2H_5)_2Se \longrightarrow (C_2H_5)_3Sn{-}Se{-}Sn(C_2H_5)_3 + 2\,C_2H_6$$

The same authors were able to use an analogous method for the synthesis of organogermyl organostannyl selenides (*64b*, *64c*, *226*, *227*):

$$(C_2H_5)_3GeSeH + (C_2H_5)_3SnH \longrightarrow (C_2H_5)_3Ge{-}Se{-}Sn(C_2H_5)_3 + H_2$$

Triethyltin isobutoxide reacts exothermically with benzylselenol, with the formation of triethyltin benzylselenide and isobutyl alcohol, which is removed by distillation (*138*):

$$(C_2H_5)_3SnOCH_2CH(CH_3)_2 + C_6H_5CH_2SeH \longrightarrow (C_2H_5)_3Sn{-}SeCH_2C_6H_5 + i\text{-}C_4H_9OH$$

Cleavage of tin-carbon bonds in tetraphenyltin or tetrabutyltin by elemental selenium occurs above 200°C in a sealed tube; the products are triphenyltin selenophenoxide and trimeric dibutyltin selenide, respectively (*195*). The successful isolation of triphenyltin selenophenoxide as a stable intermediate from nucleophilic attack by selenium on organotin compounds serves to substantiate the mechanism proposed for the analogous reaction with sulfur (*191*) (see page 321):

$$8\,(C_6H_5)_4Sn + Se_8 \longrightarrow 8\,(C_6H_5)_3SnSeC_6H_5$$

$$4\,(C_4H_9)_4Sn + Se_8 \longrightarrow \tfrac{4}{3}\,[(C_4H_9)_2SnSe]_3 + 4\,(C_4H_9)_2Se$$

Bis(triethyltin) telluride reacts with selenium forming bis(triethyltin) selenide and tellurium (*227b*):

$$[(C_2H_5)_3Sn]_2Te + Se \longrightarrow [(C_2H_5)_3Sn]_2Se + Te$$

N,*N*-Dimethylthioselenocarbamate complexes of dimethyltin have been prepared from *N*,*N*-dimethylthiocarbamoyl chloride and dimethyltin selenide in benzene (*98a*):

$$3\,(CH_3)_2NC(S)Cl + [(CH_3)_2SnSe]_3 \longrightarrow 3\ (H_3C)_2(Cl)Sn\langle{}^{S}_{Se}\rangle C{=}N(CH_3)_2$$

B. Chemical Properties and Reactions

Organotin selenides are colorless or light yellow liquids or crystalline solids, which usually dissolve readily in organic solvents such as alcohols, ethers, or aromatic hydrocarbons without decomposition. The alkyl derivatives possess a disagreeable, lingering odor. Almost all compounds are surprisingly stable toward hydrolysis and oxygen; only the lithium triorganotin selenides are decomposed rapidly by water and oxygen.

Dilute mineral acids in the cold cleave bis(triorganotin) selenides, triorganotin triorganometal selenides, and diorganotin selenides with the formation of organotin salts and hydrogen selenide, which is immediately oxidized to selenium by oxygen from the air. Bis(triphenyltin) selenide is decomposed in the same way, but only after prolonged heating in dilute hydrochloric acid.

The reactions of lithium triorganotin selenides with organometallic halides, which lead to new organotin selenides, have already been discussed in the previous section. Trimethylstannyl trimethylgermyl selenide, formed from lithium trimethylgermanium selenide and trimethyltin chloride, is thermally very unstable. It disproportionates on distillation, even below 50°C, to bis(trimethylgermyl) selenide and bis(trimethyltin) selenide (*174*). All such unsymmetrical hexaorganodimetal selenides undergo this disproportionation to some extent on long standing at room temperature (*197*, *204a*).

$$2\,(CH_3)_3Ge{-}Se{-}Sn(CH_3)_3 \longrightarrow (CH_3)_3Ge{-}Se{-}Ge(CH_3)_3 + (CH_3)_3Sn{-}Se{-}Sn(CH_3)_3$$

The tin-selenium bond in bis(tributyltin) selenide is not measurably cleaved by organotin hydrides at 160°C (*139*). By contrast, triethyltin benzylselenide, heated with acetic anhydride, forms triethyltin acetate and selenobenzyl acetate (*138*) and bis(triethyltin) selenide reacts with benzoylperoxide forming triethyltin benzoate and selenium (*227a*):

$$(C_2H_5)_3SnSeCH_2C_6H_5 + (CH_3CO)_2O \longrightarrow (C_2H_5)_3SnOC(O)CH_3 + CH_3C(O)SeCH_2C_6H_5$$

$$[(C_2H_5)_3Sn]_2Se + [C_6H_5C(O)O]_2 \longrightarrow 2\,(C_2H_5)_3SnOC(O)C_6H_5 + Se$$

Chlorine, bromine and iodine split the Sn—Se bond in bis(triethyltin) selenide with the formation of triethyltin halogenides and elemental selenium (*36b*, *227g*), and the same tin-selenium compound reacts with sulfur yielding bis(triethyltin) sulfide (*36b*, *227b*):

$$[(C_2H_5)_3Sn]_2Se + X_2 \longrightarrow 2\,(C_2H_5)_3SnX + Se \quad (X = Cl, Br, I)$$

$$8\,[(C_2H_5)_3Sn]_2Se + S_8 \longrightarrow 8\,[(C_2H_5)_3Sn]_2S + 8\,Se$$

Abel et al. (*15*, *16*) succeeded in synthesizing polynuclear, selenium-bridged manganese and rhenium carbonyls by the reaction of trimethyltin selenophenoxide, dimethyltin bis(selenophenoxide), and analogous organotin methyl- and ethylselenides with manganese and rhenium bromopentacarbonyls:

$$2\,R_3SnSeR' + 2\,M(CO)_5X \longrightarrow$$

$$R_2Sn(SeR')_2 + 2\,M(CO)_5X \longrightarrow$$

R′
OC CO Se CO OO + $2\,R_3SnX + 2\,CO$
M M
OC CO Se CO CO + $R_2SnX_2 + 2\,CO$
R′

($R = CH_3$; $R' = CH_3$, C_2H_5, C_6H_5; M = Mn, Re; X = Cl, Br)

By pyrolysis in boiling cyclohexane or xylene, these complexes lose carbon monoxide and form polymeric alkyl- and arylselenomanganese- and rhenium tricarbonyl complexes with the following unit clusters:

M—SeR′
SeR′ M SeR′
M SeR′ M

M—SeR′
R′Se M
SeR′ M
M—SeR′

Bis(trimethyltin) selenide pentacarbonyl-chromium, -molybdenum, and -tungsten complexes are obtained from $M(CO)_6$ and bis(trimethyltin) selenide by uv-irradiation (*210a*):

$$[(CH_3)_3Sn]_2Se + M(CO)_6 \xrightarrow{h\nu} [(CH_3)_3Sn]_2Se{-}M(CO)_5 + CO$$

(M = Cr, Mo, W)

C. Physical Properties

Details on specific physical properties of organotin selenides have rarely appeared in the literature. Backer and Hurenkamp (*27*) determined the unit cell constants for the rhombic, pseudotetragonal tetra(seleno-*t*-butoxy)-tin. Complete x-ray structure analyses for organotin selenides have not been reported.

The infrared spectra and, in part, Raman spectra of several organotin selenides were reported by Kriegsmann (*108*, *109*, *109a*), Schumann (*197*, *204*), and Vyazankin (*64c*). A careful and complete analysis of a Raman spectrum is available only for bis(trimethyltin) selenide (*109*), dimethyltin selenide (*109a*) and diethyltin selenide (*109a*). In this case the following bands were found:

For $[(CH_3)_3Sn_2Se]$: ν_{as}Sn—Se—Sn = 238 cm^{-1}, ν_sSn—Se—Sn = 224 cm^{-1} and δSn—Se—Sn = 57 cm^{-1}; for $[(CH_3)_2SnSe]_3$: ν_{as}Sn—Se—Sn = 267 cm^{-1}, ν_sSn—Se—Sn = 255 cm^{-1}; for $[(C_2H_5)_2SnSe]_3$: ν_{as}Sn—Se—Sn = 266 cm^{-1}, ν_sSn—Se—Sn = 255 cm^{-1}.

From these values Kriegsmann estimated the angle at selenium to be 98° and the tin-selenium force constant to be 1.5 mdyn/Å, a value consistent for a single bond.

^{1}H-nmr spectra have been measured for the organotin selenides listed in Table 10 (*8, 65, 171, 174, 188*).

TABLE 10

^{1}H-NMR SPECTRA OF ORGANOTIN SELENIDES

Compound	τ_{CH_3}	τ_{CH_2}	$J(^{117}$Sn—CH) $J(^{119}$Sn—CH)	$J(^{117}$SnSeCH) $J(^{119}$SnSeCH)	$J(^1H^{13}C)$	References
$[(C_2H_5)_3Sn]_2Se$	8.78	8.93				(*65*)
$(CH_3)_3SnSeCH_3$			53.4 55.9	30.0 31.4		(*8*)
$[(CH_3)_3Sn]_2Se$	9.43		53.5 56.0		132.0	(*171, 174, 188*)

Dessy et al. (*63*) investigated the electrochemical behavior of various organometallic compounds, among them bis(triphenyltin) selenide and triphenylstannyl triphenylgermyl selenide. It could be determined polarographically that both compounds are reduced in two one-electron processes. In the first step of the reduction of the symmetrical compound, hexaphenylditin is formed; but in the second case only triphenylgermanium hydride could be detected. This is taken as clear evidence for the preferential cleavage of the germanium-selenium bond over the tin-selenium bond.

$$(C_6H_5)_3SnSeSn(C_6H_5)_3 \xrightarrow[-2.4\text{ V}]{e} (C_6H_5)_3Sn^{\bullet} + (C_6H_5)_3SnSe^-$$

$$\xrightarrow[-2.9\text{ V}]{2e} (C_6H_5)_3Sn^- + (C_6H_5)_3SnSe^-$$

$$(C_6H_5)_3SnSeGe(C_6H_5)_3 \xrightarrow[-2.4\text{ V}]{e} (C_6H_5)_3SnSe^- + (C_6H_5)_3Ge^{\bullet}$$

$$\xrightarrow[-2.9\text{ V}]{2e} (C_6H_5)_3SnSe^- + (C_6H_5)_3Ge^-$$

The known organotin selenides with their physical constants are assembled in Appendix 11.

IV. Organotin Tellurides

In contrast to the large number of known organotin sulfides, the list of organotin tellurides shown in Appendix 12 is pitifully small. Apart from any mention in review articles, only twenty-one publications and two patents concerning the sythesis or properties of organotin tellurides have appeared.

The first was a description by Weinberg (*P122*, *P130*) of the synthesis of dibutyltin telluride. This compound is formed by the introduction of hydrogen telluride into a suspension of dibutyltin oxide in toluene. As described by the authors, it is isolated in solid form by heating to remove excess hydrogen telluride and solvent:

$$(C_4H_9)_2SnO + H_2Te \longrightarrow (C_4H_9)_2SnTe + H_2O$$

Physical data were unfortunately not given. In 1963 bis(triphenyltin) telluride was prepared from triphenyltin chloride and lithium triphenyltin telluride in tetrahydrofuran, with strict exclusion of moisture and atmospheric oxygen. The lithium triphenyltin telluride needed for the synthesis could be obtained, by analogy with the corresponding sulfur and selenium compounds, by nucleophilic attack of triphenyltin lithium on tellurium. The lithium salt could not be isolated in pure form because of its instability toward oxygen and moisture (*205*, *208*):

$$n\,(C_6H_5)_3SnLi + Te_n \longrightarrow n\,(C_6H_5)_3SnTeLi$$

$$(C_6H_5)_3SnTeLi + (C_6H_5)_3SnCl \longrightarrow (C_6H_5)_3Sn\text{—}Te\text{—}Sn(C_6H_5)_3 + LiCl$$

Bis(triphenyltin) telluride decomposes slowly in air with precipitation of tellurium, as does triphenylgermyl triphenylstannyl telluride (*209*) (obtained from lithium triphenylgermanium telluride and triphenyltin chloride). Triphenylplumbyl triphenylstannyl telluride (*208*), on the other hand, is essentially stable toward atmospheric oxygen and hydrolysis.

Vyazankin et al. (*225*, *226*, *227c*, *227g*) obtained bis(triethyltin) telluride in high yield as a distillable liquid from the reaction of triethylstannane with tellurium or bis(triethylgermyl) telluride (*36b*) or by elimination of ethane from triethylstannane and diethyltelluride. The same authors synthesized unsymmetrical organotin tellurides from triethylsilyl or triethylgermyl ethyl telluride and triethylstannane (*64a*, *64b*, *64c*, *226*, *227c*):

$$2\,(C_2H_5)_3SnH + Te \longrightarrow (C_2H_5)_3Sn\text{—}Te\text{—}Sn(C_2H_5)_3 + H_2$$

$$2\,(C_2H_5)_3SnH + (C_2H_5)_2Te \longrightarrow (C_2H_5)_3Sn\text{—}Te\text{—}Sn(C_2H_5)_3 + 2\,C_2H_6$$

$$(C_2H_5)_3M\text{—}TeC_2H_5 + (C_2H_5)_3SnH \longrightarrow (C_2H_5)_3M\text{—}Te\text{—}Sn(C_2H_5)_3 + C_2H_6$$

$$(M = Si, Ge)$$

Very little is known about the chemical behavior of organotin tellurides. They resemble the analogous selenium derivatives in general, but their thermal stability is considerably lower (*197*).

Chlorine, bromine and iodine split the Sn—Te bond in bis(triethyltin) telluride (*227g*), as well as benzoyl peroxide (*227a*) or sulfur and selenium (*36b*, *227b*) in the following manner:

$$[(C_2H_5)_3Sn]_2Te \xrightarrow{+X_2} 2(C_2H_5)_3SnX + Te$$

$$[(C_2H_5)_3Sn]_2Te \xrightarrow{+(C_6H_5CO_2)_2} 2(C_2H_5)_3SnOCOC_6H_5 + Te \quad (X = Cl, Br, I; Y = S, Se)$$

$$[(C_2H_5)_3Sn]_2Te \xrightarrow{+Y} [(C_2H_5)_3Sn]_2Y + Te$$

Bis(trimethyltin) telluride pentacarbonyl-chromium and -tungsten complexes are obtained from $M(CO)_6$ and bis(trimethyltin) telluride by uv-irradiation (*210a*):

$$\underset{(M=Cr,\ W)}{[(CH_3)_3Sn]_2Te + M(CO)_6} \xrightarrow{h\nu} [(CH_3)_3Sn]_2Te{-}M(CO)_5 + CO$$

Although the infrared spectra of the triphenyltin tellurides have been recorded (*36b*, *64a*, *64c*, *204*), no assignment was made for the tin-tellurium stretching frequency. The following signals for the ethyl protons in the ^{1}H-nmr spectrum of bis(triethyltin) telluride were reported (*65*): $\tau_{CH_2} = 8.91$; $\tau_{CH_3} = 8.80$. In the same publication the ^{1}H-nmr spectra of unsymmetrical triethyltin tellurides were compared with the analogs containing oxygen, sulfur, and selenium or silicon and germanium.

V. Comparative Summary

In the organotin chalcogens, a covalent bond exists between tin on the one hand and sulfur, selenium, or tellurium on the other. The tin atom is sp^3-hybridized; the chalcogens, with two free electron pairs, show the tendency for one of these pairs to be in the *s*-orbital. Thus the bonds in question possesses considerable *p*-character.

The strength of such a bond can be influenced either by a change in the *s*-character or by contribution of $(p \rightarrow d)\pi$ double bond interactions. By introduction of various substituents at tin, the sp^3-hybridization, and therefore the bond distance and bond angle, is altered. This effect is dependent on the electronegativity of the substituent. For molecules with a large chalcogen bridging atom and low polarity in the Sn—X—Sn bond, the angle should approach 90°. A spreading of this value should arise by *sp*-hybridization with participation of the two unshared electron pairs. At the same time the empty *d*-orbitals of the tin atom can become involved in a partial $(p\text{-}d)\pi$ double bond. The x-ray structure analysis of $[(C_6H_5)_2SnS]_3$ and $(C_6H_5)_3SnSPb(C_6H_5)_3$ indicates that the angle at sulfur in the former is nearly the same as that in dimethylsulfide (where no double bond character is possible); in the latter case the angle is only 2° larger. This is easily explained in terms of a larger spatial requirement of triphenyltin or triphenyllead groups as compared to methyl groups. The bond angles at tin and lead agree within experimental error with the expected tetrahedral angle. The measured Sn—S and Pb—S bond distances are in accord with single bonds.

There is likewise no evidence for a double bond character from infrared or Raman spectra of organotin sulfides and selenides. The uniform decrease in ν_{as} and ν_s frequencies for M—S—M in the order $(C_6H_5)_3Ge—S—Ge(C_6H_5)_3$, $(C_6H_5)_3Sn—S—Sn(C_6H_5)_3$, $(C_6H_5)_3Pb—S—Pb(C_6H_5)_3$ (*204*) can be explained by the increasing size of the bridgehead atoms. This also explains the increase in the difference $\nu_{as} - \nu_s$ in the same order, an observation which indicates an opening up of the sulfur bond angle. All of the force constants calculated by Kriegsmann (*109*) for various bis(triorganotin) sulfides and for bis(trimethyltin) selenide indicate a bond order of 1.0. Abel et al. (*2, 5*) investigated the donor properties of group IV sulfur compounds (Table 11).

TABLE 11

Frequency Shifts (cm^{-1}) for ν_{CD} Relative to ν_{CD} in Chloroform Vapor in $CDCl_3$ Solutions of Some Organotin Sulfides

M	$(CH_3)_3MSCH_3$	$(CH_3)_3MSC_2H_5$	$(CH_3)_3MSM(CH_3)_3$
C	33	36	40
Si	29	32	29
Ge	34	36	28
Sn	36	41	43
Pb	49	53	51

From the electronegativities of the group IV elements, it is thought that silicon, germanium, and tin should form the strongest bases. In spite of the small differences in electronegativity among these three elements, the donor properties in the sulfides investigated varied significantly. It thus seems justified to assume that, in the case of silicon and germanium compounds, π-bonds play an important role. With organotin sulfides this possibility is apparently negligible. The large increase in donor strength with organolead sulfides is attributed to d_π-d_π sharing between filled lead *d*-orbitals and empty sulfur orbitals.

The same conclusions have been reached by Cumper et al. (*53, 54*) from uv spectra and dipole moment measurements of similar compounds. Likewise, investigations of Mössbauer spectra (*66, 89*) of organotin sulfides indicate pure sp^3-hybridization at the tin atom and the absence of $(p \rightarrow d)\pi$ double bond character. In their chemical behavior organogermanium, organotin, and organolead sulfides are practically indistinguishable. Differences appear, however, in the acid hydrolysis of selenium compounds; the germanium-selenium bond, unlike the stable tin-selenium and lead-selenium bond, is easily cleaved. In the case of tellurium compounds, only the lead-tellurium bond is stable. Thus a group IV-group VI bond is only stable to solvolysis if the covalent radius of the group VI element is significantly smaller than that of the group IV element. The highest stability is always found in the lead compounds, provided no opposing steric effect is operating. These comparisons are given in Table 12.

TABLE 12

Comparison of Covalent Radii (Å) in Compounds Between Group IV and Group VI Elements (Elements below the lines form compounds stable toward solvolysis)

		O 0.74	S 1.04	Se 1.17	Te 1.37
Si	1.17	0.43	0.13	0.00	−0.20
Ge	1.22	0.48	0.18	0.05	−0.15
Sn	1.41	0.67	0.37	0.24	0.04
Pb	1.54	0.80	0.50	0.37	0.17

In going from oxygen to tellurium as the bridging atom in such bridged compounds, the bond angle should approach 90°. If the silicon atom is replaced by the heavier group IV elements as bridgehead atoms, however, the bond angle should become larger on steric grounds. In Table 13 is a compilation of IV–VI–IV compounds for which determinations have been made. It is noteworthy, however, that the substituents on the group IV element are not all the same.

The lead compounds, whose bond angle is largest in every case, should be best suited for overlapping of the filled *p*-orbitals of the group VI atom with the empty *d*-levels of the group IV atom. This might explain the singular stability of all such lead compounds. This overlapping, however, can scarcely account entirely for bond strengths as is clear from measurements of bond lengths, angles, and energies. In order to consider further the problem of

TABLE 13

BOND ANGLES FOR THE GROUP IV–VI–IV SKELETON (GROUPS ATTACHED TO THE GROUP IV ATOM ARE SHOWN IN PARENTHESES)

	O	S	Se	Te
C	111 (H)	105 (H)	98 (H)	—
Si	155 (H)	100 (H)	—	—
Ge	111 (H)	116 (H)	—	—
Sn	128 (CH_3)	104 (CH_3)	98 (CH_3)	—
	139 (C_6H_5)	104 (C_6H_5)	—	—
Pb	—	107 (C_6H_5)*	—	—

* $(C_6H_5)_3Sn{-}S{-}Pb(C_6H_5)_3$.

bonding in organotin sulfides, selenides, and tellurides, more x-ray structure determinations must be carried out and additional spectroscopic and chemical information be assembled.

APPENDIX 1

Compounds of the Type $R_nSn(SR')_{4-n}$

Compound	mp(°C)	bp(°C/mmHg)	n_D^{20}	d_4^{20}	References
R_3SnSR'					
$(CH_3)_3SnSCH_3$		161–3(*3, 7*)	1.5285(*3*)	1.453(*7*)	(*2, 3, 5, 7, 8, 9, 12, 17b, 219, 221a*)
			1.5303(*7*)		
$(CH_3)_3SnSC_2H_5$		177(*7*)	1.5205(*7*)	1.394(*7*)	(*2, 5, 7, 9, 10, 43*)
		105/100(*43*)	1.5215(*43*)	1.3943(*43*)	
$(CH_3)_3SnS\text{-}n\text{-}C_3H_7$		54/3(*7*)	1.5178(*7*)	1.352(*7*)	(*7, 10*)
$(CH_3)_3SnS\text{-}i\text{-}C_3H_7$		24–5/0.01(*3*)	1.5108(*3*)	1.318(*7*)	(*2, 3, 5, 7*)
		182(*7*)	1.5123(*7*)		
$(CH_3)_3SnS\text{-}n\text{-}C_4H_9$		40–2/0.01(*3*)	1.5090(*3*)	1.281(*7*)	(*3, 7, 9, 10, 79b*)
		44/0.05(*7*)	1.5098(*7*)		
$(CH_3)_3SnS\text{-}t\text{-}C_4H_9$		42/0.1(*7*)	1.5083(*7*)	1.267(*7*)	(*2, 5, 7*)
$(CH_3)_3SnS\text{-}n\text{-}C_8H_{17}$		94/0.1(*7*)	1.5000(*7*)	1.175(*7*)	(*7, 10*)
$(CH_3)_3SnSC_{12}H_{25}$					(*P3, P124*)
$(CH_3)_3SnSCH_2C_6H_5$		52/0.4			(*170a*)
$(CH_3)_3SnSC_6H_5$		69/0.01(*7*)	1.5934(*7*)	1.418(*7*)	(*7, 9*)
$(CH_3)_3SnSC_6F_5$		62/0.001(*3*)	1.5244(*3*)	1.71(*3*)	(*3, 90a, 144b*)
$(CH_3)_3SnS\text{-}p\text{-}C_6H_4I$					(*P1*)
2,3-bis[$(CH_3)_3SnS$]quinoxaline [structure: $(CH_3)_3SnS$, $SSn(CH_3)_3$, N, N]	115(*180*) (*P31*)				(*180, P31*)
$(C_2H_5)_3SnSCH_3$		94/2(*7*)	1.5274(*7*)	1.375(*7*)	(*7, 22, 24, 48, 181*)
		224(*22, 24*)	1.529(*22, 24*)	1.319(*24*)	

$(C_2H_5)_3SnSC_2H_5$	125–6/12(*176*)	1.5150(*176*)	1.278(*176*)	(*7, 43, 176, 177, 181*)
	68/0.7(*7*)	1.5153(*7*)	1.359(*7*)	
	78/3(*43*)	1.5272(*43*)	1.2689(*43*)	
$(C_2H_5)_3SnS\text{-}i\text{-}C_3H_7$	78–9/1(*176*)	1.5132(*176*)	1.236(*176*)	(*176, 177, 181*)
$(C_2H_5)_3SnS\text{-}n\text{-}C_4H_9$	88–91/1(*176*)	1.5133(*176*)	1.234(*176*)	(*48, 176, 181*)
$(C_2H_5)_3SnS\text{-}i\text{-}C_4H_9$	86–8/1(*176*)	1.5122(*176*)	1.244(*176*)	(*176, 177, 181*)
$(C_2H_5)_3SnS\text{-}t\text{-}C_4H_9$	47/0.02(*7*)	1.5130(*7*)	1.240(*7*)	(*7, 176, 181*)
	84–6/1(*176*)	1.5051(*176*)	1.240(*176*)	
$(C_2H_5)_3SnSCH_2CH_2CH(CH_3)_2$	96–7/1(*176*)	1.5060(*176*)	1.188(*176*)	(*176, 177, 181*)
$(C_2H_5)_3SnS(CH_2)_5CH_3$	126–7/1(*179*)	1.5032(*179*)	1.668(*179*)	(*179, 181*)
$(C_2H_5)_3SnS(CH_2)_6CH_3$	134–5/1(*179*)	1.5006(*179*)	1.1473(*179*)	(*179, 181*)
$(C_2H_5)_3SnSCH_2C_6H_5$	136–8/1(*176*)	1.5682(*176*)	1.304(*176*)	(*176, 179, 181*)
	138–40/1(*179*)	1.5675(*179*)		
$(C_2H_5)_3SnSC_6H_5$	98–100/0.2(*48*)	1.5812(*48*)	1.3163(*179*)	(*48, 135a, 179, 181*)
	138–40/1(*179*)	1.5828(*179*)		
$(C_2H_5)_3SnS\text{-}o\text{-}C_6H_4CH_3$	132–5/1(*176*)	1.5740(*176*)	1.295(*176*)	(*176, 179, 181*)
	132–6/1(*179*)	1.5720(*179*)		
$(C_2H_5)_3SnS\text{-}m\text{-}C_6H_4CH_3$	132–4/1(*176*)	1.5705(*176*)	1.283(*176*)	(*176, 181*)
$(C_2H_5)_3SnS\text{-}p\text{-}C_6H_4CH_3$	128–30/1(*176*)	1.5712(*176*)	1.288(*176*)	(*135a, 176, 177, 179, 181*)
	125–6/1(*177*)			
	129–32/1(*179*)			
$(C_2H_5)_3SnS\text{—}p\text{-}C_6H_4\text{—}t\text{-}C_4H_9$	$140/10^{-4}$			(*138a*)
$(C_2H_5)_3SnS\text{—}p\text{-}C_6H_4F$	100–5/1			(*135a*)
$(C_2H_5)_3SnS\text{—}p\text{-}C_6H_4Cl$	132–3/1			(*135a*)
$(C_2H_5)_3SnS\text{—}p\text{-}C_6H_4OCH_3$	136–9/1			(*135a*)
$(C_2H_5)_3SnS\text{—}p\text{-}C_6H_4N(CH_3)_2$	175–7/1			(*135a*)
$(C_2H_5)_3SnS\text{—}p\text{-}C_6H_4NO_2$	169–71/1			(*135a*)
$(C_2H_5)_3SnS\text{-}2\text{-}C_{10}H_7$	189–90/1	1.5308	1.3231	(*179*)
$(C_2H_5)_3SnSC_5H_4N$	88–9			(*107a*)

(continued)

APPENDIX 1 (*continued*)

COMPOUNDS OF THE TYPE $R_nSn(SR')_{4-n}$

Compound	mp(°C)	bp(°C/mmHg)	n_D^{20}	d_4^{20}	References
$(C_2H_5)_3SnS$–(1,3,4-thiadiazole, N—N, S)–SH	47–9				(*P82*)
$(C_2H_5)_3SnS$, $SSn(C_2H_5)_3$ (2,3-quinoxaline)					(*180, P31*)
$(C_2H_5)_3SnS$, $SSn(C_2H_5)_3$ (2,3-quinoxaline, H_3C)	91				(*180, P31*)
$(C_2H_5)_3SnS$, $SSn(C_2H_5)_3$ (2,3-quinoxaline, H_3CO)					(*180, P31*)

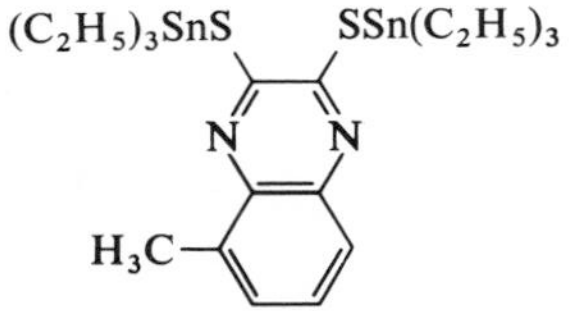 $(C_2H_5)_3SnS$ — $SSn(C_2H_5)_3$ (quinoxaline, H_3C)					(*P31*)
$(n\text{-}C_3H_7)_3SnS(CH_2)_6CH_3$		158–60/1	1.4981	1.1033	(*178*)
$(i\text{-}C_3H_7)_3SnS(CH_2)_6CH_3$		155–7/1	1.5045	1.0940	(*178*)
$(n\text{-}C_3H_7)_3SnS(CH_2)_9CH_3$		180–3/1	1.4998	1.0688	(*178*)
$(i\text{-}C_3H_7)_3SnS(CH_2)_9CH_3$		192–5/1	1.5010	1.0494	(*178*)
$(n\text{-}C_3H_7)_3SnSCH_2C_6H_5$		165–7/1	1.5558	1.2318	(*178*)
$(i\text{-}C_3H_7)_3SnSCH_2C_6H_5$		167–70/1(*178*)	1.5497(*178*)	1.2053(*178*)	(*178*, *P1*)
$(n\text{-}C_3H_7)_3SnSC_6H_5$		157–9/1(*177*)	1.5626(*177*)	1.2373(*177*)	(*177*, *178*)
$(i\text{-}C_3H_7)_3SnSC_6H_5$		138–9/1	1.5676	1.2327	(*178*)
$(n\text{-}C_3H_7)_3SnS\text{-}o\text{-}C_6H_4CH_3$					(*178*)
$(n\text{-}C_3H_7)_3SnS\text{-}p\text{-}C_6H_4CH_3$		157–9/1(*177*)	1.5516(*177*)	1.2125(*177*)	(*177*, *178*, *227i*)
		159–60/1(*178*)	1.5602(*178*)	1.2127(*178*)	
$(i\text{-}C_3H_7)_3SnS\text{-}p\text{-}C_6H_4CH_3$		157–8/1	1.5648	1.2191	(*178*)
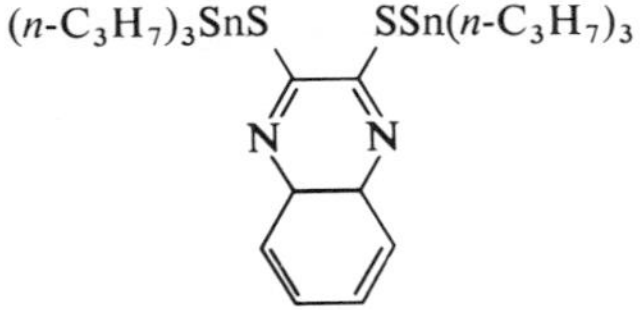 $(n\text{-}C_3H_7)_3SnS$ — $SSn(n\text{-}C_3H_7)_3$ (N, N ring)	108–10				(*180*, *P31*)
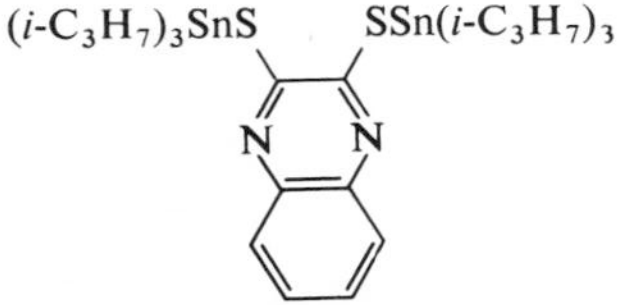 $(i\text{-}C_3H_7)_3SnS$ — $SSn(i\text{-}C_3H_7)_3$ (quinoxaline)	174				(*180*, *P31*)
$(i\text{-}C_3H_7)_3SnS$ — $SSn(i\text{-}C_3H_7)_3$ (N—N, S thiadiazole)					(*P82*)

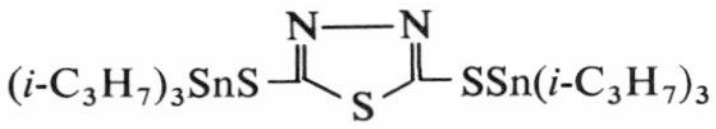

(*continued*)

APPENDIX 1 (*continued*)

Compounds of the Type $R_nSn(SR')_{4-n}$

Compound	mp(°C)	bp(°C/mmHg)	n_D^{20}	d_4^{20}	References
$(C_3H_7)_3SnSCH_2P(S)(OC_3H_7)_2$					(*P38e*)
$(C_3H_5)_3SnSC_5H_4N$		67–8/0.08			(*P160*)
$(n\text{-}C_4H_9)_3SnSCH_3$		104/0.2	1.5110(n_D^{22})		(*145a*)
$(n\text{-}C_4H_9)_3SnSC_2H_5$		119/0.4			(*128b*)
$(n\text{-}C_4H_9)_3SnSC_3H_7$		132–3/0.6			(*128b*)
$(n\text{-}C_4H_9)_3SnS\text{-}n\text{-}C_4H_9$		160–2/2(*83a*)	1.4981(*83a*)	1.102(*83a*)	(*48, 83a, 83b, 128b*)
$(n\text{-}C_4H_9)_3SnS\text{—}t\text{-}C_4H_9$		120–1/0.4			(*128b*)
$(n\text{-}C_4H_9)_3SnSC_{12}H_{25}$		190–5/0.3(*128b*)			(*P3, P39, P124, 128b*)
$(n\text{-}C_4H_9)_3SnSCH_2C_6H_5$		158–9/0.3			(*128b*)
$(n\text{-}C_4H_9)_3SnSC_6H_5$		137–44/0.26(*48*) 147/0.3(*145a*)	1.5452(*48*) 1.5479(n_D^{22})(*145a*)		(*30a, 48, 128b, 145a, P1, P158h*)
$(n\text{-}C_4H_9)_3SnS\text{-}p\text{-}C_6H_4CH_3$		170–2/0.8(*227i*)	1.5424(*277i*)		(*12, 227i*)
$(n\text{-}C_4H_9)_3SnS\text{-}p\text{-}C_6H_4\text{-}t\text{-}C_4H_9$		192–3/0.8	1.5351		(*227i*)
$(n\text{-}C_4H_9)_3SnS\text{-}p\text{-}C_6H_4Cl$		180–4/0.8	1.5505		(*227i*)
$(n\text{-}C_4H_9)_3SnSC_6F_5$		118/0.03	1.5109(n_D^{22})		(*145a*)
$(n\text{-}C_4H_9)_3SnSC_6Cl_5$		195/0.4			(*123a*)
$(C_4H_9)SnS$– (tetramethylthiazoline ring: S, N)					(*214a*)
$(C_4H_9)_3SnS$– (benzothiazole ring: S, N)					(*214a*)

$(n\text{-}C_4H_9)_3SnS$–(N—N, S ring)–SH	52–5			(*P82*, *P83*)
$(n\text{-}C_4H_9)_3SnS$—N (ring, S, H_5C_6)				(*P1*)
$(n\text{-}C_4H_9)_3SnS$–(N—N, S ring)–$SSn(n\text{-}C_4H_9)_3$		180/0.02		(*P82*)
$(n\text{-}C_4H_9)_3SnS$ $SSn(n\text{-}C_4H_9)_3$ (N, N ring)				(*180*, *P31*)
$(C_4H_9)_3SnSCH_2CH_2N(C_2H_5)_2$		165–70/3		(*P76e*, *P159*)
$(C_4H_9)_3SnSCH(CH_3)CH_2P(H)C_6H_5$		210–20/0.3		(*P76e*, *P159*)
$(C_4H_9)_3SnSC_5H_4N(2)$		183–6	$1.5480(n_D^{25})$	(*P160*)
$(C_4H_9)_3SnSC_5H_4N(4)$		120/0.01	$1.5275(n_D^{25})$	(*P160*)
$(C_4H_9)_3SnS$– (I, N, I ring)	65–6			(*P160*)
$(C_4H_9)_3SnS$– (N, S ring)		100/0.01	$1.5473(n_D^{25})$	(*P160*)
$(C_4H_9)_3SnS$– (N, N ring)		197/1.6	$1.5444(n_D^{25})$	(*P160*)

(*continued*)

APPENDIX 1 (*continued*)

COMPOUNDS OF THE TYPE $R_nSn(SR')_{4-n}$

Compound	mp(°C)	bp(°C/mmHg)	n_D^{20}	d_4^{20}	References
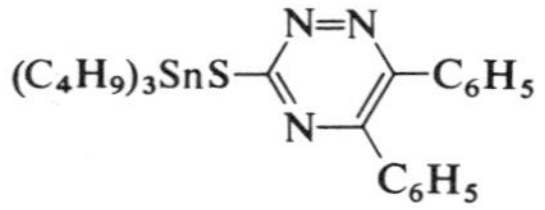 $(C_4H_9)_3SnS$–(N=N, N ring)–C_6H_5, C_6H_5			1.5897(n_D^{25})		(*P160*)
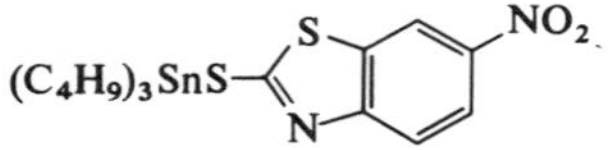 $(C_4H_9)_3SnS$–(S, N ring)–NO_2			1.5598(n_D^{25})		(*P160*)
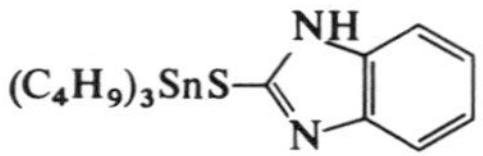 $(C_4H_9)_3SnS$–(NH, N ring)			1.5588(n_D^{25})		(*P160*)
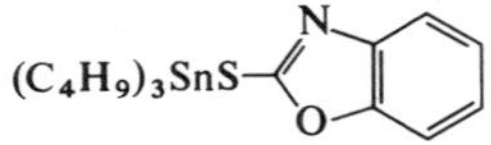 $(C_4H_9)_3SnS$–(N, O ring)		204–5/14	1.5608(n_D^{25})		(*P160*)
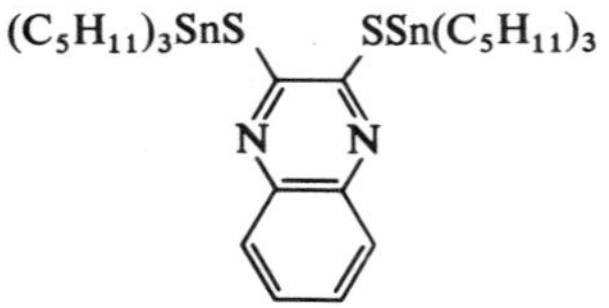 $(C_5H_{11})_3SnS$, $SSn(C_5H_{11})_3$ (N, N ring)					(*180, P31*)
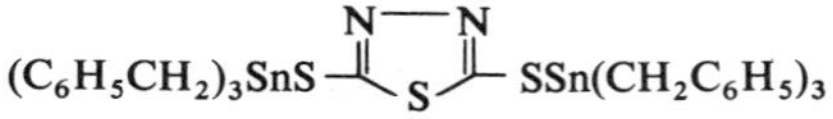 $(C_6H_5CH_2)_3SnS$–(N—N, S ring)–$SSn(CH_2C_6H_5)_3$	96–100				(*P85*)
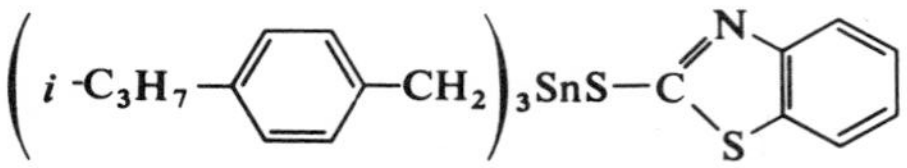 $(i\text{-}C_3H_7$–(ring)–$CH_2)_3SnS$–C(N, S ring)					(*55*)

$(C_6H_5)_3SnSCH_3$	94.6–95.4	(*145*)
$(C_6H_5)_3SnSC_{10}H_{21}$		(*P1*)
$(C_6H_5)_3SnSC_{12}H_{25}$		(*P3, P124*)
$(C_6H_5)_3SnSCH_2C_6H_5$	84	(*56*)
$(C_6H_5)_3SnS—C{\equiv}C—C_6H_5$	158	(*161*)
$(C_6H_5)_3SnSC_6H_5$	102–3(*56, 191*) 103(*32*) 98–99.5(*145*) 99–100(*48*)	(*32, 48, 56, 145, 152, 191*)
$(C_6H_5)_3SnS$-*p*-C_6H_4Cl	96–7	(*56*)
$(C_6H_5)_3SnS$—*p*-C_6H_4—*t*-C_4H_9		(*227i*)
$(C_6H_5)_3SnSC_6F_5$	76	(*90a, 144b, 145a*)
$(C_6H_5)_3SnSC_6Cl_5$	218	(*123a*)
$(C_6H_5)_3SnS$— 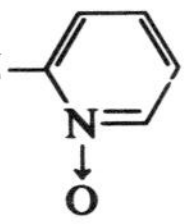 [pyridine ring, N→O]	110–2	(*P32*)
$(C_6H_5)_3SnS$— 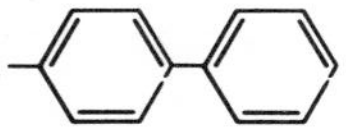 [biphenyl-4-yl]		(*P1*)
$(C_6H_5)_3SnS$-2-$C_{10}H_7$	74.5–76	(*145*)
$(C_6H_5)_3SnS$—$4C_5H_4N$	129–30	(*107a*)
$(C_6H_5)_3SnS$— 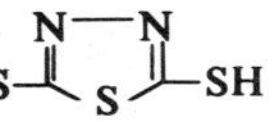 [N—N, S thiadiazole ring] —SH	114	(*P85*)
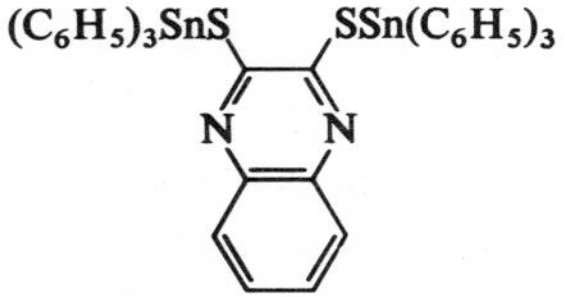 $(C_6H_5)_3SnS$, $SSn(C_6H_5)_3$ [on quinoxaline ring, N N]	220–1	(*180, P31*)

(*continued*)

APPENDIX 1 (*continued*)

Compounds of the Type $R_nSn(SR')_{4-n}$

Compound	mp(°C)	bp(°C/mmHg)	n_D^{20}	d_4^{20}	References
$(C_6H_5)_3SnS$–(1,3,4-thiadiazole-2,5-diyl)–$SSn(o\text{-}C_6H_4CH_3)_3$					(*P85*)
6-[$S-Sn(C_6H_5)_3$]-9-[$Sn(C_6H_5)_3$]purine · 6-[$S-Sn(C_6H_5)_3$]-9H-purine	216–7(decomp.)				(*117a*)
$(C_6H_5)_3SnSC(Cl)_2P(C_6H_5)_2$	93				(*199, 201*)
$(C_6H_5)_3SnSC(NH_2)_2P(C_6H_5)_2$	115				(*199, 201*)
$(o\text{-}CH_3C_6H_4)_3SnS$–(1,3,4-thiadiazole-2,5-diyl)–$SSn(o\text{-}C_6H_4CH_3)_3$	67–68.5				(*P85*)
$(p\text{-}CH_3C_6H_4)_3SnS$–(1,3,4-thiadiazole-2,5-diyl)–SH	181–6(decomp.)				(*P85*)
6-[$S-Sn(C_6H_4CH_3)$]-9-[$Sn(C_6H_4CH_3)_3$]purine · 6-[$S-Sn(C_6H_4CH_3)_3$]-9H-purine	203–4				(*117a*)
6-[$S-Sn(C_6H_4F)_3$]-9-[$Sn(C_6H_4F)_3$]purine · 6-[$S-Sn(C_6H_4F)_3$]-9H-purine	226–7				(*117a*)

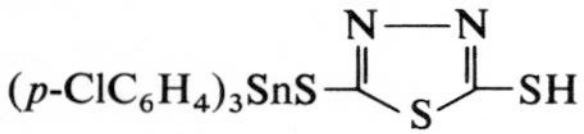 $(p\text{-}ClC_6H_4)_3SnS$–(1,3,4-thiadiazole-2-yl)–SH [structure: $N—N$ ring with S, bearing SH]	120–3				(*P85*)
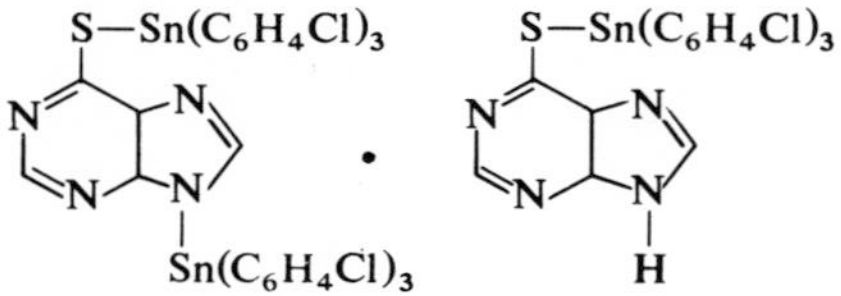 6-[$S—Sn(C_6H_4Cl)_3$]-9-[$Sn(C_6H_4Cl)_3$]purine · 6-[$S—Sn(C_6H_4Cl)_3$]-9H-purine [structures]	210–1				(*117a*)
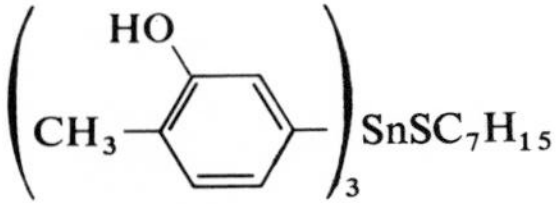 $(2\text{-}HO\text{-}4\text{-}CH_3C_6H_3)_3SnSC_7H_{15}$ [structure]					(*P1*)
$(C_5H_5)_3SnSC_6H_5$		150/1.7	1.5794	1.314	(*7*)
$Cl_3SnSC_2H_5$					(*28*)
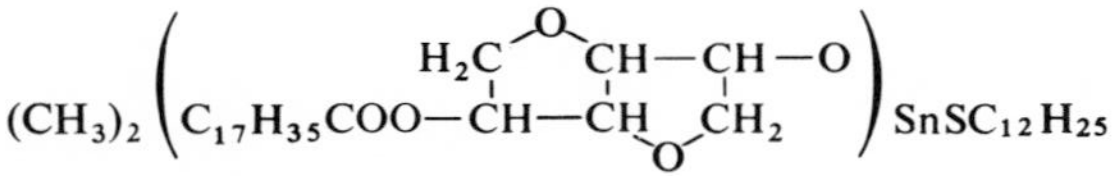 $(CH_3)_2\left(C_{17}H_{35}COO—CH—CH(—CH_2—O—)CH—CH—O\right)SnSC_{12}H_{25}$ [structure: H_2C, O, CH, CH_2 ring]					(*P89b*)
$(C_2H_5)_2(CH_2{=}CH—)SnSC_{12}H_{25}$			1.498	1.06	(*P59*)
$(C_2H_5)_2(CH_3COO)SnS\text{-}iso\text{-}C_6H_{13}$					(*P158b*)
$(C_4H_9)_2(CH_3O—)SnSC_4H_9$					(*P84*)
$(C_4H_9)_2(CH_3O—)SnSC_{12}H_{25}$					(*P39, P84, P140*)
$(C_4H_9)_2(C_2H_5O—)SnSC_2H_5$					(*P103, P104, P105, P106*)
$(C_4H_9)_2(C_2H_5O—)SnSC_4H_9$					(*215*)
$(C_4H_9)_2(C_2H_5O—)SnS\text{-}t\text{-}C_4H_9$					(*215*)
$(C_4H_9)_2(C_4H_9O—)SnSC_8H_{17}$					(*P21*)
$(C_4H_9)_2(C_{12}H_{25}O—)SnSC_{12}H_{25}$					(*P31, P84*)

(*continued*)

APPENDIX 1 (*continued*)

COMPOUNDS OF THE TYPE $R_nSn(SR')_{4-n}$

Compound	mp(°C)	bp(°C/mmHg)	n_D^{20}	d_4^{20}	References
$(C_4H_9)_2(C_8H_{17}-C_6H_4-O)-SnS-$(2-benzothiazolyl)					(*P153*)
$(n\text{-}C_4H_9)_2(CH_3COO)SnSC_4H_9$		140/0.3			(*P158b*)
$(n\text{-}C_4H_9)_2(CH_3COO)SnSC_8H_{17}$					(*P158b*)
$(n\text{-}C_4H_9)_2(CH_3COO)SnSC_6H_5$					(*P158b*)
$(C_4H_9)_2(HSCH_2COO)SnSC_8H_{17}$					(*P158k*)
$(C_4H_9)_2(C_4H_9(C_2H_5)CHCH_2C(=O)O)-SnS-$(2-benzothiazolyl)					(*P153*)
$(C_4H_9)_2(CH_3(C_2H_5)CH(CH_2)_3C(=O)O)-SnS-$(2-benzothiazolyl)					(*P63*)
$(C_4H_9)_2(C_{11}H_{23}C(=O)O-)SnSC_{12}H_{25}$			1.4834	1.036	(*P21, P39, P140, P145*)
$(C_4H_9)_2(C_{11}H_{23}C(=O)O-)SnSC_{16}H_{33}$					(*P67*)

Compound	Ref.
$(C_4H_9)_2\left(C_{11}H_{23}\overset{O}{\overset{\Vert}{C}}O\text{—}\right)SnSCH_2C_6H_5$	(*P67*)
$(C_4H_9)_2\left(CH_2{=}CH\overset{O}{\overset{\Vert}{C}}O\text{—}\right)SnSC_8H_{17}$	(*P67*)
$(C_4H_9)_2\left(CH_2{=}CH\overset{O}{\overset{\Vert}{C}}O\text{—}\right)SnSC_{12}H_{25}$	(*P21*)
$(C_4H_9)_2\left(CH_2{=}\underset{\underset{CH_3}{\vert}}{C}\text{—}\overset{O}{\overset{\Vert}{C}}O\text{—}\right)SnSC_{12}H_{25}$	(*P21*)
$(C_4H_9)_2\left(CH_3CH{=}CH\text{—}\overset{O}{\overset{\Vert}{C}}O\text{—}\right)SnSC_{12}H_{25}$	(*P21*)
$(C_4H_9)_2\left[CH_3(CH_2)_7CH{=}CH(CH_2)_7\underset{\underset{O}{\Vert}}{C}O\text{—}\right]SnSC_{12}H_{25}$	(*P21*)
$(C_4H_9)_2(C_{12}H_{25}OOCCH_2S)SnSC_{12}H_{25}$	(*P76g*)
$(C_4H_9)_2(C_4H_9OOCCH{=}CHCOO)SnSC_{12}H_{25}$	(*P89e*)
$(C_4H_9)_2(CH_3OSn[(C_4H_9)_2]O)SnSC_{12}H_{25}$	(*P89e*)
$(C_4H_9)_2\left(C_6H_5CH{=}CHCH_2\overset{O}{\overset{\Vert}{C}}O\text{—}\right)SnSC_{16}H_{33}$	(*P67*)
$(C_4H_9)_2\left(C_6H_5CH{=}CHCH_2\overset{O}{\overset{\Vert}{C}}O\text{—}\right)SnSCH_2C_6H_5$	(*P67*)

(continued)

APPENDIX 1 (*continued*)

COMPOUNDS OF THE TYPE $R_nSn(SR')_{4-n}$

Compound	mp(°C)	bp(°C/mmHg)	n_D^{20}	d_4^{20}	References
$(C_4H_9)_2(C_6H_5C(=O)O\text{—})SnSC_8H_{17}$					(*P67*)
$(C_4H_9)_2(C_6H_5C(=O)O\text{—})SnSC_{16}H_{33}$					(*P67*)
$(C_4H_9)_2(C_6H_5C(=O)O\text{—})SnSCH_2C_6H_5$					(*P67*)
$(C_4H_9)_2(CH_3OC(=O)CH{=}CHC(=O)O\text{—})SnSC_{12}H_{25}$					(*P21*)
$(C_4H_9)_2(C_8H_{17}OC(=O)CH{=}CHC(=O)O\text{—})SnSC_8H_{17}$					(*P67*)
$(n\text{-}C_4H_9)_2(C_6H_5COS)SnSC_{12}H_{25}$					(*P76g*)
$(C_4H_9)_2(C_{18}H_{35}OC(=O)CH{=}CHC(=O)O\text{—})SnSC_{12}H_{25}$			1.4890	1.038	(*P39, P140, P145*)
$(C_4H_9)_2(C_4H_9\text{—}C_6H_4\text{—}C(=O)\text{—}O\text{—}Sb(\text{—}O\text{—}C_6H_4\text{—}C_9H_{19})\text{—}SCH_2C(=O)O\text{—})SnS\text{—}o\text{—}C_6H_4CH_3$					(*P142*)

$(C_4H_9)_2ClSnSC_4H_9$	116(*58*)	96–8/1(*202*)	1.5209(*83a*)	1.225(*83a*)	(*58*, *83a*, *83b*, *202*)
$(C_4H_9)_2(C_8H_{17}O\overset{O}{\overset{\parallel}{C}}CH_2S-)SnS\text{-}2\text{-}C_{10}H_7$			1.5796	1.2006	(*P39*, *P140*)
$(C_4H_9)_2(C_6H_5\overset{O}{\overset{\parallel}{C}}S-)SnSC_{12}H_{25}$					(*P33*, *P34*, *P64*)
$(C_4H_9)_2(C_6H_5\overset{O}{\overset{\parallel}{C}}S-)SnSCH_2CH(C_2H_5)(CH_2)_3OH$					(*P64*)
$(C_4H_9)_2(C_6H_5\overset{O}{\overset{\parallel}{C}}S-)SnSCH_2C_6H_5$					(*P67*)
$(C_4H_9)_2(H_3C\text{-}C_6H_4\text{-}SO_2N(CH_2\text{-}CH\text{-}CH_2\text{(epoxide, O)})\text{-})SnSC_{12}H_{25}$					(*P152*)
$(C_4H_9)(C_6H_{13})(C_6H_5COS)SnSC_{12}H_{25}$					(*P76g*)
$(C_8H_{17})_2(C_7H_{15}\overset{O}{\overset{\parallel}{C}}O-)SnSC_8H_{17}$					(*P67*)
$(C_8H_{17})_2(C_6H_5CH{=}CHCH_2\overset{O}{\overset{\parallel}{C}}O-)SnSC_8H_{17}$					(*P67*)
$(C_8H_{17})_2(C_6H_5\overset{O}{\overset{\parallel}{C}}S-)SnSC_8H_{17}$					(*P67*)

(*continued*)

APPENDIX 1 (*continued*)

COMPOUNDS OF THE TYPE $R_nSn(SR')_{4-n}$

Compound	mp(°C)	bp(°C/mmHg)	n_D^{20}	d_4^{20}	References
$(C_8H_{17})_2(HSCH_2COO)SnSC_{12}H_{25}$					(*P158k*)
$(C_6H_5CH_2)_2(CH_3O-)SnSC_{12}H_{25}$					(*P84*)
$(C_6H_5)_2(HSCH_2COO)SnSC_8H_{17}$					(*P158k*)
$(C_6H_5)_2(CH_3O-)SnSC_{12}H_{25}$					(*P84*)
$(2,3\text{-}(CH_3)_2C_6H_3)(4,2\text{-}CH_3(CH_3)C_6H_3)((C_{10}H_{21}O)(C_{18}H_{35}COO)Sb-O-C(=O)(CH_2)_3C(=O)O)SnSC_{12}H_{25}$					(*P142*)
$(C_4H_9)(CH_3O-)_2SnSC_{12}H_{25}$					(*P84*)
$C_8H_{17}OC(=O)CH_2S-Sn(CH_3)_2-(O-Sn(CH_3)_2)_2-S-1-C_{10}H_7$					(*P50*)
$C_{12}H_{25}S-(Sn(C_4H_9)_2-O)_3-Sn(C_4H_9)_2-SC_{12}H_{25}$					(*P20, P125*)
$C_8H_{17}OC(=O)CH_2S-Sn(C_4H_9)_2-(O-Sn(C_4H_9)_2)_2-S-1-C_{10}H_7$					(*P50*)

$[(C_4H_9)_2(C_{11}H_{23}\overset{O}{\overset{\|}{C}}O-)SnS(CH_2)_4]_2O$					(*P21*)
$[(C_4H_9)_2(CH_3CH{=}CH\overset{O}{\overset{\|}{C}}O-)SnS(CH_2)_4-]_2O$					(*P21*)
$(C_6H_5)_2(C_2H_5O)SnSC_4H_9$					(*P89e*)
$\left[C_{12}H_{25}-S-\underset{C_8H_{17}}{\overset{C_8H_{17}}{Sn}}-O\right]B-O-\underset{C_8H_{17}}{\overset{C_8H_{17}}{Sn}}-SCH_2\overset{O}{\overset{\|}{C}}O(CH_2)_2O\overset{O}{\overset{\|}{C}}CH_2SH$					(*P92*)
$(CH_3)_3Sn(H_5C_6)S-Cr(CO)_5$	45(decomp.)				(*181b*)
$R_2Sn(SR')_2$					
$(CH_3)_2Sn(SCH_3)_2$		40/0.03(*3*)	1.5953(*3*)	1.547(*7*)	(*3*, *7–12*, *14*, *219*)
		44/0.05(*7*)	1.6003(*7*)		
$(CH_3)_2Sn(SC_2H_5)_2$		58/0.07(*7*)	1.5713(*7*)	1.440(*7*)	(*7*, *9–12*)
$(CH_3)_2Sn(-S\text{-}n\text{-}C_3H_7)_2$		74/0.1(*7*)	1.5498(*7*)	1.323(*7*)	(*3*, *7*, *10*)
$(CH_3)_2Sn(-S\text{-}n\text{-}C_4H_9)_2$		81/0.1(*7*)	1.5400(*7*)	1.280(*7*)	(*7*, *10*, *P6*, *P38f*, *P60*, *P95b*, *P115*)
		110/0.7(*P95b*)	1.5371(*P6*, *P60*, *P115*)		
			1.537(*P95b*)		
$(CH_3)_2Sn(-SC_8H_{17})_2$		166/0.2(*7*)	1.5129(*7*)	1.092(*7*)	(*7*, *10*)
$(CH_3)_2Sn(-SC_{12}H_{25})_2$					(*P3*, *P79g*, *P124*)
$(CH_3)_2Sn(-SCH_2C_6H_5)_2$		175–85/0.25			(*162*, *P38f*, *P95b*)
$(CH_3)_2Sn(SC_6H_5)_2$	38–9	130–5/0.001			(*3*)
$(CH_3)_2Sn(-S\text{-}p\text{-}C_6H_4CH_3)_2$	60–1(*P6*, *P60*)				(*P6*, *P60*, *P115*)
$(CH_3)_2Sn(-SCH_2CHOH-CH_2OH)_2$			1.6026	1.6596	(*P39*, *P140*)
$(CH_3)_2Sn(-SCH_2CH_2SC_8H_{17})_2$					(*P68*)

(*continued*)

APPENDIX 1 (*continued*)

COMPOUNDS OF THE TYPE $R_nSn(SR')_{4-n}$

Compound	mp(°C)	bp(°C/mmHg)	n_D^{20}	d_4^{20}	References
$(CH_3)_2Sn-S-[\text{2-benzothiazolyl}]_2$	106				(*P113*)
$(CH_3)_2Sn-S-[\text{tetramethyl-tetrahydrobenzothiazolyl}]_2$	122				(*P113*)
$(C_2H_5)_2Sn(—SCH_3)_2$		61/0.1(*7*)	1.5793(*7*)	1.440(*7*)	(*7, 9*)
$(C_2H_5)_2Sn(—SC_2H_5)_2$		94/0.05	1.5572	1.319	(*7*)
$(C_2H_5)_2Sn(SC_4H_9)_2$					(*P38d*)
$(C_2H_5)_2Sn[S-C_6H_4-C(CH_3)_3]_2$					(*P66*)
$(n\text{-}C_3H_7)_2Sn(—SC_6H_5)_2$		226–30/1	1.6298		(*177*)
$(n\text{-}C_4H_9)_2Sn(SCH_3)_2$		94/0.02	1.5538(n_D^{22})		(*145a*)
$(n\text{-}C_4H_9)_2Sn(—SC_2H_5)_2$		97–100/0.4(*215*)			(*215, P3, P124, P137*)
$(n\text{-}C_4H_9)_2Sn(—S\text{-}n\text{-}C_3H_7)_2$		121–3/0.15			(*128b, 215*)
$(n\text{-}C_4H_9)_2Sn(—S\text{-}n\text{-}C_4H_9)_2$		170/1.5(*215*) 158–60/2.5(*83a*)	1.5238(*83a*)	1.145(*83a*)	(*83a, 83b, 128b, 215, P1, P3, P34, P124, P137*)
$(n\text{-}C_4H_9)_2Sn(—S\text{-}sec\text{-}C_4H_9)_2$					(*P3, P124, P137*)
$(n\text{-}C_4H_9)_2Sn(—S\text{-}iso\text{-}C_4H_9)_2$		148/0.6(*215*)			(*215, P3, P124, P137*)
$(n\text{-}C_4H_9)_2Sn(—S\text{-}t\text{-}C_4H_9)_2$		118/0.1			(*215*)
$(n\text{-}C_4H_9)_2Sn(—SC_5H_{11})_2$			1.5200(*P6, P115*)		(*P3, P6, P115, P124, P137*)

$(n\text{-}C_4H_9)_2Sn(\text{—}SC_6H_{13})_2$					(*P3*, *P76g*, *P124*, *P137*)
$(n\text{-}C_4H_9)_2Sn(\text{—}S\text{-}t\text{-}C_6H_{13})_2$					(*P3*, *P124*, *P137*)
$(n\text{-}C_4H_9)_2Sn(\text{—}SC_7H_{15})_2$					(*P3*, *P124*, *P137*)
$(n\text{-}C_4H_9)_2Sn(\text{—}SC_8H_{17})_2$					(*175*, *P3*, *P58a*, *P111*, *P124*, *P137*, *P143*)
$(n\text{-}C_4H_9)_2Sn(\text{—}S\text{-}n\text{-}C_9H_{19})_2$					(*P3*, *P124*, *P137*)
$(n\text{-}C_4H_9)_2Sn(\text{—}S\text{-}n\text{-}C_{10}H_{21})_2$					(*P3*, *P124*, *P137*)
$(n\text{-}C_4H_9)_2Sn(\text{—}SC_{12}H_{25})_2$		160/10(*P6*, *P115*)	1.4992(*P90*, *P65*, *P53*) 1.5177(*P39*, *P140*) 1.5011(*P6*, *P115*)	1.045(*P39*, *P140*)	(*215*, *P3*, *P6*, *P34*, *P39*, *P53*, *P65*, *P76g*, *P89a*, *P90*, *P92*, *P100*, *P110*, *P111*, *P115*, *P124*, *P140*, *P143*, *P146*,)
$(n\text{-}C_4H_9)_2Sn(\text{—}S\text{-}t\text{-}C_{12}H_{25})_2$			1.5168(*P6*, *P60*, *P115*)		(*P3*, *P6*, *P60*, *P115*, *P137*)
$(n\text{-}C_4H_9)_2Sn(\text{—}SC_{13}H_{27})_2$			1.4982		(*P6*, *P115*)
$(n\text{-}C_4H_9)_2Sn(\text{—}SC_{14}H_{29})_2$					(*P3*, *P124*, *P137*)
$(n\text{-}C_4H_9)_2Sn(\text{—}SC_{16}H_{33})_2$			1.4975(*P6*, *P60*, *P115*)		(*P3*, *P6*, *P60*, *P115*, *P124*, *P137*)
$(n\text{-}C_4H_9)_2Sn(\text{—}S\text{-}t\text{-}C_{16}H_{33})_2$			1.5064(*P6*, *P60*, *P115*)		(*P3*, *P6*, *P60*, *P115*, *P137*)
$(n\text{-}C_4H_9)_2Sn(\text{—}SC_{18}H_{37})_2$	26(*P6*, *P60*, *P115*)				(*P6*, *P60*, *P115*, *P143*)
$(n\text{-}C_4H_9)_2Sn(\text{—}SCH_2CH{=}CH_2)_2$		134–5/0.075			(*118*)
$(n\text{-}C_4H_9)_2Sn[S(CH_2)_8CH{=}CH(CH_2)_7CH_3]_2$					(*P76g*)
$(n\text{-}C_4H_9)_2Sn(\text{—}SCH_2C_6H_5)_2$			1.5989(*P6*, *P60*, *P115*)		(*215*, *P6*, *P60*, *P34*, *P76g*, *P115*)

(*continued*)

COMPOUNDS OF THE TYPE $R_nSn(SR')_{4-n}$

Compound	mp(°C)	bp(°C/mmHg)	n_D^{20}	d_4^{20}	References
$(n-C_4H_9)_2Sn(-SCH_2-C_6H_4-Cl)_2$			1.6031		(*P6, P60, P115*)
$(n\text{-}C_4H_9)_2Sn(-SC_2H_4OH)_2$			1.5663(*P60, P115*)		(*P14, P60, P76g, P115, P136*)
$(n\text{-}C_4H_9)_2Sn(-SCH_2CHOH-CH_2OH)_2$			1.5703	1.3702	(*P39, P76g, P140*)
$(n\text{-}C_4H_9)_2Sn(SCH_2CHOHCHOHCH_2OH)_2$					(*P76g*)
$(n\text{-}C_4H_9)_2Sn(SCH_2CH_2OCH_2CH_2OH)_2$					(*P76g*)
$(n\text{-}C_4H_9)_2Sn(SCH_2CH_2SCH_2CH_2OH)_2$					(*P76g*)
$(n\text{-}C_4H_9)_2Sn(-SCH_2CH_2OCH_2CH(C_4H_9)(C_2H_5))_2$			1.5073	1.1313	(*P39*)
$(n\text{-}C_4H_9)_2Sn(-SCH_2CH_2OCH_2CH_2O-C_6H_4-C_8H_{17})_2$			1.5429		(*P6*)
$(C_4H_9)_2Sn(-SCH_2CH_2-S-C_8H_{17})_2$					(*P68*)
$(C_4H_9)_2Sn(-SCH_2CH_2SC_{18}H_{37})_2$					(*P68*)
$(C_4H_9)_2Sn(-SCH_2CH_2SC_{22}H_{45})_2$					(*P68*)
$(C_4H_9)_2Sn(-SC_6H_5)_2$		179/0.1(*215*) 162/0.05(*145a*)	1.6105(*P6, P115, P60*) 1.6158(n_D^{22})(*145a*)		(*145a, 215, P3, P6, P58a, P60, P111, P115, P124, P137*)

Compound				
$(C_4H_9)_2Sn(SC_6F_5)_2$	30	139/0.02		(*145a*)
$(C_4H_9)_2Sn(SC_6Cl_5)_2$	105(*123a*)			(*123a*, *P1*)
$(C_4H_9)_2Sn(—S\text{-}o\text{-}C_6H_4CH_3)_2$		210/0.5		(*215*)
$(C_4H_9)_2Sn(—S\text{-}p\text{-}C_6H_4CH_3)_2$			1.6029(*P6*, *P60*, *P115*)	(*P3*, *P6*, *P60*, *P76g*, *P115*, *P124*, *P137*)
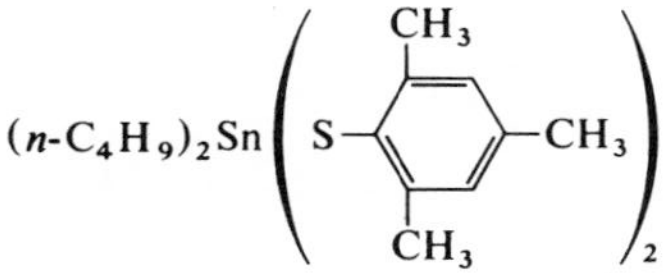 $(n\text{-}C_4H_9)_2Sn(S\text{-}C_6H_2(CH_3)_3)_2$ [S–2,4,6-trimethylphenyl]				(*P76g*)
$(C_4H_9)_2Sn(—S\text{-}o\text{-}C_6H_4NH_2)_2$				(*P3*, *P124*, *P137*)
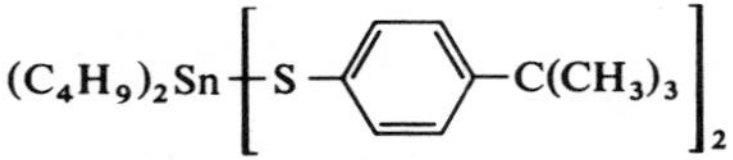 $(C_4H_9)_2Sn[S\text{-}C_6H_4\text{-}C(CH_3)_3]_2$		191–200/0.1(*47*)	1.5744(*47*)	(*47*, *P66*)
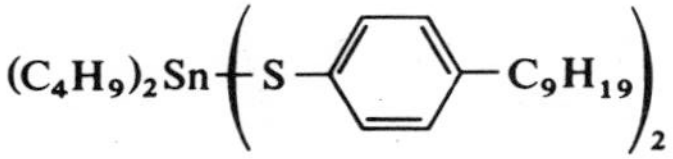 $(C_4H_9)_2Sn(S\text{-}C_6H_4\text{-}C_9H_{19})_2$				(*P66*)
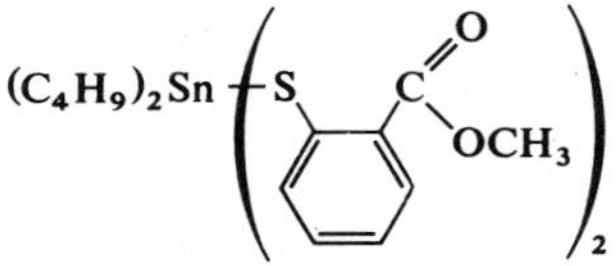 $(C_4H_9)_2Sn(S\text{-}C_6H_4\text{-}C(=O)OCH_3)_2$ [ortho]				(*P4*, *P15*, *P60*, *P102*, *P121*)
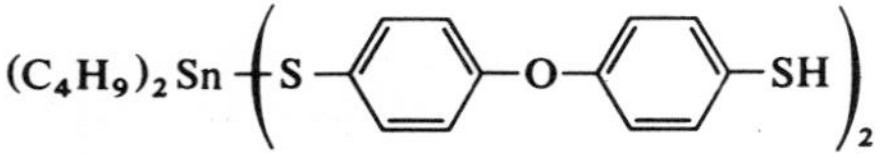 $(C_4H_9)_2Sn(S\text{-}C_6H_4\text{-}O\text{-}C_6H_4\text{-}SH)_2$				(*P6*, *P60*, *P115*)
$(C_4H_9)_2Sn(—S\text{-}2\text{-}C_{10}H_7)_2$			1.6630	(*P6*, *P60*, *P115*)
$(C_4H_9)_2Sn(—S\text{-}1\text{-}C_{10}H_7)_2$			1.6770	(*P6*, *P60*, *P115*)

(*continued*)

APPENDIX 1 (*continued*)

COMPOUNDS OF THE TYPE $R_nSn(SR')_{4-n}$

Compound	mp(°C)	bp(°C/mmHg)	n_D^{20}	d_4^{20}	References
$(C_4H_9)_2Sn[S-C_4H_3S]_2$ (2-thienyl)			1.6280		(*P60, P115*)
$(C_4H_9)_2Sn(S-C_3H_2NS)_2$ (2-thiazolyl)					(*P113*)
$(C_4H_9)_2Sn(S-C_2N_2S-SH)_2$ (5-mercapto-1,3,4-thiadiazol-2-yl)	176				*P*(*83*)
$(C_4H_9)_2Sn(S-C_2N_2S(=S)CH_3)_2$ (4-methyl-5-thioxo-1,3,4-thiadiazolin-2-yl)	120				(*P154*)
$(C_4H_9)_2Sn(S-C_7H_4NS)_2$ (2-benzothiazolyl)	185(*P113*)				(*52, P3, P113, P124, P137*)
$(C_4H_9)_2Sn(S-N(C_6H_5)-C_3H_2NS)_2$					(*P1*)
$(C_4H_9)_2Sn[S-C_7H_3NS(C_6H_5)]_2$ (7-phenyl-2-benzothiazolyl)					(*P113*)

Compound			
$(CH_3CH{=}CHCH_2)_2Sn(SC_{18}H_{37})_2$			(*P76g*)
$(C_5H_{11})_2Sn(—SC_{12}H_{25})_2$			(*P3, P124*)
$(C_6H_{13})_2Sn(SC_{18}H_{37})_2$			(*P76g*)
$(C_7H_{15})_2Sn(—SC_8H_{17})_2$			(*P2*)
$(C_8H_{17})_2Sn(—SC_6H_{13})_2$			(*P143*)
$(C_8H_{17})_2Sn(—SC_8H_{17})_2$			(*P143*)
$(C_8H_{17})_2Sn(—SC_{12}H_{25})_2$			(*P143*)
$(C_8H_{17})_2Sn(—SC_6H_5)_2$		1.5688	(*P53, P65, P90*)
$(C_8H_{17})_2Sn(—S—C_6H_4—C_4H_9)_2$			(*P2*)
$(C_8H_{17})_2Sn[—S—C_6H_4—C(CH_3)_3]_2$			(*P66*)
$(C_8H_{17})_2Sn(—S\text{-}iso\text{-}C_8H_{16}—OH)_2$			(*P95*)
$(C_8H_{17})_2Sn[—S—C_2N_2S(=S)—C_6H_5]_2$ (3-phenyl-2-thioxo-1,3,4-thiadiazol-5-yl)	95		(*P154*)
$(C_{10}H_{21})_2Sn(—SC_8H_{17})_2$			(*P143*)
$(C_{10}H_{21})_2Sn[—S—C_2N_2S(=S)—CH_2CH_2CH_3]_2$ (3-propyl-2-thioxo-1,3,4-thiadiazol-5-yl)			(*P154*)
$(C_{12}H_{25})_2Sn(—SC_5H_{11})_2$			(*P3, P137*)
$(C_{12}H_{25})_2Sn(—SC_6H_{13})_2$			(*P143*)
$(C_{12}H_{25})_2Sn(—SC_{12}H_{25})_2$			(*P3, P124, P143*)

(*continued*)

APPENDIX 1 (*continued*)

Compounds of the Type $R_nSn(SR')_{4-n}$

Compound	mp(°C)	bp(°C/mmHg)	n_D^{20}	d_4^{20}	References
$(C_{12}H_{25})_2Sn(—SCH_2C_6H_5)_2$					(*P3, P124*)
$(C_{18}H_{37})_2Sn(SC_6H_5)_2$					(*P76g*)
$(C_6H_5CH_2)_2Sn(SC_{12}H_{25})_2$					(*P76g*)
$(C_6H_5CH_2)_2Sn[S-C_6H_4-C(CH_3)_3]_2$					(*P66*)
$(C_6H_5CH_2)_2Sn(S-\text{benzimidazol-2-yl})_2$					(*P96*)
$(C_6H_5CH_2)_2Sn(S-\text{benzothiazol-2-yl})_2$					(*P96*)
$[(CH_3)_2CH-C_6H_4-CH_2]_2Sn[S-\text{benzothiazol-2-yl}]_2$	93–5				(*55*)
$(C_6H_5)_2Sn(—SC_4H_9)_2$			1.4967(*P6, P115*) 1.5001(*P60*)		(*P6, P60, P111, P115*)
$(C_6H_5)_2Sn(—SC_{12}H_{25})_2$					(*P3, P76g, P124, P137*)
$(C_6H_5)_2Sn(—S—C{\equiv}C—C_6H_5)_2$	137–9				(*161*)
$(C_6H_5)_2Sn(—SCH_2C_6H_5)_2$			1.6610(*113*)		(*113, P1, P3, P124*)

$(C_6H_5)_2Sn(\text{—}SC_6H_5)_2$	65–65.5(*29, 191*) 66–7(*32*)		(*29, 32, 56, 191, 203, P1, P111*)
$(C_6H_5)_2Sn(\text{—}S\text{-}p\text{-}C_6H_4CH_3)_2$		1.6835	(*P6, P60, P115*)
$(C_6H_5)_2Sn[S-C_6H_4-C(CH_3)_3]_2$			(*P66*)
$(C_6H_5)_2Sn(S\text{-}(1,3,4\text{-thiadiazol-2-yl})\text{-}SH)_2$	176		(*P83*)
$(C_6H_5)_2Sn(S\text{-}(2\text{-thiazolyl}))_2$			(*P113*)
$(C_6H_5)_2Sn(S\text{-}(2\text{-benzothiazolyl}))_2$	158		(*P113*)
$(C_6H_5)_2Sn(S-CH_2CH_2-O-CH_2CH_2-O-C_6H_4(C_8H_{17}))_2$		1.5695	(*P6*)
$(2,4,6\text{-}(CH_3)_3-C_6H_2)_2Sn(SC_{12}H_{25})_2$			(*P76g*)
$(m\text{-}Cl-C_6H_4)_2Sn(S\text{-}(4\text{-}CH_2CH_2CH_3\text{-}5\text{-thioxo-}1,3,4\text{-thiadiazol-2-yl}))_2$			(*P154*)

(*continued*)

APPENDIX 1 (*continued*)

Compounds of the Type $R_nSn(SR')_{4-n}$

Compound	mp(°C)	bp(°C/mmHg)	n_D^{20}	d_4^{20}	References
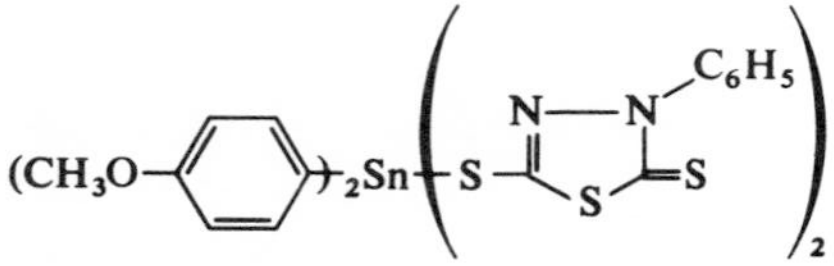 $(CH_3O-C_6H_4)_2Sn(-S-C=N-N(C_6H_5)-C(=S)-S-)_2$ (1,3,4-thiadiazole-2-thione ring)					(*P154*)
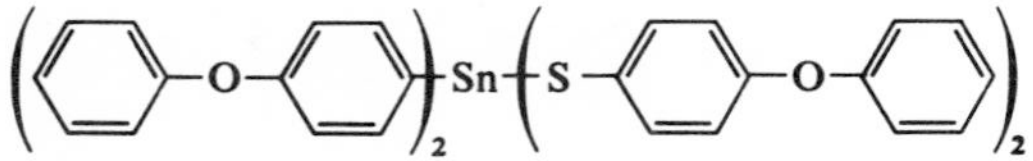 $(C_6H_5-O-C_6H_4)_2Sn(-S-C_6H_4-O-C_6H_5)_2$					(*P1*)
$(C_4H_9)(CH_3O-)Sn(-SC_{12}H_{25})_2$					(*P84*)
$(C_4H_9)_2Sn(-SC_{12}H_{25})(-SC_8H_{17})$					(*P67*)
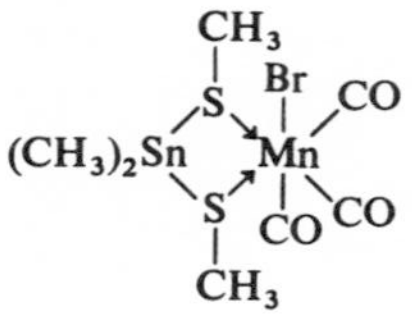 $(CH_3)_2Sn(\mu\text{-}SCH_3)_2Mn(Br)(CO)_3$					(*14*)
$[\pi\text{-}C_5H_5Fe(CO)_2]_2Sn(-SC_2H_5)_2$	129–31				(*24a*, *135*)
$[\pi\text{-}C_5H_5Fe(CO)_2]_2Sn(-SC_6H_5)_2$	132–4				(*161a*)
$[\pi\text{-}C_5H_5-(CO)_3Mo]_2Sn(SC_2H_5)_2$	115–6				(*136a*)
$[\pi\text{-}C_5H_5-(CO)_3W]_2Sn(SC_2H_5)_2$	136.5–8.5				(*136a*)
RSn(SR')₃					
$CH_3Sn(-SCH_3)_3$		75/0.01(*3*)	1.6352(*3*)	1.63(*3*)	(*3*, *8*, *219*)
$CH_3Sn(-SC_2H_5)_3$		90/0.05	1.5972	1.469	(*7*)

$CH_3Sn(—S\text{-}n\text{-}C_3H_7)_3$		95/0.001	1.5684	1.337	(7)
$CH_3Sn(—S\text{-}n\text{-}C_4H_9)_3$			1.5541(*P7, P114, P109*)		(*P7, P109, P112, P114*)
$CH_3Sn(—S\text{-}sec\text{-}C_5H_{11})_3$			1.5452(*P7, P109, P114*)		(*P7, P109, P112, P114*)
$CH_3Sn(—SC_{12}H_{25})_3$					(*P3, P124*)
$CH_3Sn(—S\text{-}t\text{-}C_{12}H_{25})_3$					(*P7, P109, P112, P114*)
$CH_3Sn(—SC_{18}H_{37})_3$	50–60(*P7, P109, P114*)				(*P7, P109, P112, P114*)
$CH_3Sn(—SCH_2—C_6H_4—Cl)_3$			1.6523		(*P7, P109, P112, P114*)
$CH_3Sn(—S—C_6H_4—CH_3)_3$	143–5				(*P7, P109, P112, P114*)
$CH_3Sn(—S\text{-}1\text{-}C_{10}H_7)_3$			1.71		(*P7, P109, P112, P114*)
$CH_3Sn(—SCH_2CH_2OH)_3$			1.6168		(*P7, P109, P112, P114*)
$C_2H_5Sn(—SCH_3)_3$		66/0.001	1.6232	1.548	(7)
$C_2H_5Sn(—SC_{12}H_{25})_3$					(*P3, P124*)
$n\text{-}C_4H_9Sn(—S\text{-}n\text{-}C_4H_9)_3$			1.5420		(*P7, P109, P112, P114*)
$n\text{-}C_4H_9Sn(—SC_{12}H_{25})_3$					(*P2, P3, P95*)
$n\text{-}C_4H_9Sn(—S\text{-}t\text{-}C_{12}H_{25})_3$					(*P7, P109, P112, P114*)

(*continued*)

APPENDIX 1 (*continued*)

COMPOUNDS OF THE TYPE $R_nSn(SR')_{4-n}$

Compound	mp(°C)	bp(°C/mmHg)	n_D^{20}	d_4^{20}	References
$n\text{-}C_4H_9Sn(S-C_6H_4-CH_3)_3$			1.6540		(*P7*, *P109*, *P112*, *P114*)
$n\text{-}C_4H_9Sn(S-C_7H_4NS)_3$					(*P108*, *P113*)
$C_6H_5Sn(-S\text{-}n\text{-}C_4H_9)_3$			1.5710		(*P7*, *P109*, *P112*, *P114*)
$C_6H_5Sn(S-C_6H_4-CH_3)_3$			1.6890		(*P7*, *P109*, *P112*, *P114*)
$ClSn[-S-C(CH_3)_2-CH_2CH_3]_3$					(*28*)
$\pi\text{-}C_5H_5-(CO)_3MoSn(SC_2H_5)_3$	89–90				(*136a*)
Sn(SR)$_4$					
$Sn(-SCH_3)_4$	31(*25*, *28*, *30*)	81/0.001(*28*)			(*8*, *25*, *28*, *30*)
$Sn(-SC_2H_5)_4$		105/0.001(*28*) 200(*45*) 152/0.8(*128*)	1.6188(28)		(*28*, *30*, *45*, *128*, *P5*)
$Sn(-S\text{-}n\text{-}C_3H_7)_4$		123/0.001(*28*) 160/0.1(*128*)	1.5851(*28*)		(*28*, *128*)
$Sn(-S\text{-}iso\text{-}C_3H_7)_4$		92/0.001(*28*) 145/0.1(*128*)	1.5789(*28*)		(*28*, *30*, *128*)
$Sn(-S\text{-}n\text{-}C_4H_9)_4$		136/0.001(*28*) 200/1(*128*)	1.5639(*28*)		(*28*, *128*)

$Sn\left(-S-CH\begin{matrix}C_2H_5\\CH_3\end{matrix}\right)_4$		111/0.001	1.5668		(*28*)
$Sn\left(-S-CH_2-CH\begin{matrix}CH_3\\CH_3\end{matrix}\right)_4$		126/0.0005(*28*) 166–7/0.1(*128*)	1.5599(*28*)		(*28*, *128*)
$Sn(-S\text{-}t\text{-}C_4H_9)_4$	188(*26*, *28*, *30*)	180–90/0.8(*128*)			(*26*, *27*, *28*, *30*, *99*, *128*)
$Sn(-S\text{-}n\text{-}C_5H_{11})_4$		162/0.004	1.5475		(*29*)
$Sn[-S-C(CH_3)_2CH_2CH_3]_4$	44				(*28*)
$Sn(-SC_6H_{13})_4$					(*29*)
$Sn(-SC_{12}H_{25})_4$	35.5(*29*) 28–30(*P110*)				(*29*, *128*, *P5*, *P110*)
$Sn(-SC_{16}H_{33})_4$	53–53.5				(*29*)
$Sn(-SC_{20}H_{41})_4$					(*P5*)
$Sn(-SCH_2CH{=}CH_2)_4$					(*29*)
$Sn\left(-SCH_2\underset{\displaystyle CH_3}{\underset{\vert}{C}}{=}CH_2\right)_4$					(*29*)
$Sn(-S\text{-}cyclo\text{-}C_6H_{11})_4$	53–4				(*29*, *30*)
$Sn(-SCH_2-C_6H_5)_4$				1.384	(*29*, *128*, *233*)
$Sn(-SC_6H_5)_4$	67(*29*, *30*) 67.5(*233*) 66–7(*158*) 62–4(*P110*)	120–4/0.5(*128*)			(*29*, *30*, *128*, *158*, *233*, *P110*)

(*continued*)

APPENDIX 1 (*continued*)

Compounds of the Type $R_nSn(SR')_{4-n}$

Compound	mp(°C)	bp(°C/mmHg)	n_D^{20}	d_4^{20}	References
$Sn(—S\text{-}p\text{-}C_6H_4CH_3)_4$	100(*29, 30*)				(*29, 30, P5*)
$Sn[S\text{-}C_6H_4\text{-}C(CH_3)_3]_4$	106				(*29, 30*)
$Sn(—S\text{-}p\text{-}C_6H_4Cl)_4$	189				(*29*)
$Sn(—S\text{-}p\text{-}C_6H_4Br)_4$	217				(*29*)
$Sn(—S\text{-}p\text{-}C_6H_4NH_2)_4$	166				(*233*)
$Sn[S\text{-}C_6H_4\text{-}N(CH_3)_2]_4$	159				(*233*)
$Sn(SC_6F_5)_4$	128–30(*105a*) 144(*145b*)				(*105a, 145b*)
$Sn(—S\text{-}1\text{-}C_{10}H_7)_4$	103–4				(*233*)
$(C_6H_5S)_4Sn$ ← 2,2'-bipyridyl (N,N')	135–6				(*158*)

APPENDIX 2

CHEMICAL SHIFTS AND COUPLING CONSTANTS FOR COMPOUNDS $(CH_3)_nSn(SCH_3)_{4-n}$

	$(CH_3)_4Sn$	$(CH_3)_3SnSCH_3$	$(CH_3)_2Sn(SCH_3)_2$	$CH_3Sn(SCH_3)_3$	$Sn(SCH_3)_4$
$\tau(M{-}CH_3)$	9.92	9.61	9.29	9.06	
$\tau(S{-}CH_3)$		8.00	7.84	7.73	7.68
$J(^{119}Sn{-}C{-}H)$	54.0	56.9	61.2	66.6	
$J(^{117}Sn{-}C{-}H)$	51.5	54.5	58.5	63.5	
$J(^{119}Sn{-}S{-}C{-}H)$		37.5	44.2	56.7	66.0
$J(^{117}Sn{-}S{-}C{-}H)$		36.0	42.6	53.2	66.0
$J(H{-}^{13}C(Sn))$		135	135	135	
$J(H{-}^{13}C(S))$		142.5	142.5	142.5	

APPENDIX 3

Compounds of the Type $R_nSn(—SR'COOR'')_{4-n}$ and $R_nSn(—SR'OR'')_{4-n}$

Compound	mp(°C)	bp(°C/mmHg)	n_D^{20}	d_4^{20}	References
$R_3SnSR'COOR''$					
$(C_2H_5)_3SnSCH_2COOH$					(*P4, P127, P128*)
$(C_2H_5)_3SnSCH_2COOC_2H_5$		107–9/1	1.5101		(*183*)
$(C_2H_5)_3SnSCH_2COO\text{-}iso\text{-}C_8H_{17}$					(*P15, P121*)
$(C_2H_5)_3SnSCH_2COOC_{10}H_{21}$					(*P4, P127, P128*)
$(C_2H_5)_3SnSCH_2CH_2O-B\langle O-CH_2-CH_2 / O-CH_2-CH_2 \rangle$ (cyclic)					(*P138*)
$(C_2H_5)_3SnSCH_2CH_2O-B\langle O, O \rangle(CH_2)_5$ (cyclic)					(*P138*)
$(C_3H_7)_3SnSCH_2CH_2C(O)N(H)C_5H_{11}$					(*P56*)
$(C_4H_9)_3SnSCH_2COOH$					(*P4, P127, P128*)
$(C_4H_9)_3SnSCH_2COOC_4H_9$					(*P100, P101, P123*)
$(C_4H_9)_3SnSCH_2COOC_{10}H_{21}$					(*P4, P102, P127, P128*)
$(C_4H_9)_3SnSCH_2COOCH_2$-(tetrahydrofuryl ring, O)					(*P89*)
$(C_4H_9)_3SnSCH_2CONHC_5H_{11}$					(*P107*)
$(C_4H_9)_3SnSCH_2CH_2COO\text{-}iso\text{-}C_8H_{17}$					(*P15, P121*)

$(C_4H_9)_3SnSCH_2CH_2CONHC_5H_{11}$		(*P9, P41, P107*)
$(C_4H_9)_3SnS(CH_2)_3CON(C_4H_9)_2$		(*P51, P146*)
$(C_4H_9)_3SnS(CH_2)_4COO(CH_2)_2OOC(CH_2)_4SH$	125/2(*P91*)	(*P27, P91*)
$(C_4H_9)_3SCH_2HC(CH_2OCOC_{11}H_{23})(OCOC_{11}H_{23})$		(*P10, P42, P132, P133*)
$(C_8H_{17})_3SnS(CH_2)_4COO(CH_2)_2OOC(CH_2)_4SH$		(*P27, P91*)
$(CH_2{=}CH)_3SnSCH_2CH_2OB(OC_{20}H_{33})_2$		(*P138*)
$(C_6H_5CH_2)_3SnSCH_2COO(CH_2)_2OOCCH_2SH$	150/2(*P91*)	(*P27, P91*)
$(C_6H_5CH_2)_3SnS{-}C_6H_4{-}O{-}B(O{-}C_6H_4{-}CH_3)_2$		(*P138*)
$(C_6H_5CH_2)_3SnSCH(COO{-}C_4H_7O)(CH_2COO{-}C_4H_7O)$ (tetrahydrofurfuryl)		(*P45*)
$(C_6H_5)_3SnSCH_2COOC_{10}H_{21}$		(*P4, P102, P127, P128*)
$(C_6H_5)_3SnSCH_2COOCH_2C_6H_5$		(*P40, P100, P101*)
$(C_6H_5)_3SnSCH_2COO(CH_2)_2OOCCH_2SH$	150/2(*P91*)	(*P27, P91*)
$(C_6H_5)_3SnS{-}C_6H_4{-}CH_2OB(OC_6H_{11})_2$		(*P138*)
$(C_6H_5)_3SnSCH(COOC_4H_9)(COOC_4H_9)$		(*P15, P121*)

(*continued*)

APPENDIX 3 (*continued*)

COMPOUNDS OF THE TYPE $R_nSn(—SR'COOR'')_{4-n}$ AND $R_nSn(—SR'OR'')_{4-n}$

Compound	mp(°C)	bp(°C/mmHg)	n_D^{20}	d_4^{20}	References
$(C_4H_9)_2(CH_3O—)SnSCH_2COO(CH_2)_2OCH_3$					(*P84*)
$(C_4H_9)_2(CH_3O—)SnS(CH_2)_2OOCC_{11}H_{23}$					(*P84*)
$(C_4H_9)_2(C_2H_5O—)SnSCH_2OOCC_2H_5$					(*P84*)
$(C_4H_9)_2(C_8H_{17}O—)SnSCH_2COOC_8H_{17}$			1.4896	1.0892	(*P39*, *P140*)
$(n\text{-}C_4H_9)_2(C_6H_5COS)SnSCH_2COO—iso\text{-}C_8H_{17}$					(*P76g*)
$(C_4H_9)_2ClSnSCH_2COOC_4H_9$					(*P77*)
$(C_4H_9)_2(Cl)SnSCH_2CH_2COO—iso\text{-}C_8H_{17}$					(*P58f*)
$C_4H_9(Cl)_2SnSCH_2CH_2COO—iso\text{-}C_8H_{17}$					(*P58f*)
$(C_4H_9)_2(2—C_{10}H_7S—)SnSCH_2COOC_8H_{17}$			1.5796	1.2006	(*P39*, *P140*)
$(C_4H_9)_2[C_6H_5C(O)S—]SnSCH_2COOCH_2CH(C_4H_9)(C_2H_5)$					(*P33*)
$(C_4H_9)_2[C_6H_5C(O)S—]SnSCH_2CH_2COOH$					(*P33*)
$(C_4H_9)_2[H_3C—C_6H_4—SO_2N(CH_2CH(O)CH_2)]SnSCH_2COO\text{-}iso\text{-}C_8H_{17}$ (epoxide ring: $H_2C—CH$ bridged by O)					(*P152*)
$(iso\text{-}C_3H_7)(C_7H_{15}COO—)_2SnS(CH_2)_2COOCH_2C(CH_3)_2CH_2OH$					(*P2*)
$CH_2—CH_2—C(O)O(CH_2)_2SSn(C_4H_9)_3$ | $CH_2—CH_2—C(O)O(CH_2)_2SSn(C_4H_9)_3$					(*P13*)
$H_2C[(CH_2)_3—C(O)O(CH_2)_4SSn(—CH_2C_6H_5)_3]_2$					(*P13*, *P46*, *P134*, *P135*)

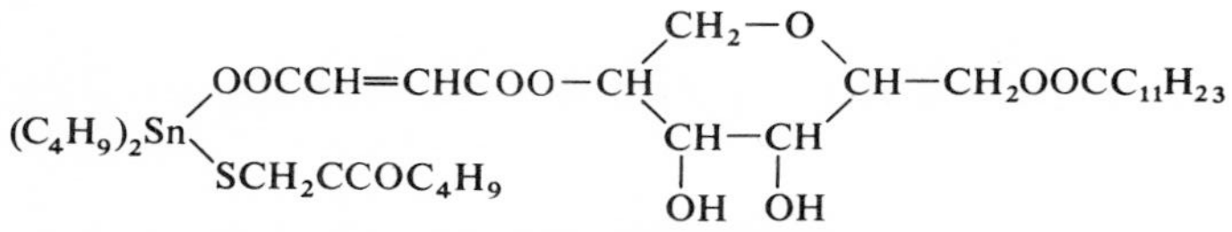 $(C_4H_9)_2Sn(OOCCH{=}CHCOO{-}CH\langle CH_2{-}O{-}CH(CH_2OOCC_{11}H_{23}){-}CH(OH){-}CH(OH)\rangle)(SCH_2CCOC_4H_9)$			(*P89b*)
$CH{-}COOSn(C_4H_9)_2({-}OOCCH{=}CHCOOC_8H_{17})$ ‖ $CH{-}COOSn(C_4H_9)_2({-}SCH_2COOC_8H_{17})$		$1.5110(n_D^{25})$	(*P38*)
$CH{-}COOSn(C_4H_9)_2({-}SCH_2COOC_8H_{17})$ ‖ $CH{-}COOSn(C_4H_9)_2({-}SCH_2COOC_8H_{17})$		$1.5110(n_D^{25})$	(*P38*)
$CH{-}COOSn(C_4H_9)_2({-}SCH_2COOC_6H_5)$ ‖ $CH{-}COOSn(C_4H_9)_2({-}OOCCH{=}CHCOOC_8H_{17})$		$1.5235(n_D^{25})$	(*P38*)
$CH_2{-}S{-}Sn(C_4H_9)_2{-}(OOCCH_2COOCH_2CH(C_2H_5)(C_4H_9))$ | $CH_2{-}COOSn(C_4H_9)_2{-}(SCH_2COOC_{12}H_{25})$		$1.4959(n_D^{25})$	(*P37*)
$CH_2{-}S{-}Sn(C_4H_9)_2{-}(OOCCH_2COOCH_2CH(C_2H_5)(C_4H_9))$ | $CH_2{-}COOSn(C_4H_9)_2{-}(SCH_2COOCH_2C_6H_5)$		$1.5116(n_D^{25})$	(*P37*)
$CH_2{-}S{-}Sn(C_4H_9)_2{-}(OOCCH_2COOCH_2CH(C_2H_5)(C_4H_9))$ | $CH_2{-}COOSn(C_4H_9)_2{-}(SCH_2COOC_6H_5)$	89–110		(*P37*)

(*continued*)

APPENDIX 3 (*continued*)

COMPOUNDS OF THE TYPE $R_nSn(\text{—}SR'COOR'')_{4-n}$ AND $R_nSn(\text{—}SR'OR'')_{4-n}$

Compound	mp(°C)	bp(°C/mmHg)	n_D^{20}	d_4^{20}	References
$CH_2\text{—}S\text{—}Sn(C_4H_9)_2\left(\text{—}OOCCH_2COOCH_2CH(C_2H_5)(C_4H_9)\right)$ / $CH_2\text{—}COOSn(C_4H_9)_2\left(\text{—}SCH_2COOCH_2CH(C_2H_5)(C_4H_9)\right)$			$1.5200(n_D^{25})$		(*P37*)
$CH_2\text{—}S\text{—}Sn(C_4H_9)_2(\text{—}OOCCH{=}CHCOOC_8H_{17})$ / $CH_2\text{—}COOSn(C_4H_9)_2(\text{—}OOCCH{=}CHCOOC_8H_{17})$					(*P37*)
$CH_2SCH_2COOSn(C_4H_9)_2\left(\text{—}OOCCH_2COOCH_2CH(C_2H_5)(C_4H_9)\right)$ / $CH_2SCH_2COOSn(C_4H_9)_2\left(\text{—}SCH_2COOCH_2CH(C_2H_5)(C_4H_9)\right)$	129–32				(*P37*)
$CH_2CH_2SCH_2COOSn(C_4H_9)_2\left(\text{—}OOCCH_2COOCH_2CH(C_2H_5)(C_4H_9)\right)$ / $CH_2CH_2SCH_2COOSn(C_4H_9)_2\left(\text{—}SCH_2COOCH_2CH(C_2H_5)(C_4H_9)\right)$	115–22				(*P37*)

$H_2C\begin{cases} SCH_2COOSn(C_4H_9)_2\left(-OOCCH_2COOCH_2CH\begin{matrix} C_2H_5 \\ C_4H_9 \end{matrix}\right) \\ SCH_2COOSn(C_4H_9)_2\left(-SCH_2COOCH_2CH\begin{matrix} C_2H_5 \\ C_4H_9 \end{matrix}\right) \end{cases}$	113–21		(*P37*)
$O\begin{cases} CH_2CH_2SCH_2COOSn(C_4H_9)_2\left(-OOCCH_2COOCH_2CH\begin{matrix} C_2H_5 \\ C_4H_9 \end{matrix}\right) \\ CH_2CH_2SCH_2COOSn(C_4H_9)_2\left(-SCH_2COOCH_2CH\begin{matrix} C_2H_5 \\ C_4H_9 \end{matrix}\right) \end{cases}$		$1.5086(n_D^{25})$	(*P37*)
$\begin{matrix} CH_2OOCCH_2SSn(C_4H_9)_2\left(-OOC-C_6H_4(OH)\right) \\ \vert \\ CH_2OOCCH_2SSn(C_4H_9)_2(-OOCC_{11}H_{23}) \end{matrix}$ (HO on the ring ortho to OOC)			(*P78*)
$(CH_2)_4\begin{cases} COO(CH_2)_2S-Sn(C_4H_9)\left(\begin{matrix} SCH_2CH_2OOC \\ SCH_2CH_2OOC \end{matrix}(CH_2)_4\right) \\ COO(CH_2)_2S-Sn(C_4H_9)\left(\begin{matrix} SCH_2CH_2OOC \\ SCH_2CH_2OOC \end{matrix}(CH_2)_4\right) \end{cases}$			(*P13*)

(*continued*)

APPENDIX 3 (*continued*)

COMPOUNDS OF THE TYPE $R_nSn(—SR'COOR'')_{4-n}$ AND $R_nSn(—SR'OR'')_{4-n}$

Compound	mp(°C)	bp(°C/mmHg)	n_D^{20}	d_4^{20}	References
$[(C_6H_5CH_2—)_3SnSCH_2CH_2COOCH_2CH_2—]_2$					(*P16*, *P43*, *P116*)
$(C_4H_9)_2Sn(OOCCH{=}CHCOOCH_2CH_2OOCCH{=}CHCOOSn(C_4H_9)_2{+}OOCC_{11}H_{23}))(SCH_2COOCH_2CH_2OOCCH_2SSn(C_4H_9)_2{+}OOCC_{11}H_{23}))$					*P*(*78*)
$C_{12}H_{25}OOCCH{=}CHCOO[Sn(C_4H_9)_2SCH_2COOCH_2CH_2COOCH_2S—]_{1.8}Sn(C_4H_9)_2OOCCH{=}CHCOOC_{12}H_{25}$					*P*(*78*)
$[C_8H_{17}OOCCH_2SSn(C_2H_5)_2]_2O$					(*P20*)
$[(C_2H_5)(C_4H_9)CHCH_2OOCCH_2SSn(C_4H_9)_2]_2O$			1.5128	1.245	(*P20*, *P50*, *P125*)
$O[Sn(C_4H_9)_2{+}OOCCH{=}CHCOOC_{18}H_{35})][Sn(C_4H_9)_2{+}SCH_2COOCH_2CH(C_2H_5)(C_4H_9))]$			1.5037	1.164(d_4^{25})	(*P20*, *P125*)
$C_{10}H_7—S{+}[Sn(CH_3)_2—O]_2{+}Sn(CH_3)_2—SCH_2COOC_8H_{17}$ (1-naphthyl)					(*P50*)
$C_{10}H_7—S{+}[Sn(C_4H_9)_2—O]_2{+}Sn(C_4H_9)_2—SCH_2COOC_8H_{17}$ (1-naphthyl)					(*P50*)

Compound			
$\left[(C_2H_5)(C_4H_9)CHCH_2OOCCH_2S{-}\left[Sn(C_4H_9)_2{-}O\right]_3{-}OCCH{=}CHCOOCH_2CH(OH)CH_3\right]$			(*P20, P125*)
$\left[(C_2H_5)(C_4H_9)CHCH_2OOCCH_2S{-}\left[Sn(C_4H_9)_2{-}O\right]_2{-}Sn(C_4H_9)_2\right]$	1.5166	1.286	(*P20, P125*)
$(C_2H_5)(C_4H_9)CHCH_2OOCCH_2S{-}\left[Sn(C_4H_9)_2{-}O\right]_3{-}OCC_{11}H_{25}$	1.5069	1.256	(*P20, P50, P125*)
$\left[(C_2H_5)(C_4H_9)CHCH_2OOCCH_2SSn(C_4H_9)_2{-}O{-}Sn(C_4H_9)_2{-}\right]_2O$	1.5220(*P20, P125*)	1.330(*P20, P125*)	(*P20, P50, P125*)
$S[Sn(C_4H_9)_2(-SCH_2COOC_4H_9)][Sn(C_4H_9)_2(-OOCCH{=}CHCOOC_4H_9)]$			(*P77*)
$C_6H_4\left[COO(CH_2)_2SSn(i\text{-}C_3H_7)\left[S(CH_2)_2OOC\right]_2C_6H_4\right]_2$			(*P13, P46, P134, P135*)

(continued)

APPENDIX 3 (*continued*)

COMPOUNDS OF THE TYPE $R_nSn(—SR'COOR'')_{4-n}$ AND $R_nSn(—SR'OR'')_{4-n}$

Compound	mp(°C)	bp(°C/mmHg)	n_D^{20}	d_4^{20}	References
$iso\text{-}C_8H_{17}OOCCH{=}CHCOO(C_4H_9)_2Sn{-}O{-}Sn(C_4H_9)_2SCH_2COOC_8H_{17}\text{-}iso$ (written as: iso-$C_8H_{17}OOCCH{=}CHCOO(C_4H_9)_2Sn$ \ O / iso-$C_8H_{17}OOCCH_2S(C_4H_9)_2Sn$)					(*P58c*)
$[C_3H_7OOCCH{=}CHCOS(C_4H_9)_2Sn]_2O$					(*P58c*)
$[iso\text{-}C_8H_{17}OOCCH_2S(C_4H_9)_2Sn]_2O$					(*P58c*)
B with: $O{-}Sn(CH_3)_2SCH_2COOCH(CH_3)CH_2OOCCH_2SH$; $O{-}Sn(CH_3)_2OOCC_{11}H_{23}$; $O{-}Sn(CH_3)_2OOCCH{=}CHCOOCH_2C_6H_5$					*P*(*92*)
$B[—O—Sn(C_4H_9)_2(—SCH_2COOCH_2CH_2OOCCH_2SH)]_3$					(*P92*)
$B[—O—Sn(C_4H_9)_2(—SCH_2CH_2COOCH_2CH_2OOCCH_2CH_2SH)]_3$					(*P92*)
B with: $O{-}Sn(C_4H_9)_2(—OOCCH{=}CHCOOCH_2C_6H_5)$; $[—O—Sn(C_4H_9)_2(—SCH_2CH_2OOCCH{=}CHCOOCH_2CH_2SH)]_2$					(*P92*)
B with: $O{-}Sn(C_4H_9)_2(—OOCC_{11}H_{23})$; $[—O—Sn(C_4H_9)_2(—SCH_2CH_2OOCCH{=}CHCOOCH_2CH_2SH)]_2$					(*P89a, P92*)
$B[—O—Sn(C_8H_{17})_2(—SCH_2COOCH_2CH_2OOCCH_2SH)]_3$					(*P92*)
B with: $O{-}Sn(C_8H_{17})_2(—SCH_2COOCH_2CH_2OOCCH_2SH)$; $[—O—Sn(C_8H_{17})_2(—SC_{12}H_{25})]_2$					(*P89a, P92*)

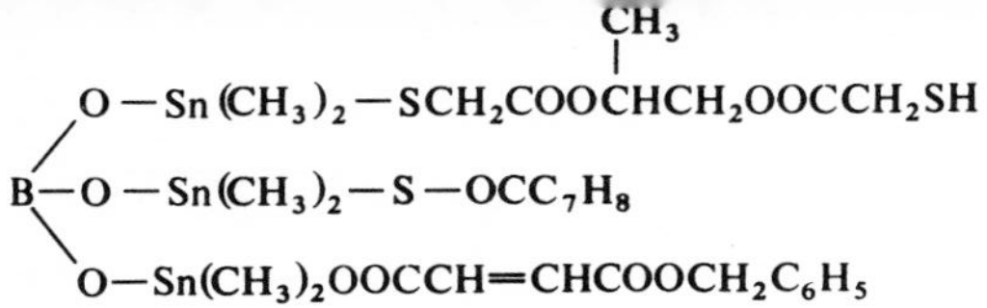

Compound	Ref.
$B(-O-Sn(CH_3)_2-SCH_2COOCH(CH_3)CH_2OOCCH_2SH)(-O-Sn(CH_3)_2-S-OCC_7H_8)(-O-Sn(CH_3)_2OOCCH{=}CHCOOCH_2C_6H_5)$	(*P89a*)
$B(-O-CH_2CH(CH_3)-S-Sn(C_6H_{11})_3)(-O-CH_2CH(CH_3)-S-Sn(C_6H_{11})_3)(-O-CH_2CH(CH_3)-S-Sn(C_6H_{11})_3)$	(*P139*)
$B(-O-C_6H_4-CH_2S-Sn(-CH_2CH{=}CHCH_3)_3)_3$	(*P139*)
$B(-O-CH_2CH(C_6H_{13})-S-Sn(C_6H_4-CH_3)_3)_3$	(*P139*)

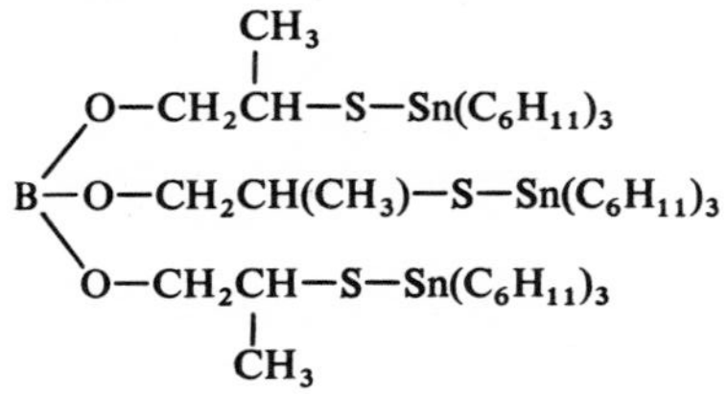

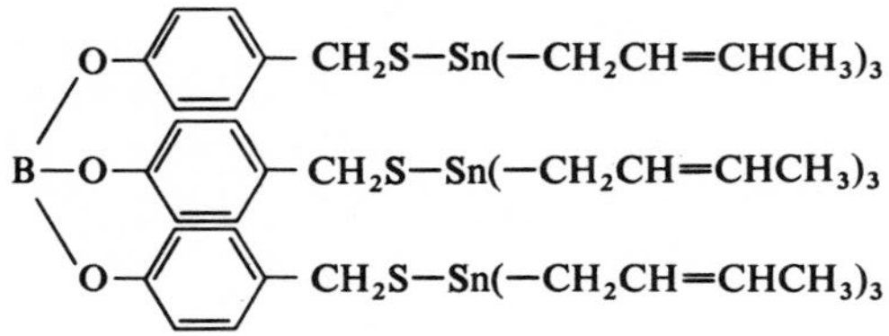

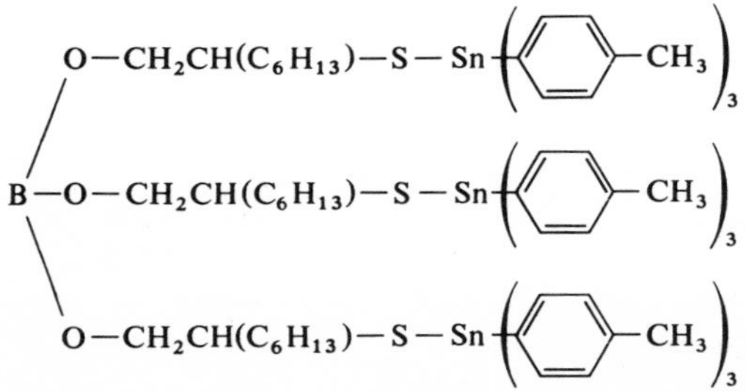

(*continued*)

APPENDIX 3 (*continued*)

COMPOUNDS OF THE TYPE $R_nSn(—SR'COOR'')_{4-n}$ AND $R_nSn(—SR'OR'')_{4-n}$

Compound	mp(°C)	bp(°C/mmHg)	n_D^{20}	d_4^{20}	References
$B[-O-(CH_2)_4-S-Sn(CH_2-C_6H_4-Cl)_3]_3$					(*P139*)
$R_2Sn(SR'COOR'')_2$					
$(CH_3)_2Sn(—SCH_2COO\text{-}iso\text{-}C_8H_{17})_2$					(*P52*, *P58*)
$(CH_3)_2Sn(—SCH_2COOC_{10}H_{21})_2$					(*P4*, *P102*)
$(CH_3)_2Sn(SCH_2COO-C_6H_4-CH_3)_2$					(*P15*, *P121*)
$(CH_3)_2Sn(—SCH_2CH_2COOC_{18}H_{37})_2$					(*P15*, *P121*)
$(CH_3)_2Sn(—SCH(C_6H_5)—CH_2OOCC_3H_7)_2$					(*P19*)
$(C_2H_5)_2Sn(SCHOHCOOC_2H_5)_2$					(*P38d*)
$(C_3H_7)_2Sn(SCH(CH_3)CH_2CH_2COO-C_4H_7O)_2$					(*P100*)

$(C_4H_9)_2Sn(—SCH_2COOH)_2$				(*P76g*, *P102*)
$(n\text{-}C_4H_9)_2Sn(SCH_2COOCH_3)_2$				(*P76g*)
$(C_4H_9)_2Sn(—SCH_2COOC_2H_5)_2$				(*P34*)
$(C_4H_9)_2Sn(—SCH_2COOCH_2CH_2Cl)_2$				(*P79*)
$(C_4H_9)_2Sn(—SCH_2COOC_4H_9)_2$				(*P38d*, *P100*, *P101*, *P103*, *P104*, *P105*, *P106*, *P123*)
$(C_4H_9)_2Sn\left(—SCH_2COOCH_2\overset{\overset{CH_3}{\vert}}{C}{=}CH_2\right)_2$				(*P80*)
$(C_4H_9)_2Sn(—SCH_2COOC_6H_{11})_2$				(*P4*, *P40*, *P100*, *P101*, *P102*, *P123*, *P127*, *P128*)
$(C_4H_9)_2Sn\left(—SCH_2COOCH_2\overset{\overset{C_2H_5}{\vert}}{C}H—C_4H_9\right)_2$	75–80(*72*) 124/25(*P95b*)	1.5075(*P53*, *P65*) 1.5073(*P140*)	1.1313(*P140*)	(*72*, *74*, *76*, *130*, *P15*, *P38a*, *P38f*, *P52*, *P53*, *P58*, *P58a,b,d*, *P59a*, *P65*, *P76a*, *P89a*, *P92*, *P95b*, *P103*, *P104*, *P105*, *P106*, *P121*, *P140*, *P146*)
$(iso\text{-}C_4H_9)_2Sn\left(—SCH_2COOCH_2\overset{\overset{C_2H_5}{\vert}}{C}H—C_4H_9\right)_2$		1.5033(*P65*) 1.5073(*P125*)	1.131(*P125*)	(*P65*, *P125*)

(*continued*)

APPENDIX 3 (*continued*)

COMPOUNDS OF THE TYPE $R_nSn(—SR'COOR'')_{4-n}$ AND $R_nSn(—SR'OR'')_{4-n}$

Compound	mp(°C)	bp(°C/mmHg)	n_D^{20}	d_4^{20}	References
$(C_4H_9)_2Sn(—SCH_2COOCH_2CH_2—C(CH_3)(CH_3)—CH_2CH(CH_3)—CH_3)_2$					(*P4*, *P15*, *P102*, *P121*, *P127*, *P128*)
$(C_4H_9)_2Sn[—SCH_2COO(CH_2)_8C(CH_3)_3]_2$					(*P4*, *P102*, *P127*, *P128*)
$(n\text{-}C_4H_9)_2Sn(SCH_2COO—n\text{-}C_{16}H_{33})_2$					(*P76g*)
$(C_4H_9)_2Sn(—SCH_2COOC_{18}H_{37})_2$					(*P4*, *P15*, *P102*, *P103*, *P104*, *P105*, *P106*, *P121*)
$(C_4H_9)_2Sn(—SCH_2COO—C_4H_7O)_2$ (tetrahydrofuryl ester)					(*P81*)
$(C_4H_9)_2Sn(—SCH_2COOCH_2—C_4H_7O)_2$ (tetrahydrofurfuryl ester)					(*P89*)
$(C_4H_9)_2Sn(—SCH_2COOC_{20}H_{30})_2$					(*P40*, *P100*, *P101*)
$(C_4H_9)_2Sn(—SCH_2COOC_{20}H_{32})_2$					(*P4*, *P102*, *P127*, *P128*)
$(C_4H_9)_2Sn(—SCH_2COOCH_2CH_2OC_6H_5)_2$					(*P4*, *P102*, *P127*, *P128*)
$(C_4H_9)_2Sn(—SCH_2COOCH_2CH_2OOCCH_2SH)_2$					(*P27*, *P91*)
$(C_4H_9)_2Sn[—SCH_2COOCH_2CH_2OOC(CH_2)_7CH{=}CHCH_2CH(OH)(CH_2)_5CH_3]_2$					(*P4*)

Compound	References
$(C_4H_9)_2Sn\left(-SCH_2COOCH(CH_3)CH_2OOCCH_2SH\right)_2$	(*P89a*, *P92*)
$(C_4H_9)_2Sn[-SCH_2COO(CH_2)_4OOCCH_2SH]_2$	(*P27*, *P91*)
$(C_4H_9)_2Sn[-SCH_2COO(CH_2)_4OOCC_{11}H_{23}]_2$	(*P4*, *P102*, *P127*, *P128*)
$(C_4H_9)_2Sn[-SCH_2COO(CH_2)_5OOCCH_2SH]_2$	(*P27*)
$(C_4H_9)_2Sn[-SCH_2COO(CH_2)_8OOCCH_2SH]_2$	(*P27*, *P91*)
$(C_4H_9)_2Sn\left(-SCH_2COOCH(C_2H_5)-CH_2-CH(C_3H_7)-OOCCH_2SH\right)_2$	(*P27*)
$(C_4H_9)_2Sn(-SCH_2CONHC_5H_{11})_2$	(*P41*, *P107*)
$(C_4H_9)_2Sn(-SCH_2CONHC_6H_{11})_2$	(*P9*)
$(C_4H_9)_2Sn\left[SCH_2CON\left(CH_2CH(C_4H_9)(C_2H_5)\right)_2\right]_2$	(*P9*)
$(C_4H_9)_2Sn\left[SCH_2CON\left(CH(C_5H_{11})(C_2H_5)\right)_2\right]_2$	(*P107*)
$(C_4H_9)_2Sn\left(SCH_2CO-N(CH_2-CH_2)_2O\right)_2$	(*P9*, *P28*, *P107*)
$(C_4H_9)_2Sn\left[-SCH(CH_3)-COO(CH_2)_5C(CH_3)_3\right]_2$	(*P4*, *P102*)

(*continued*)

APPENDIX 3 *(continued)*

Compounds of the Type $R_nSn(-SR'COOR'')_{4-n}$ and $R_nSn(-SR'OR'')_{4-n}$

Compound	mp(°C)	bp(°C/mmHg)	n_D^{20}	d_4^{20}	References
$(C_4H_9)_2Sn(-SCH(C_2H_5)-COOC_5H_{11})_2$					(*P100*)
$(C_4H_9)_2Sn[-SCH(C_2H_5)-COO(CH_2)_5C(CH_3)_3]_2$					(*P4*, *P102*, *P127*, *P128*)
$(C_4H_9)_2Sn[-SCH(C_3H_7)-COO(CH_2)_5C(CH_3)_3]_2$					(*P4*, *P102*, *P127*, *P128*)
$(C_4H_9)_2Sn(-SCH(C_4H_9)-COOC_4H_9)_2$					(*P4*, *P102*, *P127*, *P128*)
$(C_4H_9)_2Sn[-SCH(C_4H_9)-COO(CH_2)_5C(CH_3)_3]_2$					(*P4*, *P102*, *P127*, *P128*)
$(C_4H_9)_2Sn(-SCH(C_6H_{13})-COOC_4H_9)_2$					(*P4*, *P127*, *P128*)
$(C_4H_9)_2Sn(-SCH(C_7H_{15})-COOC_4H_9)_2$					(*P4*, *P102*, *P127*, *P128*)
$(C_4H_9)_2Sn(-SCH(C_8H_{17})-COOC_4H_9)_2$					(*P102*)

Compound	Ref.
$(C_4H_9)_2Sn\left(-S\overset{C_{10}H_{21}}{\overset{\mid}{C}}H-COOC_4H_9\right)_2$	(*P4, P102, P127, P128*)
$(C_4H_9)_2Sn\left(-S\overset{C_{14}H_{29}}{\overset{\mid}{C}}H-COOC_4H_9\right)_2$	(*P4, P102, P127, P128*)
$(n\text{-}C_4H_9)_2Sn\left(S\underset{C_{16}H_{33}}{\underset{\mid}{C}}HCOOH\right)_2$	(*P76g*)
$(C_4H_9)_2Sn\left(-S\overset{C_{16}H_{33}}{\overset{\mid}{C}}H-COOC_4H_9\right)_2$	(*P4, P102, P127, P128*)
$(C_4H_9)_2Sn\left(-S\overset{C_6H_5}{\overset{\mid}{C}}H-COOC_2H_5\right)_2$	(*P4, P102, P127, P128*)
$(C_4H_9)_2Sn\left(-S-\overset{C_2H_5}{\underset{C_2H_5}{C}}-COOCH_2CH_2-\overset{CH_3}{\underset{CH_3}{C}}-CH_2-\overset{CH_3}{\overset{\mid}{C}H}-CH_3\right)_2$	(*P4, P127, P128*)
$(C_4H_9)_2Sn\left(-S-\overset{C_4H_9}{\underset{C_2H_5}{C}}-COOC_2H_5\right)_2$	(*P4, P102, P127, P128*)

(*continued*)

APPENDIX 3 (*continued*)

COMPOUNDS OF THE TYPE $R_nSn(—SR'COOR'')_{4-n}$ AND $R_nSn(—SR'OR'')_{4-n}$

Compound	mp(°C)	bp(°C/mmHg)	n_D^{20}	d_4^{20}	References
$(C_4H_9)_2Sn\left(—S—C(C_6H_5)_2—COOC_2H_5\right)_2$					(*P4*, *P102*, *P127*, *P128*)
$(C_4H_9)_2Sn\left[—S—C(C_6H_5)_2—COO(CH_2)_5C(CH_3)_3\right]_2$					(*P102*)
$(C_4H_9)_2Sn(—SCH_2CH_2COOC_4H_9)_2$					(*P15*, *P121*)
$(C_4H_9)_2Sn(—SCH_2CH_2OOCC_5H_{11})_2$					(*P19*)
$(C_4H_9)_2Sn\left[—SCH_2CH_2OOCCH_2(CH(CH_3)—)_3CH_3\right]_2$					(*P10*, *P132*, *P133*)
$(C_4H_9)_2Sn\left[SCH_2CH_2COOCH_2CH(C_2H_5)(C_4H_9)\right]_2$			1.5046(*P65*)		(*P42*, *P53*, *P65*)
$(C_4H_9)_2Sn\left[—SCH_2CH_2COOCH_2(CH(CH_3)—)_3C_2H_5\right]_2$					(*P127*, *P128*)
$(C_4H_9)_2Sn(—SCH_2CH_2OOCC_9H_{19})_2$					(*P19*)
$(C_4H_9)_2Sn(—SCH_2CH_2OOCC_{11}H_{23})_2$					(*P10*, *P19*, *P42*, *P132*, *P133*)

$(C_4H_9)_2Sn\left(-SCH_2CH_2COOCH_2CH(C_4H_9)(C_6H_{13})\right)_2$	(*P4*, *P102*, *P127*, *P128*)
$(C_4H_9)_2Sn(-SCH_2CH_2OOCC_{19}H_{33})_2$	(*P10*, *P132*, *P133*)
$(C_4H_9)_2Sn\left(-SCH_2CH_2COOCH_2CH(OH)CH_2OH\right)_2$	(*P2*)
$(C_4H_9)_2Sn[-SCH_2CH_2OOC(CH_2)_8COOH]_2$	(*P19*)
$(C_4H_9)_2Sn(-SCH_2CH_2OOCC_6H_5)_2$	(*P19*)
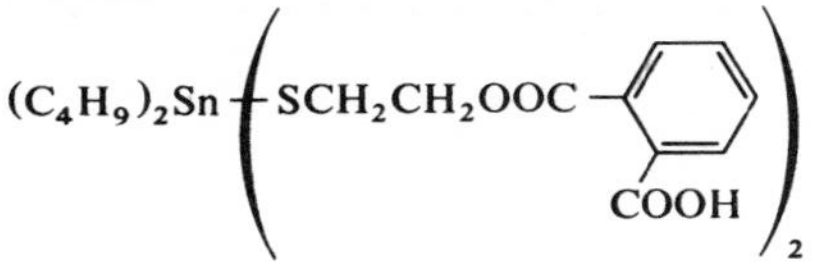 $(C_4H_9)_2Sn\left(-SCH_2CH_2OOC-C_6H_4-COOH\right)_2$ (ortho)	(*P19*)
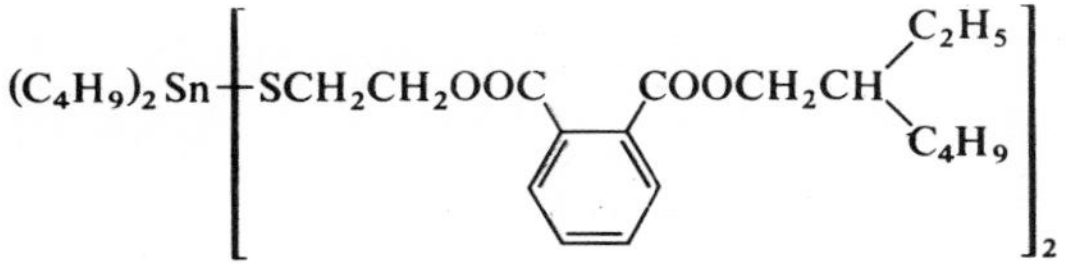 $(C_4H_9)_2Sn\left[-SCH_2CH_2OOC-C_6H_4-COOCH_2CH(C_2H_5)(C_4H_9)\right]_2$ (ortho)	(*P19*)
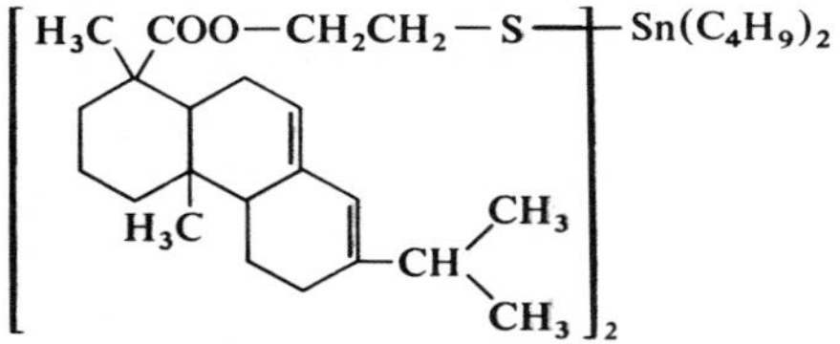 $[H_3C(COO-CH_2CH_2-S-)$ hydrophenanthrene ring with H_3C and $-CH(CH_3)_2$ substituents$]_2Sn(C_4H_9)_2$	(*P19*)
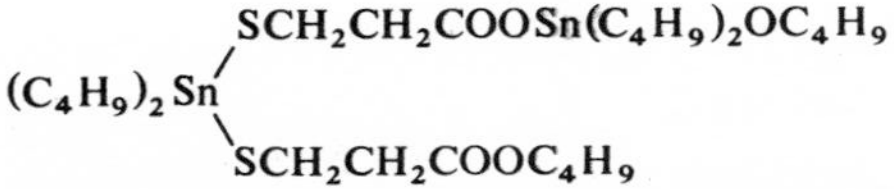 $(C_4H_9)_2Sn(SCH_2CH_2COOSn(C_4H_9)_2OC_4H_9)(SCH_2CH_2COOC_4H_9)$	(*P76c*)

(*continued*)

APPENDIX 3 (*continued*)

COMPOUNDS OF THE TYPE $R_nSn(\text{—}SR'COOR'')_{4-n}$ AND $R_nSn(\text{—}SR'OR'')_{4-n}$

Compound	mp(°C)	bp(°C/mmHg)	n_D^{20}	d_4^{20}	References
$(C_4H_9)_2Sn[SCH_2CH_2COOSn(C_4H_9)_2OC_4H_9]_2$					(*P76c*)
$(C_4H_9)_2Sn(SCH_2CH_2COOSn(C_4H_9)_2OC_8H_{17})(SCH_2CH_2COOC_8H_{17})$					(*P76c*)
$(C_4H_9)_2Sn[SCH_2CH_2COOSn(C_4H_9)_2OC_8H_{17}]_2$					(*P76c*)
$(C_4H_9)_2Sn(SCH(CH_3)CH_2COOSn(C_4H_9)_2OC_8H_{17})(SCH(CH_3)CH_2COOC_8H_{17})$					(*P76c*)
$(C_4H_9)_2Sn(SCH_2C(CH_3)_2COOSn(C_4H_9)_2OC_8H_{17})(SCH_2C(CH_3)_2COOC_8H_{17})$					(*P76c*)
$(C_4H_9)_2Sn\left[SCH_2CH_2O\text{—}B\left(OCH_2CH(C_2H_5)C_4H_9\right)_2\right]_2$					(*P18, P54, P138*)
$(C_4H_9)_2Sn\left[SCH_2CH_2O\text{—}B\left(OCH_2CH(C_6H_{13})C_4H_9\right)_2\right]_2$					(*P54, P138*)
$(C_4H_9)_2Sn\left[SCH_2CH_2O\text{—}B\begin{matrix}O\text{—}CH_2\\ \vert \\ O\text{—}CH_2\end{matrix}\right]_2$					(*P18, P54, P138*)

Compound			Ref.
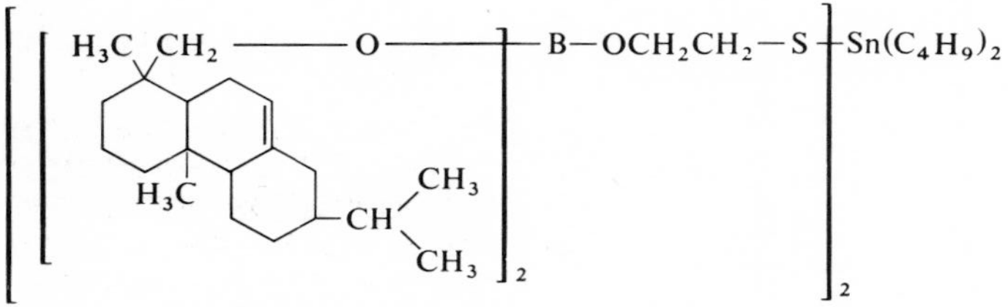 $\left[\left[\text{(abietyl: } H_3C, CH_2 — O\text{; ring } H_3C\text{; } CH(CH_3)_2)\right]_2 B—OCH_2CH_2—S\right]_2 Sn(C_4H_9)_2$			(*P18*, *P138*)
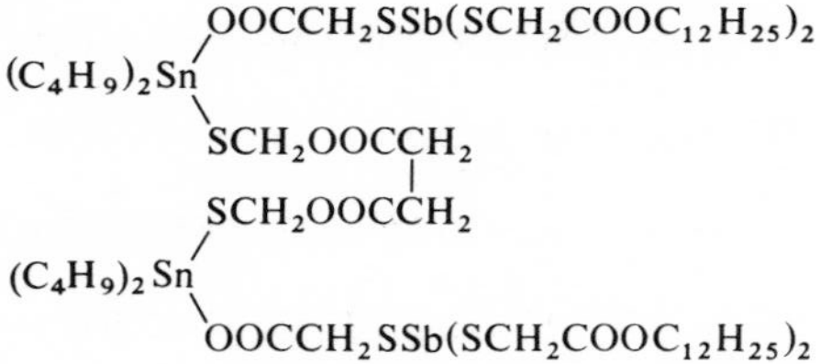 $(C_4H_9)_2Sn(OOCCH_2SSb(SCH_2COOC_{12}H_{25})_2)(SCH_2OOCCH_2—)$ / $(C_4H_9)_2Sn(SCH_2OOCCH_2—)(OOCCH_2SSb(SCH_2COOC_{12}H_{25})_2)$			(*P76d*)
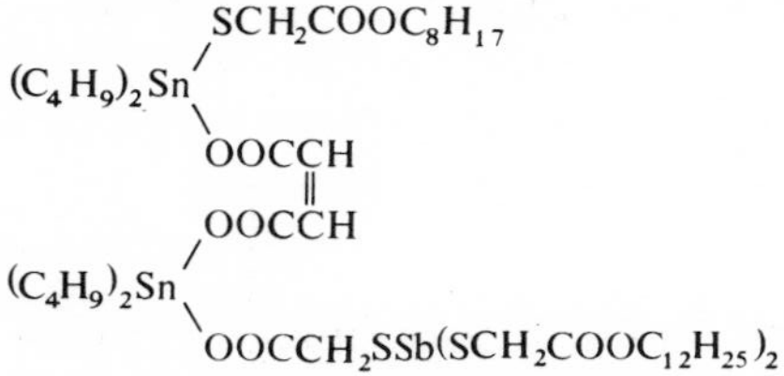 $(C_4H_9)_2Sn(SCH_2COOC_8H_{17})(OOCCH=)$ / $(C_4H_9)_2Sn(=CHCOO)(OOCCH_2SSb(SCH_2COOC_{12}H_{25})_2)$			(*P76d*)
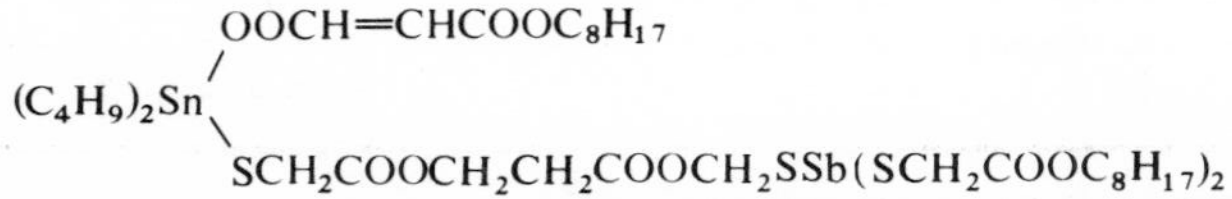 $(C_4H_9)_2Sn(OOCH=CHCOOC_8H_{17})(SCH_2COOCH_2CH_2COOCH_2SSb(SCH_2COOC_8H_{17})_2)$	1.5290	1.246	(*P76d*)

(*continued*)

APPENDIX 3 (*continued*)

COMPOUNDS OF THE TYPE $R_nSn(—SR'COOR'')_{4-n}$ AND $R_nSn(—SR'OR'')_{4-n}$

Compound	mp(°C)	bp(°C/mmHg)	n_D^{20}	d_4^{20}	References
$(C_8H_{17})_2Sn[—SCH_2OOCCH_2CH_2COOCH_2SSb(SCH_2COOC_8H_{17})—SCH_2OOCCH_2CH_2COOCH_2SSn(C_8H_{17})_2—OOC—R]_2$					(*P76d*)
$(C_4H_9)_2Sn[—S—CH(CH_3)—CH_2COO(CH_2)_5C(CH_3)_3]_2$					(*P4*, *P102*, *P127*, *P128*)
$(C_4H_9)_2Sn[—S(CH_2)_4OOCC_8H_{17}]_2$					(*P10*, *P42*, *P132*, *P133*)
$(C_4H_9)_2Sn[—S(CH_2)_4COO(CH_2)_5C(CH_3)_3]_2$					(*P102*, *P127*, *P128*)
$(C_4H_9)_2Sn[—S(CH_2)_5COOC_5H_{10}C(CH_3)_3]_2$					(*P4*, *P102*, *P127*, *P128*)
$(C_4H_9)_2Sn[—S(CH_2)_6COOC_5H_{10}C(CH_3)_3]_2$					(*P4*, *P102*, *P127*, *P128*)
$(C_4H_9)_2Sn[—S(CH_2)_7COOC_4H_9]_2$					(*P4*, *P127*, *P128*)
$(C_4H_9)_2Sn[—S(CH_2)_7COOC_5H_{10}C(CH_3)_3]_2$					(*P4*, *P102*)
$(n\text{-}C_4H_9)_2Sn(—SC_{11}H_{22}COOH)_2$					(*P76g*)
$(C_4H_9)_2Sn[—S(CH_2)_{11}COOC_4H_9]_2$					(*P4*, *P127*, *P128*)
$(C_4H_9)_2Sn[—S(CH_2)_{11}COOC_5H_{10}C(CH_3)_3]_2$					(*P102*)
$(C_4H_9)_2Sn(—S—HC(COOH)COOH)_2$					(*P127*, *P128*, *P129*)

Compound	Ref.
$(C_4H_9)_2Sn\left(S-CH(COOC_4H_9)(COOC_4H_9)\right)_2$	(*P129*)
$(C_4H_9)_2Sn\left[-S-CH(COOCH_2CH(C_2H_5)(C_4H_9))(COOCH_2CH(C_2H_5)(C_4H_9))\right]_2$	(*P15, P121*)
$(C_4H_9)_2Sn\left[S-CH(COOC_4H_9)(CH_2COOC_4H_9)\right]_2$	(*P4, P45, P57*)
$(C_4H_9)_2Sn\left[S-CH(COO-CH(C_5H_{11})(C_2H_5))(CH(CH_3)-COO-CH(C_5H_{11})(C_2H_5))\right]_2$	(*P45*)
$(n\text{-}C_4H_9)_2Sn\left(-SC(COOH)=CHCOOH\right)_2$	(*P76g*)
$(C_4H_9)ClSn(-SCH_2CH_2COO-i\text{-}C_8H_{17})_2$	(*P58f*)
$[C_4H_9Sn(-SCH_2CH_2COOC_8H_{17})_2]_2O$	(*P76c*)

(*continued*)

APPENDIX 3 (*continued*)

COMPOUNDS OF THE TYPE $R_nSn(—SR'COOR'')_{4-n}$ AND $R_nSn(—SR'OR'')_{4-n}$

Compound	mp(°C)	bp(°C/mmHg)	n_D^{20}	d_4^{20}	References
$[C_4H_9Sn(—SCH_2CH_2COOSn(C_4H_9)_2OC_8H_{17})_2]_2O$					(*P76c*)
$[C_4H_9Sn(—SCH_2CH_2COOSn(C_8H_{17})_2OC_8H_{17})_2]_2O$					(*P76c*)
$(C_5H_{11})_2Sn(S—CH(COOC_{10}H_{21})_2)_2$					(*P15, P121*)
$(n\text{-}C_8H_{17})_2Sn(—SCH_2COOC_4H_9)_2$			1.5070(*P53, P65*)		(*124, 232, P53, P65*)
$(C_8H_{17})_2Sn(SCH_2COOCH_2CH(C_2H_5)C_4H_9)_2$			1.4992(*P53, P65*)		(*124, 211a, P38a, P53, P58, P65, P76a*)
$(C_8H_{17})_2Sn(—SCH_2COOC_{18}H_{37})_2$			1.4992(*P53, P65*)		(*P53, P58a, P65*)
$(iso\text{-}C_8H_{17})_2Sn(S—CH(COOH)CH_3)_2$					(*P2*)
$(iso—C_8H_{17})_2Sn[—SCH_2COO(CH_2)_2OOCCH_2SH]_2$					(*P27*)
$(C_8H_{17})_2Sn[—SCH_2COO(CH_2)_4OOCCH_2SH]_2$					(*P27, P91*)

$(C_8H_{17})_2Sn\left[S-CH\begin{matrix}COOCH_2CH(C_2H_5)(C_4H_9)\\COOCH_2CH(C_2H_5)(C_4H_9)\end{matrix}\right]_2$	1.4892	(*P90*)
$(C_8H_{17})_2Sn\left[S-CH\begin{matrix}COOCH_2HC(C_2H_5)(C_4H_9)\\CH_2COOCH_2HC(C_2H_5)(C_4H_9)\end{matrix}\right]_2$	1.4892	(*P53, P65*)
$(C_{10}H_{21})_2Sn(—SCH_2COO\text{-}iso\text{-}C_8H_{17})_2$		(*P2*)
$(C_{12}H_{25})_2Sn(—SCH_2COOC_4H_9)_2$		(*P4, P15, P102, P121, P127, P128*)
$(C_{12}H_{25})_2Sn[—S(CH_2)_3COOC_4H_9]_2$		(*P40, P100, P101*)
$(C_{12}H_{25})_2Sn\left(S-CH\begin{matrix}CH_2COOC_2H_5\\COOC_2H_5\end{matrix}\right)_2$		(*P2*)
$(C_{12}H_{25})_2Sn\left(S-C_6H_4-CH_2OOCC_5H_{11}\right)_2$		(*P19*)
$(c\text{-}C_6H_{11})_2Sn\left(SCH_2COOCH_2CH_2-C(CH_3)_2-CH(CH_3)_2\right)_2$		(*P15, P121*)

(*continued*)

APPENDIX 3 (*continued*)

COMPOUNDS OF THE TYPE $R_nSn(—SR'COOR'')_{4-n}$ AND $R_nSn(—SR'OR'')_{4-n}$

Compound	mp(°C)	bp(°C/mmHg)	n_D^{20}	d_4^{20}	References
$(C_6H_5CH_2)_2Sn(—SCH_2CH_2COO\text{-}iso\text{-}C_8H_{17})_2$					(*P15, P121*)
$(NCCH_2CH_2)_2Sn(SCH_2COO\text{-}iso\text{-}C_8H_{17})_2$			1.5208(n_D^{28})	1.2084(d_4^{25})	(*P95a*)
$(C_6H_5)_2Sn(—SCH_2COOC_{12}H_{25})_2$					(*P15, P121*)
$(C_6H_5)_2Sn(—SCH_2CH_2OOCC_{11}H_{23})_2$					(*P19*)
$(C_6H_5)_2Sn\left(—S—CH\begin{matrix}CH_2COOC_4H_9\\COOC_4H_9\end{matrix}\right)_2$			1.2290	1.5400	(*P39, P140*)
$(C_6H_5)_2Sn[—SCH_2CH_2O—B(—OC_6H_5)_2]_2$					(*P12, P138*)
$(o\text{-}CH_3C_6H_4)_2Sn\left(—S—CH\begin{matrix}COOC_4H_9\\CH_2COOC_4H_9\end{matrix}\right)_2$					(*P121*)
$(o\text{-}CH_3C_6H_4)_2Sn\left[—S—CH\begin{matrix}COOCH_2CH_2C(CH_3)_2—CH_2CH(CH_3)_2\\COOCH_2CH_2C(CH_3)_2—CH_2CH(CH_3)_2\end{matrix}\right]_2$					(*P15*)
$(NC—CH_2CH_2)_2Sn(—SCH_2COO\text{-}iso\text{-}C_8H_{17})_2$			1.5215(n_D^{25})		(*166*)

Compound	Property	Reference
$(C_2H_5)(C_6H_5)Sn{-}(SCH_2COO\text{-}iso\text{-}C_8H_{17})_2$		(*P15, P121*)
$CH_2{-}S{-}Sn(C_4H_9)_2{-}(SCH_2COOCH_2CH(C_2H_5)(C_4H_9))$ / $CH_2{-}COO{-}Sn(C_4H_9)_2{-}(SCH_2COOCH_2CH(C_2H_5)(C_4H_9))$	1.5103(n_D^{25})	(*P37*)
$O[Sn(C_4H_9){-}(SCH_2COO\text{-}iso\text{-}C_8H_{17})_2]_2$		(*P93*)
$O[Sn(C_4H_9){-}(SCH_2CH_2COO\text{-}iso\text{-}C_8H_{17})_2]_2$		(*P93*)
$(C_4H_9)_2Sn\left[S(CH_2)_2COOCH_2(HOCH_2)C(CH_2OOC(CH_2)_2S)_2Sn(C_4H_9)_2\right]_2$		(*P35*)
$(C_4H_9)_2Sn\left[S(CH_2)_2COOCH_2(C_{11}H_{23}COOCH_2)C(CH_2OOC(CH_2)_2S)_2Sn(C_4H_9)_2\right]_2$		(*P35*)

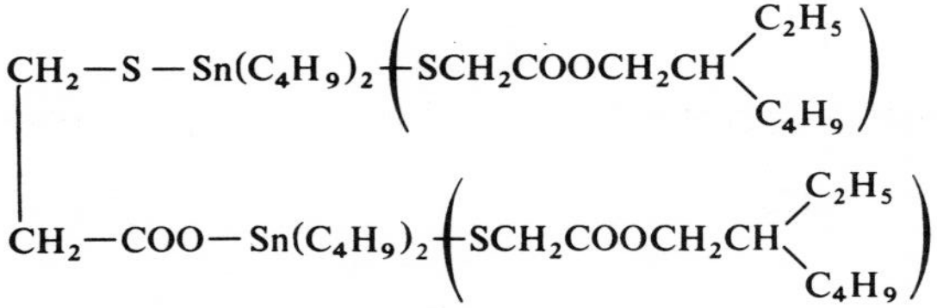

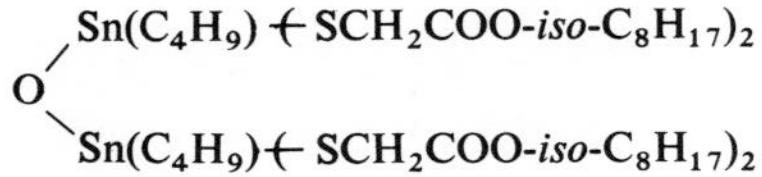

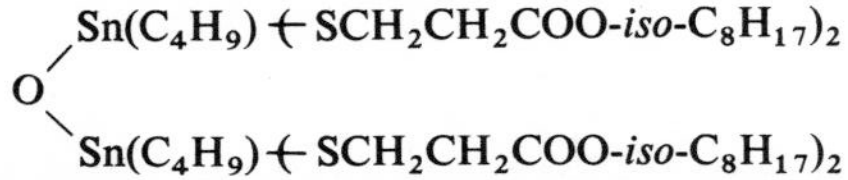

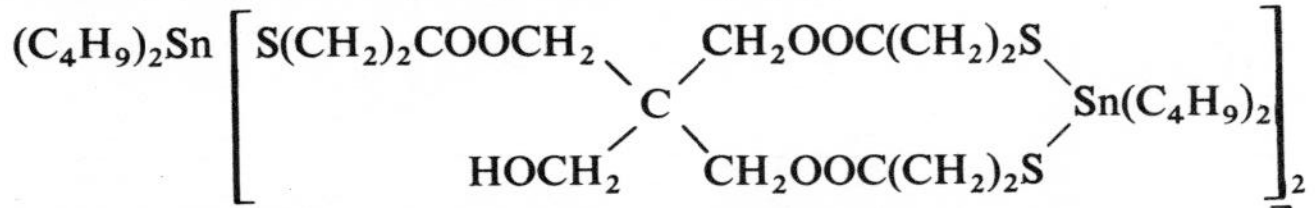

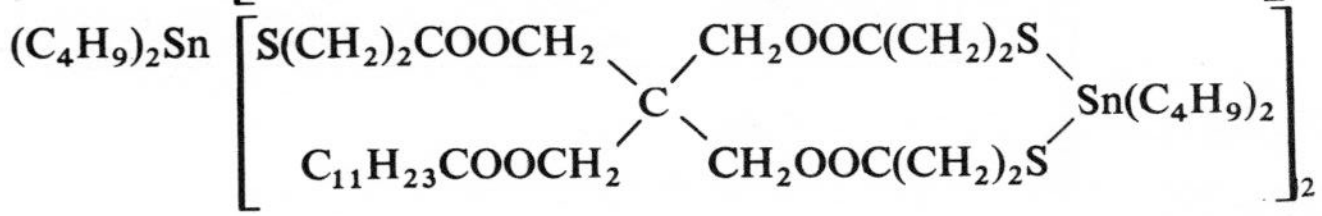

(*continued*)

APPENDIX 3 (*continued*)

COMPOUNDS OF THE TYPE $R_nSn(—SR'COOR'')_{4-n}$ AND $R_nSn(—SR'OR'')_{4-n}$

Compound	mp(°C)	bp(°C/mmHg)	n_D^{20}	d_4^{20}	References
$HO(CH_2)_2OOCCH_2S\left[\overset{C_4H_9}{\underset{C_4H_9}{Sn}}—SCH_2COO(CH_2)_2OOCCH_2S\right]_{2.8}\overset{C_4H_9}{\underset{C_4H_9}{Sn}}—SCH_2COO(CH_2)_2OH$			$1.566(n_D^{28})$	1.31	(*P78*)
$C_8H_{17}OOCCH_2S\left[\overset{C_4H_9}{\underset{C_4H_9}{Sn}}—SCH_2COO(CH_2)_2OOCCH_2S\right]_{0.9}\overset{C_4H_9}{\underset{C_4H_9}{Sn}}—SCH_2COOC_8H_{17}$			$1.520(n_D^{28})$	1.225	(*P78*)
$(iso\text{-}C_3H_{17}COO)_2SbOOC—\underset{C_{12}H_{23}}{CH}—CH_2—COO\overset{C_4H_9}{\underset{C_4H_9}{Sn}}—SCH_2COO\text{-}iso\text{-}C_8H_{17}$					(*P142*)
$[(3,5\text{-}(H_3C)_2C_6H_3)O]_2SbOOC—\underset{C_8H_{17}}{CH}—CH_2COO\overset{C_6H_5}{\underset{C_6H_5}{Sn}}—S—CH(COOC_4H_9)(CH_2COOC_4H_9)$					(*P142*)

Structure	References
$B(O-C_6H_4-S-)_3$ with each pair of S atoms bridged by $Sn(CH_3)_2$ (B—O—C₆H₄—S / $Sn(CH_3)_2$ units)	(*P12, P54, P139*)
$B(O-CH_2CH_2-S-)_3$ with each pair of S atoms bridged by $Sn(C_4H_9)_2$	(*P12, P54, P139*)

(*continued*)

APPENDIX 3 (*continued*)

COMPOUNDS OF THE TYPE $R_nSn(—SR'COOR'')_{4-n}$ AND $R_nSn(—SR'OR'')_{4-n}$

Compound	mp(°C)	bp(°C/mmHg)	n_D^{20}	d_4^{20}	References
Structure: B—O and two O, each O on a benzene ring bearing CH_2CH_2S; the S atoms pair up on $Sn(C_{12}H_{25})_2$ groups. Drawn as two B(O-aryl) units with three $Sn(C_{12}H_{25})_2$ bridges (labels: O, CH_2CH_2S, $Sn(C_{12}H_{25})_2$, B—O)					(*P12*, *P54*, *P139*)

Structure	References
$B[-O-CH_2CH(C_6H_5)-S-Sn[CH_2CH(C_6H_5)_2]_2-S-CH(C_6H_5)CH_2-O-]_3B$ (cyclic: two B atoms, each bonded to three O; each O–$CH_2CH(C_6H_5)$–S pair bridged by $Sn[CH_2CH(C_6H_5)_2]_2$)	(*P12, P54, P139*)
$B[-O-CH_2CH_2-S-Sn(C_6H_5)_2-S-CH_2CH_2-O-]_3B$ (cyclic: two B atoms, each bonded to three O; each O–CH_2CH_2–S pair bridged by $Sn(C_6H_5)_2$)	(*P21, P139*)

(*continued*)

APPENDIX 3 (*continued*)

COMPOUNDS OF THE TYPE $R_nSn(—SR'COOR'')_{4-n}$ AND $R_nSn(—SR'OR'')_{4-n}$

Compound	mp(°C)	bp(°C/mmHg)	n_D^{20}	d_4^{20}	References
$RSn(SR'COOR'')_3$					
$CH_3Sn[—S—C_6H_{10}—O—B(—OC_{10}H_{21})_2]_3$ (S and O on adjacent ring carbons of the cyclohexane ring)					(*P138*)
$C_2H_5Sn(—SCH_2COO\text{-}iso\text{-}C_8H_{17})_3$					(*P15*, *P121*)
$i\text{-}C_3H_7Sn(—SCH_2CH_2OOCC_8H_{17})_3$					(*P10*, *P42*, *P132*, *P133*)
$i\text{-}C_3H_7Sn(—SCH_2CH_2CH_2CONHC_4H_9)_3$					(*P9*, *P41*, *P107*)
$i\text{-}C_3H_7Sn(—S—CH(COOCH_2C_6H_5)(CH_2COOCH_2C_6H_5))_3$					(*P45*)
$C_4H_9Sn(—SCH_2COOH)_3$					(*P127*, *P128*)
$C_4H_9Sn(—SCH_2COOC_4H_9)_3$					(*P100*, *P101*, *P123*)
$C_4H_9Sn(—SCH_2COOC_8H_{17})_3$			1.5020(n_D^{30})		(*P38f*, *P95b*)
$C_4H_9Sn(—SCH_2COOC_{10}H_{21})_3$					(*P4*, *P102*, *P127*, *P128*)
$C_4H_9Sn[—SCH_2COOCH_2—C_4H_7O]_3$ (tetrahydrofurfuryl)					(*P89*)
$C_4H_9Sn(—SCH_2CONHC_5H_{11})_3$					(*P107*)

$i\text{-}C_4H_9Sn\left(-S-CH(CH_3)COOCH_2CH_2OCH_3\right)_3$	(*P2*)
$C_4H_9Sn(-SCH_2CH_2COOC_6H_{13})_3$	(*P40, P100, P101*)
$C_4H_9Sn(-SCH_2CH_2COOC_8H_{17})_3$	(*P76c*)
$C_4H_9Sn(-SCH_2CH_2COOCH_2C_6H_5)_3$	(*P15, P121*)
$n\text{-}C_4H_9Sn[-S(CH_2)_2COO(CH_2)_2O(CH_2)_2OH]_3$	(*P2*)
$C_4H_9Sn[-SCH_2CH_2COOSn(C_4H_9)_2OC_8H_{17}]_3$	(*P76c*)
$C_4H_9Sn[-SCH(CH_3)-CH_2COO(CH_2)_3OOCCH_2CH(CH_3)-SH]_3$	(*P27*)
$C_4H_9Sn[-S(CH_2)_3COO(CH_2)_3OOC(CH_2)_3SH]_3$	(*P91*)
$C_8H_{17}Sn[-S(CH_2)_4O-B(-OC_4H_9)_2]_3$	(*P138*)
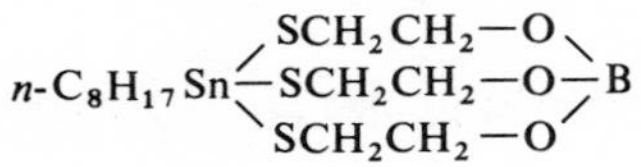 $n\text{-}C_8H_{17}Sn(-SCH_2CH_2-O-)_3B$	(*P139*)
$C_6H_5CH_2Sn\left[-S(CH_2)_{10}O-B\left(-OCH_2CH(C_2H_5)C_2H_5\right)_2\right]_3$	(*P138*)
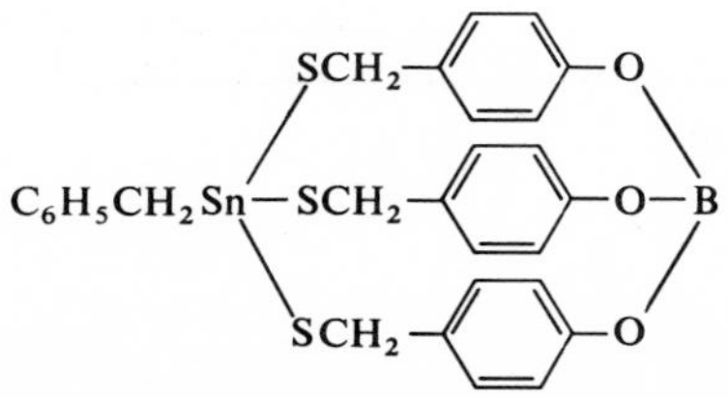 $C_6H_5CH_2Sn(-SCH_2-C_6H_4-O-)_3B$	(*P139*)

(*continued*)

APPENDIX 3 (*continued*)

COMPOUNDS OF THE TYPE $R_nSn(\text{—}SR'COOR'')_{4-n}$ AND $R_nSn(\text{—}SR'OR'')_{4-n}$

Compound	mp(°C)	bp(°C/mmHg)	n_D^{20}	d_4^{20}	References
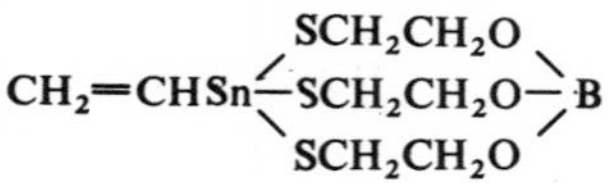 $CH_2{=}CHSn(\text{—}SCH_2CH_2O\text{—})_3B$ (three SCH_2CH_2O groups on Sn, joined to B)					(*P139*)
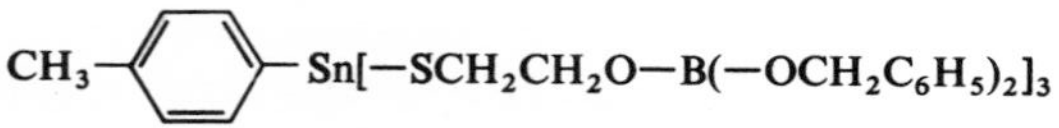 $CH_3\text{—}C_6H_4\text{—}Sn[\text{—}SCH_2CH_2O\text{—}B(\text{—}OCH_2C_6H_5)_2]_3$					(*P138*)
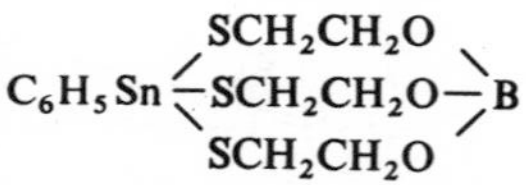 $C_6H_5Sn(\text{—}SCH_2CH_2O\text{—})_3B$ (three SCH_2CH_2O groups on Sn, joined to B)					(*P139*)
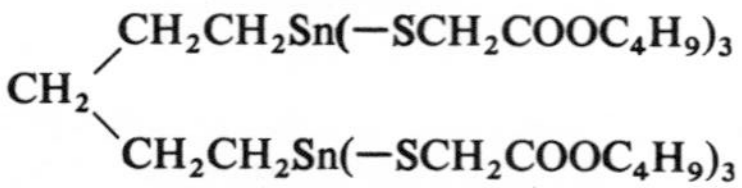 $CH_2[CH_2CH_2Sn(\text{—}SCH_2COOC_4H_9)_3]_2$					(*P2*)

APPENDIX 4

Organotin Thiocarboxylates, Thiocarbonate Esters and Related Compounds

Compound	mp(°C)	bp(°C/mmHg)	n_D^{20}	d_4^{20}	References
$>Sn-S-C(=O)-C<$					
$(CH_3)_3SnSC(O)CH_3$		86–8/0.02			(*79a*)
$(n\text{-}C_4H_9)_3SnSC(O)CH_3$					(*52*)
$(C_5H_{11})_3SnSC(O)C_9H_{19}$					(*P1*)
$(C_6H_5)_3SnSC(O)C_4H_9$					(*P1*)
$(CH_3)_3SnSC(O)C_6H_5$			1.5906(n_D^{27})		(*P38f, P95b*)
$(C_6H_5)_3SnSC(O)C_6H_5$	108–9				(*56, 205, 206*)
$(p\text{-}CH_3C_6H_4)_3SnSC(O)CH_2CH_2F$					(*P1*)
$(p\text{-}ClC_6H_4)_3SnSC(O)CCl_3$					(*P1*)
$(C_4H_9)_2(CH_2{=}CHCOO-)SnSC(O)CH_3$					(*P21*)
$(C_4H_9)_2(CH_3CH{=}CHCOO-)SnSC(O)CH_3$					(*P21*)
$(C_4H_9)_2(CH_3OOCCH{=}CHCOO-)SnSC(O)CH_3$					(*P21*)
$(C_4H_9)_2(C_{12}H_{25}S-)SnSC(O)C_6H_5$					(*P33, P34, P64*)
$(C_4H_9)_2(C_6H_5CH_2S-)SnSC(O)C_6H_5$					(*P67*)
$(C_4H_9)_2(HOOCCH_2CH_2S-)SnSC(O)C_6H_5$					(*P33*)
$(C_4H_9)_2\left(HOCH_2CH_2CH_2CH(C_2H_5)-CH_2S-\right)SnSC(O)C_6H_5$					(*P64*)
$(C_4H_9)_2\left((C_2H_5)(C_4H_9)CHCH_2OOCCH_2S-\right)SnSC(O)C_6H_5$					(*P33*)

(*continued*)

APPENDIX 4 (*continued*)

ORGANOTIN THIOCARBOXYLATES, THIOCARBONATE ESTERS AND RELATED COMPOUNDS

Compound	mp(°C)	bp(°C/mmHg)	n_D^{20}	d_4^{20}	References
$(C_8H_{17})_2(C_8H_{17}S-)SnSC(O)C_6H_5$					(*P67*)
$(n\text{-}C_4H_9)(C_3H_7COO-)(iso\text{-}C_8H_{17}COO-)SnSC(O)C_4H_9$					(*P2*)
$[(C_4H_9)_2(C_6H_5C(O)S-)Sn]_2O$					(*P58e*)
$(CH_3)_2Sn(SCOC_6H_5)_2$	134–5				(*P38f*, *P95b*)
$(CH_3)_2Sn[-SC(O)-C_6H_4(2\text{-}CH_3)]_2$			1.6468		(*P6*, *P60*, *P115*)
$(C_2H_5)_2Sn(SCOC_6H_5)_2$	86–7				(*P38f*, *P95b*)
$(C_4H_9)_2Sn[-SC(O)CH_3]_2$		130–3/1.1	1.5363(n_D^{25})		(*P76g*, *P148*)
$(C_4H_9)_2Sn[-SC(O)C_{11}H_{23}]_2$					(*P76g*, *P148*)
$(n\text{-}C_4H_9)_2Sn[-SC(O)C_{15}H_{31}]_2$					(*P76g*)
$(n\text{-}C_4H_9)_2Sn[-SC(O)CH_2C_6H_5]_2$					(*P76g*)
$(C_4H_9)_2Sn[-SC(O)C_6H_5)]_2$	46–50(*P148*) 27–29(*P26*) 51–2(*P38f*, *P95b*)				(*P26*, *P34*, *P38f*, *P76g*, *P95b*, *P148*)
$(n\text{-}C_4H_9)_2Sn[-SC(O)-C_6H_2(2,4,6\text{-}(CH_3)_3)]_2$					(*P76g*)

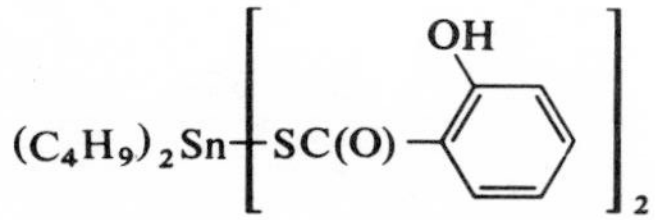			(*P115*)
$(C_4H_9)_2Sn[\text{—}SC(O)CH{=}CHC_6H_5]_2$			(*P60*, *P115*)
$(n\text{-}C_4H_9)_2Sn[\text{—}SC(O)(CH_2)_7CH{=}CH(CH_2)_7CH_3]_2$			(*P76g*)
$(C_4H_9)_2Sn[\text{—}SC(O)CH{=}CHCOOH]_2$			(*P102*)
$(C_4H_9)_2Sn[\text{—}SC(O)CH{=}CHC(O)SC_4H_9]_2$			(*P102*)
$(C_4H_9)_2Sn[\text{—}SC(O)CH(SH)\text{—}CH_2COO\text{-}iso\text{-}C_8H_{17}]_2$			(*P103*, *P104*, *P105*, *P106*)
$(C_4H_9)_2Sn[\text{—}SC(O)CH(SH)\text{—}CH_2COO\text{-}iso\text{-}C_{12}H_{25}]_2$			(*P103*, *P104*, *P105*, *P106*)
$(C_4H_9)_2Sn[\text{—}SC(O)CH(SH)\text{—}CH_2COOC_{18}H_{37}]_2$			(*P103*, *P104*, *P105*, *P106*)
$(C_8H_{17})_2Sn(SCOC_6H_5)_2$		1.5787(n_D^{26})	(*P38f*, *P95b*)
$(iso\text{-}C_8H_{17})_2Sn[\text{—}SC(O)C_{11}H_{23}]_2$			(*P2*)
$(C_{12}H_{25})_2Sn[\text{—}SC(O)C_6H_5]_2$		1.5496	(*P53*, *P65*, *P90*)
$(C_6H_5CH_2)_2Sn[\text{—}SC(O)C_{11}H_{23}]_2$			(*P1*)
$(C_6H_5CH_2)_2Sn[\text{—}SC(O)C_{17}H_{33}]_2$			(*P97*)
$(C_6H_5)_2Sn[\text{—}SC(O)C_2H_5]_2$			(*P1*)
$(C_6H_5)_2Sn[\text{—}SC(O)C_6H_5]_2$	141–2(*207*) 152–3(*P38f*, *P95b*)		(*92b*, *203*, *207*, *P38f*, *P95b*)
$(1\text{-}C_{10}H_7)_2Sn[\text{—}SC(O)CH_2Cl]_2$			(*P1*)

(*continued*)

APPENDIX 4 (*continued*)

ORGANOTIN THIOCARBOXYLATES, THIOCARBONATE ESTERS AND RELATED COMPOUNDS

Compound	mp(°C)	bp(°C/mmHg)	n_D^{20}	d_4^{20}	References
$CH_3Sn[SC(O)-C_6H_4-2-CH_3]_3$			1.6579		(*P7, P109, P112*)
n-$C_4H_9Sn[SC(O)-C_6H_4-2-CH_3]_3$					(*P7, P109, P112, P114*)
$\rangle Sn-S-C(=O)-O-$					
$(C_6H_5)_3SnSC(O)OC_2H_5$	67–8				(*83, 191a, 196a*)
$\rangle Sn-S-C(=S)-C\langle$					
$Sn[-SC(S)CH_2C_6H_5]_4$					(*24b*)
$\rangle Sn-S-C(=S)-O-$					
$(CH_3)_3SnSC(S)OC_6H_5$					(*P38e*)
$(CH_3)_3SnSC(S)O-C_6H_3-3,5-(C_2H_5)_2$					(*P38e*)

Compound		Reference
$(CH_3)_3SnSC(S)O{-}1{-}C_{10}H_7$		(*P38e*)
$CH_3(C_4H_9)_2SnSC(S)O{-}p{-}C_6H_4{-}NO_2$		(*P38e*)
$CH_3(C_6H_5)_2SnSC(S)OCH_3$		(*P38e*)
$(iso\text{-}C_3H_7)_3SnSC(S)OC_4H_9$		(*P11, P44, P117*)
$(C_4H_9)_3SnSC(S)OCH_3$		(*34, 35, 36, 57, P158h*)
$(C_4H_9)_3SnSC(S)O\text{-}iso\text{-}C_3H_7$		(*P51, P117, P146*)
$C_4H_9(C_6H_5)_2SnSC(S)OC_6H_{13}$		(*P38e*)
$(C_4H_9)_3SnSC(S)OC_{12}H_{25}$		(*P38e*)
$(C_6H_{13})_3SnSC(S)O{-}p{-}C_6H_4{-}OH$		(*P38e*)
$(C_6H_5)_3SnSC(S)OC_2H_5$	46–7(*83*)	(*83, 191a, 196a*)
$(n\text{-}C_4H_9)_2ClSnSC(S)OCH_3$		(*59*)
$(n\text{-}C_4H_9)_2(CH_3O-)SnSC(S)OCH_3$	64–7	(*59*)
$(n\text{-}C_4H_9)_2(CH_3COO-)SnSC(S)OCH_3$		(*59*)
$(C_4H_9)_3SnSC(S)OSn(C_4H_9)_3$		(*57, 158h*)
$(n\text{-}C_4H_9)_2Sn[{-}SC(S)OCH_3]_2$		(*59*)
$(n\text{-}C_4H_9)_2Sn[{-}SC(S)O\text{-}iso\text{-}C_3H_7]_2$		(*P11, P44, P117*)
$(n\text{-}C_4H_9)_2Sn[{-}SC(S)OC_4H_9]_2$		(*P8, P24*)
$(C_{12}H_{25})_2Sn\left[SC(S)OCH(CH_3)_2\right]_2$		(*P8*)
$(C_6H_5)_2Sn[{-}SC(S)OC_2H_5]_2$	86	(*83, 196a*)
$C_4H_9Sn[{-}SC(S)O\text{-}iso\text{-}C_3H_7]_3$		(*P117*)
$C_6H_5CH_2Sn[{-}SC(S)OC_4H_9]_3$		(*P11, P44, P117*)

$$>Sn{-}S{-}\overset{S}{\overset{\|}{C}}{-}S{-}$$

(continued)

APPENDIX 4 (*continued*)

Organotin Thiocarboxylates, Thiocarbonate Esters and Related Compounds

Compound	mp(°C)	bp(°C/mmHg)	n_D^{20}	d_4^{20}	References
$(C_6H_5)_3SnSC(S)SC_{12}H_{25}$					(*P8*, *P119*)
$>Sn-S-\overset{S}{\overset{\Vert}{C}}-N<$					
$C_6H_5(C_4H_9)_2SnSC(S)NH_2$					(*P38e*)
$(C_4H_9)_3SnSC(S)NH-p\text{-}C_6H_4-C_4H_9$					(*P38e*)
$(CH_3)_3SnSC(S)N(CH_3)_2$	63(*79*, *97*)				(*57*, *79*, *97*)
$Cl(CH_3)_2SnSC(S)N(CH_3)_2$	135–7				(*90c*)
$Br(CH_3)_2SnSC(S)N(CH_3)_2$	128–30				(*90c*)
$I(CH_3)_2SnSC(S)N(CH_3)_2$	141–4				(*90c*)
$Cl(CH_3)_2SnSC(S)N(C_2H_5)_2$	83.5–4.5(*60b*)				(*60a,b*)
$Br(CH_3)_2SnSC(S)N(C_2H_5)_2$	113–4(*60b*)				(*60b,a*)
$I(CH_3)_2SnSC(S)N(C_2H_5)_2$	118–8.5(*60b*)				(*60b,a*)
$(C_2H_5)_3SnSC(S)N(C_2H_5)_2$		165/2			(*39*)
$(C_4H_9)_3SnSC(S)N(C_2H_5)_2$					(*P25*)
$(C_4H_9)_3SnSC(S)-N\langle\;\rangle NH$					(*P28*)
$(C_6H_5)_3SnSC(S)N(CH_3)_2$	136–7(*117*) 135(*235*)				(*92b*, *117*, *235*)
$(C_6H_5)_3SnSC(S)N(C_2H_5)_2$	133–4(*117*) 132(*39*, *235*)				(*39*, *117*, *235*)
$(C_6H_5)_3SnSC(S)N\genfrac{}{}{0pt}{}{H}{CH_2C_6H_5}$	124–6				(*116*, *117*, *117b*)

$(C_6H_5)_3SnSC(S)N(C_6H_5)_2$	194–5(*39*) 182–3(*117*)		(*39, 117*)
$(C_6H_5)_2ClSnSC(S)N(CH_3)_2$	139–41		(*117*)
$(C_6H_5)_2ClSnSC(S)N(C_2H_5)_2$	143–5		(*117*)
$(C_6H_5)_2ClSnSC(S)N(H)(CH_2C_6H_5)$	139–140		(*117*)
$(C_6H_5)_2ClSnSC(S)N(C_6H_5)_2$	203–4		(*117*)
$(CH_3)_2Sn[—SC(S)N(CH_3)_2]_2$	198–200(*90c*)		(*90c, 91*)
$(CH_3)_2Sn[—SC(S)N(C_2H_5)_2]_2$	134–5(*60b*)		(*60a,b, 70a, 91*)
$(CH_3)_2Sn[—S—C(S)N(C_6H_5)_2]_2$			(*70a*)
$(CH_3)_2Sn[S—C(S)N(CH_2CH_2)(CH_2CH_2)]_2$ (pyrrolidine ring)			(*70a*)
$(C_2H_5)_2Sn[—SC(S)N(C_2H_5)_2]_2$	84	180/0.1(Subl.)	(*39*)
$(n\text{-}C_4H_9)_2Sn[—SC(S)N(C_5H_{11})_2]_2$			(*P8, P120*)
$(C_4H_9)_2Sn[S—C(S)N(CH_2C_6H_5)_2]_2$			(*70a*)
$(C_4H_9)_2Sn[S—C(S)N(C_6H_5)_2]_2$			(*70a*)
$(C_4H_9)_2Sn[S—C(S)N(CH_2CH_2)(CH_2CH_2)]_2$ (pyrrolidine ring)			(*70a*)
$(C_4H_9)_2Sn[SC(S)—N(CH_2CH_2)_2NH]_2$ (piperazine ring)			(*P28*)

(*continued*)

APPENDIX 4 (*continued*)

ORGANOTIN THIOCARBOXYLATES, THIOCARBONATE ESTERS AND RELATED COMPOUNDS

Compound	mp(°C)	bp(°C/mmHg)	n_D^{20}	d_4^{20}	References
$(C_6H_{13})_2Sn[—SC(S)NH_2]_2$					(*P25*)
$(C_8H_{17})_2Sn[SC(S)—N(CH_2CH_2)_2NH_2]_2$ (piperazine ring)					(*P28*)
$(C_{12}H_{25})_2Sn[—SC(S)N(C_4H_9)_2]_2$					(*P8, P120*)
$(C_6H_5)_2Sn[—SC(S)N(CH_3)_2]_2$	203–5				(*117*)
$(C_6H_5)_2Sn[—SC(S)N(C_2H_5)_2]_2$	152–4(*117*) 145–6(*39*)				(*39, 70a, 117*)
$(C_6H_5)_2Sn[S—C(S)N(CH_2C_6H_5)_2]_2$					(*70a*)
$(C_6H_5)_2Sn[S—C(S)N(H_2C—CH_2)(H_2C—CH_2)]_2$ (pyrrolidine ring)					(*70a*)
$(C_6H_5)_2Sn[SC(S)N(H)CH_2C_6H_5]_2$	92–94				(*117, 117b*)
$(C_6H_5)_2Sn[—SC(S)N(C_6H_5)_2]_2$	117–8(*39*) 216–8(*117*)				(*39, 117, 117b*)
$CH_3ClSn[—SC(S)N(CH_3)_2]_2$	201–3(*90c*)				(*90c, 91*)
$CH_3BrSn[—SC(S)N(CH_3)_2]_2$	192–3(*90c*)				(*90c, 91*)
$CH_3ISn[—SC(S)N(CH_3)_2]_2$	155(decomp.)(*90c*)				(*90c, 91*)
$Cl_2Sn[—SC(S)N(CH_3)_2]_2$	244–6				(*39a, 90c*)
$Cl_2Sn[—SC(S)N(C_2H_5)_2]_2$	220–1				(*39, 39a*)

$Br_2Sn[—SC(S)N(C_2H_5)_2]_2$	220	(*39*)
$I_2Sn[—SC(S)N(C_2H_5)_2]_2$	202	(*39*)
$\left[Cl(O)Sn—S—C(S)N(CH_2CH_2)_2CH_2 \right] \cdot 2H_2O$	330	(*39a*)
$Sn[SC(S)N(CH_3)_2]_4$	160(decomp.)	(*39a*)
$Sn[—SC(S)N(C_2H_5)_2]_4$	168–170(*P76*) 154–6(*39a*)	(*39a*, *40a*, *P76*, *P76b*)
$Sn[—SC(S)N(n\text{-}C_4H_9)_2]_4$	118–120	(*P76*, *P76b*)
$Sn[—SC(S)N(iso\text{-}C_4H_9)_2]_4$	154–5	(*P76*, *P76b*)
$Sn\left[SC(S)N(CH_2CH_2)_2CH_2 \right]_4$	180(decomp.)	(*39a*)
$Sn\left[SC(S)N(CH_2CH_2)_2CH_2 \right]_4 \cdot 2C_6H_6$	180(decomp.)	(*39a*)
$>Sn—S—C(=S)—P<$		
$(C_6H_5)_3SnSC(S)P(C_6H_5)_2$	68	(*199*, *201*)
$Sn[—SC(S)P(C_6H_5)_2]_4$	55	(*201*)
$>Sn—S—C(=N—)—C<$		

(continued)

APPENDIX 4 (*continued*)

Organotin Thiocarboxylates, Thiocarbonate Esters and Related Compounds

Compound	mp(°C)	bp(°C/mmHg)	n_D^{20}	d_4^{20}	References
$(CH_3)_3SnSC(NC_6H_5)H$					(*121*)
$(C_2H_5)_3SnSC(NC_6H_5)H$		115–6/0.2(*141*)	1.5910(*141*)		(*121, 140–142*)
$(n\text{-}C_4H_9)_3SnSC(NC_6H_5)H$					(*121*)
$(C_6H_5)_3SnSC(NC_6H_5)H$					(*121*)
$(C_2H_5)_3SnSC(NC_6H_5)CH_3$		108–9/0.2	1.5577		(*140*)
$>Sn-S-C(=N-)-O-$					
$(n\text{-}C_4H_9)_3SnSC(NC_3H_5)OCH_3$		84–5/0.02			(*36*)
$(n\text{-}C_4H_9)_2(CH_3O-)SnSC(NC_3H_5)OCH_3$					(*59*)
$(C_2H_5)_3SnSC(NC_6H_5)OC_2H_5$		62–3/0.2(*140*)	1.5740(*140*)		(*48, 140*)
$(C_4H_9)_3SnSC(NC_6H_5)OCH_3$		68/0.2(decomp.)(*36*) 60/0.01(*36*)			(*35, 36*)
$(n\text{-}C_4H_9)_2ClSnSC(NC_6H_5)OCH_3$					(*59*)
$(n\text{-}C_4H_9)_2(CH_3O-)SnSC(NC_6H_5)OCH_3$					(*59*)
$(n\text{-}C_4H_9)_2(CH_3COO-)SnSC(NC_6H_5)OCH_3$					(*59*)
$(n\text{-}C_4H_9)_2\left(\begin{matrix}CH_3O \\ Cl_3C\end{matrix}\!>\!CH-O\right)SnSC(NC_6H_5)OCH_3$					(*59*)
$(n\text{-}C_4H_9)_2Sn[-SC(NC_6H_5)OCH_3]_2$					(*59*)
$(n\text{-}C_4H_9)_2Sn[-SC(NC_3H_5)OCH_3]_2$					(*59*)

$\rangle Sn-S-C(=N-)-S-$				
$(n\text{-}C_4H_9)_3SnS$ $\rangle C=N-CN$ $(n\text{-}C_4H_9)_3SnS$		145/0.3	$1.5276(n_D^{23})$	(*P158*)
$(C_8H_{17})_3SnS$ $\rangle C=N-CN$ $(C_8H_{17})_3SnS$				(*P158*)
$(CH_3)_2Sn[-SC(NC_4H_9)SCH_2C_6H_5]_2$		120/5		(*P23*)
$(C_4H_9)_2Sn[-SC(NC_4H_9)SCH_2C_6H_5]_2$				(*P23*)
$(C_4H_9)_2Sn[-SC(NC_6H_{11})SCH_2C_6H_5]_2$				(*P23*)
$(C_4H_9)_2Sn[-SC(NCH_2C_6H_5)SCH_2C_6H_5]_2$		125/10		(*P23*)
$(C_6H_5)_3SnSC(NCN)SCH_3$	153.5–4.5			(*211b*)
$\rangle Sn-S-C(=N-)-N\langle$				
$(CH_3)_3SnSC(NC_2H_5)N(CH_3)_2$				(*79a*)

APPENDIX 5

Compounds of the Type $R_3Sn—S—MR_3$ ($M = Si, Ge, Sn$ and Pb)

Compound	mp(°C)	bp(°C/mmHg)	n_D^{20}	d_4^{20}	References
$(CH_3)_3SnSSn(CH_3)_3$	7(*84*)	50/0.05(*7*) 118/18(*84*) 63–6/3(*94*) 52–7/0.4(*94*) 107/11(*108*) 118/15(*108*)	1.5600(*7*)	1.649(d_4^{25})(*106*) 1.6392(d_4^{25})(*84*)	(*2, 5, 7, 9, 84, 93a, 94, 94b, 106, 108, 109, 109a, 188*)
$(C_2H_5)_3SnSSn(C_2H_5)_3$		122–8/0.6(*48*) 187–8/20(*84*) 165/12(*108*) 135–7/1(*109*) 162–8/12(*126*) 125–8/0.1(*142*) 89–9/0.03(*160*) 133–7/1(*176*) 127–9/1.5(*225*) 136–7/1.5(*227*)	1.5365(*22*) 1.5465(*48*) 1.5489(*109*) 1.5485(*142*) 1.5481(*160*) 1.5468(*176*) 1.5488(*225*)	1.4216(*84*) 1.4130(*160*) 1.429(*176*) 1.5160(*225*)	(*22–24, 36b, 48, 64c 65, 84, 108, 109, 109a, 114, 115, 122, 126, 142, 160, 176, 225, 227, 227a, 227b, P150*)
$(C_3H_5)_3SnSSn(C_3H_5)_3$					(*P156*)
$(n\text{-}C_3H_7)_3SnSSn(n\text{-}C_3H_7)_3$		215–9/18(*84*) 160–5/3(*P88*)	1.5290(n_D^{18})(*84*)	1.2110(d_4^{35})(*84*)	(*84, 86, P88, P150, P156*)
$(i\text{-}C_3H_7)_3SnSSn(i\text{-}C_3H_7)_3$					(*P156*)
$(C_4H_9)_3SnSSn(C_4H_9)_3$		140/0.15(*36*) 189/0.9(*53*) 208/1(*108, 109*) 185–8/0.6(*164*) 194–8/1.5(*225, 227*) 147–9/0.02(*36a*)	1.5180(n_D^{21})(*P22, P47, P61, P150*) 1.517($n_D^{19.5}$)(*P22, P47, P61, P150*) 1.518(n_D^{25})(*129*) 1.5151(*225*) 1.5153(n_D^{25})(*36a*)	1.197(*225*)	(*35, 36, 36a, 48, 51, 52, 53, 57, 108, 109, 109a, 129, 139, 164, 211b, 214d, 225, 227, P22, P47, P61, P88, P150, P156*)

Compound	M.p. (°C)	References
$(i\text{-}C_4H_9)_3SnSSn(i\text{-}C_4H_9)_3$		(*P156*)
$(sec\text{-}C_4H_9)_3SnSSn(sec\text{-}C_4H_9)_3$		(*P156*)
$(t\text{-}C_4H_9)_3SnSSn(t\text{-}C_4H_9)_3$		(*P156*)
$(n\text{-}C_8H_{17})_3SnSSn(n\text{-}C_8H_{17})_3$		(*P88*)
$\left[\left[C_6H_5\text{—}C(CH_3)_2\text{—}CH_2\text{—}\right]_3Sn\right]_2S$	97.5–8	(*165*)
$(C_6H_5CH_2\text{—})_3SnSSn(\text{—}CH_2C_6H_5)_3$	85(*144c*)	(*144c, P1*)
$\left[\left[(CH_3)_2CH\text{—}C_6H_4\text{—}CH_2\right]_3Sn\right]_2S$	65	(*55*)
$(C_6H_5)_3SnSSn(C_6H_5)_3$	145–6(*39*) 137–43(*48*) 143–4(*109, 116, 214*) 143–5(*123*) 145.5–7(*162*) 144–4.5(*145*) 141.5–3(*143, 164*) 140–1(*200, 212*) 144(*203*) 144–6(*211*) 96–7(*P88*)	(*39, 48, 56, 89, 92b, 109, 116, 117, 123, 143, 144c,d, 145, 152, 162, 164, 194, 200, 203, 204, 205, 206, 211, 211b, 212, 214, P1, P29, P88*)
$[(p\text{-}CH_3C_6H_4)_3Sn]_2S$	141–2(*211*)	(*63c, 211*)
$[(o\text{—}CH_3C_6H_4)_3Sn]_2S$		(*213b*)

(*continued*)

APPENDIX 5 (*continued*)

COMPOUNDS OF THE TYPE $R_3Sn—S—MR_3$
(M = Si, Ge, Sn AND Pb)

Compound	mp(°C)	bp(°C/mmHg)	n_D^{20}	d_4^{20}	References
$[(C_2H_5OOC—C_6H_4)_3Sn]_2S$	132.3				(*134*)
$[(1\text{-}C_{10}H_7)_3Sn]_2S$					(*P1*)
$[(NCCH_2CH_2)_3Sn]_2S$			1.5745(n_D^{28})		(*P158g*)
$[(CH_3)_2ClSn]_2S$	59–61				(*60*)
$[(CH_3)_2BrSn]_2S$	39–42				(*60*)
$[(n\text{-}C_4H_9)_2(CH_2{=}CH)Sn]_2S$					(*P156*)
$[(C_2H_5)_2(C_5H_{11})Sn]_2S$					(*P156*)
$[(n\text{-}C_4H_9)_2(n\text{-}C_3H_7)Sn]_2S$					(*P156*)
$[C_3H_7(C_6H_5)_2Sn]_2S$	52				(*P58g*)
$[C_4H_9(C_6H_5)_2Sn]_2S$			1.6316		(*P58g*)
$[(CH_3)_2(CH_3COO)Sn]_2S$					(*P94a*)
$[(C_4H_9)_2FSn]_2S$	90–3				(*60*)
$[(C_4H_9)_2ClSn]_2S$	35.5–7(*60*) 35(*202*) 34–5(*129*)	110–5/1(*202*)			(*60*, *129*, *202*)
$(C_4H_9)_2ClSnSSnCl_2C_4H_9$	28–30				(*60*)
$(C_4H_9)_2ClSnSSnCl_2C_6H_5$	38–38.5				(*60*)
$(C_4H_9)_2BrSnSSnBr_2C_4H_9$					(*60*, *60a*)
$[(C_4H_9)_2SCN—Sn]_2S$	115–7				(*60*)
$[(C_4H_9)_2(CH_3O—)Sn]_2S$	86–8				(*60*)
$[(C_4H_9)_2(CH_3COO)Sn]_2S$					(*P94a*)

Compound	Ref.
$\left[(C_4H_9)_2\left(\begin{matrix}C_2H_5\\ \\ C_2H_5\end{matrix}\!\!>\!CHCOO\right)Sn\right]_2S$	(*P94a*)
$[(C_4H_9)_2(C_7H_{15}COO-)Sn]_2S$	(*60*)
$[(C_4H_9)_2(CH_3(CH_2)_{10}COO)Sn]_2S$	(*P94a*)
$\left[(C_4H_9)_2\left((CH_3)_3CCH_2\underset{\displaystyle CH_3}{\underset{\mid}{C}H}CH_2COO\right)Sn\right]_2S$	(*P94a*)
$\left[(C_4H_9)_2\left(CH_3(CH_2)_3\underset{\displaystyle C_2H_5}{\underset{\mid}{C}H}COO\right)Sn\right]_2S$	(*P94a*)
$[(C_4H_9)_2(iso\text{-}C_8H_{17}COOCH{=}CHCOO)Sn]_2S$	(*P94a*)
$[(C_4H_9)_2(C_{11}H_{23}COO-)Sn]_2S$	(*P77*)
$(C_4H_9)_2(C_4H_9OOCCH_2S-)Sn-S-Sn(-OOCCH{=}CHCOOC_4H_9)(C_4H_9)_2$ [printed as: $(C_4H_9)_2(C_4H_9OOCCH_2S-)Sn\diagdown S \diagup Sn(C_4H_9)_2(C_4H_9OOCCH{=}CHCOO-)$]	(*P77*)
$[(C_4H_9)_2(C_4H_9OOCCH{=}CHCOO-)Sn]_2S$	(*P77*)
$(i\text{-}C_4H_9)_2SnOOC(CH_2)_{10}CH_3$ / S / $(i\text{-}C_4H_9)_2SnOOCCH{=}CHOOCCH_2\underset{\displaystyle C_2H_5}{\underset{\mid}{C}H}(CH_2)_3CH_3$	(*P94a*)
$(C_4H_9)_2(C_{11}H_{23}COO)Sn-S-Sn(C_8H_{17})_2(C_{11}H_{23}COO)$	(*P94a*)
$[(C_8H_{17})_2(C_{11}H_{23}COO)Sn]_2S$	(*P94a*)
$[(C_4H_9)_2(iso\text{-}C_8H_{17}OOCCH_2CH_2S)Sn]_2S$	(*P58f*)

(continued)

APPENDIX 5 (*continued*)

COMPOUNDS OF THE TYPE $R_3Sn—S—MR_3$
(M = Si, Ge, Sn AND Pb)

Compound	mp(°C)	bp(°C/mmHg)	n_D^{20}	d_4^{20}	References
$(C_4H_9)(iso\text{-}C_8H_{17}OOCCH_2CH_2S)_2Sn—S—Sn(iso\text{-}C_8H_{17}OOCCH_2CH_2S)(C_4H_9)_2$					(*P58f*)
$C_4H_9(iso\text{-}C_8H_{17}OOCCH_2CH_2S)_2Sn]_2S$					(*P58f*)
$[[8\text{-quinolyl-}O—]_2(C_4H_9)Sn]_2S$	225–6				(*105*)
$[(C_2H_5)_3Sn]_2S_{1.3}$					(*211*)
$[(C_2H_5)_3Sn]_2S_{1.4}$		78.6–9.5/760			(*211*)
$[(n\text{-}C_4H_9)_3Sn]_2S_{1.3}$					(*211*)
$[(n\text{-}C_4H_9)_3Sn]_2S_{2.3}$					(*211*)
$[(C_6H_5)_3Sn]_2S_2$					(*P29*)
$[(C_6H_5)_3Sn]_2S_3$	127–33(*211*)				(*211, P29*)
$[(C_6H_5)_3Sn]_2S_4$	118–25(*211*)				(*211, P29*)
$[(C_2H_5)_2ClSn]_2S_2$	50				(*129*)
$[(C_4H_9)_2ClSn]_2S_{2.5}$	26–7				(*129*)
$[(C_4H_9)_2ClSn]_2S_4$	29–30				(*129*)
$[(C_4H_9)_2(HOOC—C\equiv C—COO—)Sn]_2S_4$	105				(*129*)
$[(C_4H_9)_2(2\text{-furyl-}COO)Sn]_2S_{2.5}$	27–8				(*129*)

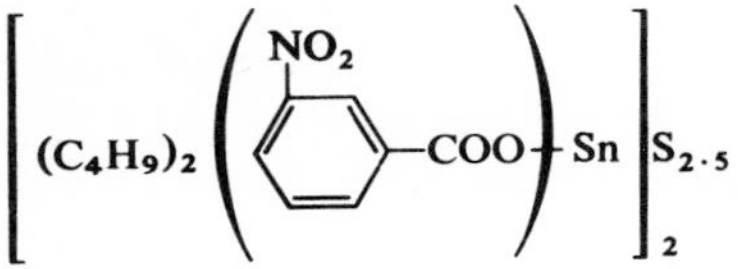 $[(C_4H_9)_2(m\text{-}NO_2C_6H_4COO)Sn]_2S_{2.5}$	47				(*129*)
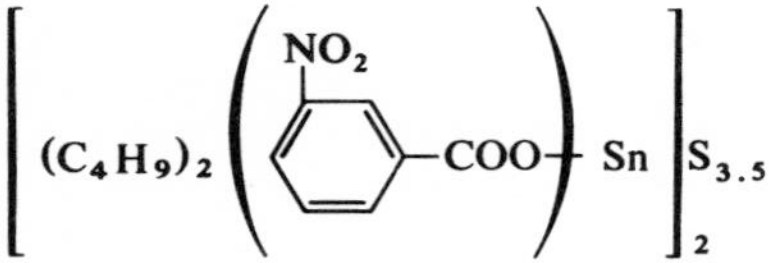 $[(C_4H_9)_2(m\text{-}NO_2C_6H_4COO)Sn]_2S_{3.5}$	35				(*129*)
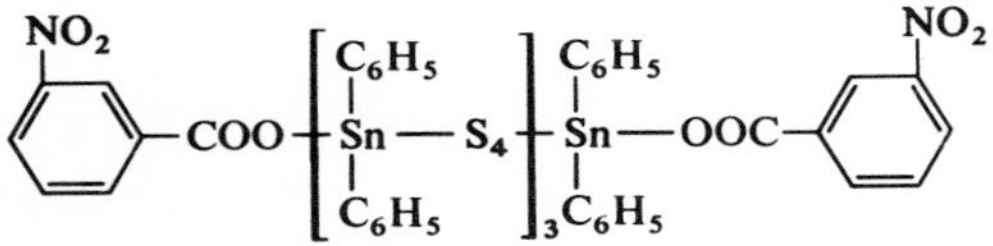 $m\text{-}NO_2C_6H_4COO[Sn(C_6H_5)_2-S_4]_3Sn(C_6H_5)_2-OOC\text{-}C_6H_4\text{-}m\text{-}NO_2$	95–100				(*129*)
$(CH_3)_3Sn-S-Si(CH_3)_3$	–36(*171*, *190*)	75/12(*171*, *190*)			(*171*, *188*, *190*)
$(CH_3)_3Sn-S-Ge(CH_3)_3$	–8(*173*)	89–90/12(*173*) 51–52/1(*173*)			(*173*, *203*)
$(C_2H_5)_3Sn-S-Ge(C_2H_5)_3$		100–3/1(*226*, *227c*) 121–8/2(*227d*)	1.5279(*226*)	1.2990(*226*)	(*65*, *226*, *227c*, *227d*)
$(C_6H_5)_3Sn-S-Ge(C_6H_5)_3$	137(*203*) 136(*206*, *209*) 135(*205*)				(*88*, *203*, *204*, *205*, *206*, *209*)
$(C_6H_5)_3Sn-S-Pb(C_6H_5)_3$	137–8				(*203*, *204*, *205*, *206*, *210*)
$[(CH_3)_3Sn]_2S-Cr(CO)_5$	95(decomp.)				(*181b*)

APPENDIX 6

Compounds of the Type $(R_2SnS)_3$

Compound	mp(°C)	bp(°C/mmHg)	n_D^{20}	d_4^{20}	References
$[(CH_3)_2SnS]_3$	142–9(*164*) 148(*42*, *100*, *108*) 149(*85*, *132*) 149–51(*162*)				(*42*, *52*, *60*, *60a*, *82*, *83a*, *83b*, *85*, *92b*, *100*, *108*, *109a*, *132*, *148*, *162*, *164*, *212*, *213*, *P122*, *P141*, *P144*)
$[(C_2H_5)_2SnS]_3$	25(*108*) 24(*85*, *129*)	202/4(*108*) 219–21/8(*85*) 192–3/2(*109a*)		1.7264(d_4^{35})	(*85*, *100*, *108*, *109a*, *122*, *129*, *P87*, *P150*)
$[(n\text{-}C_3H_7)_2SnS]_3$		254/16(*87*, *86*)			(*86*, *87*, *P150*)
$[(iso\text{-}C_3H_7)_2SnS]_3$					(*P76h*)
$[(C_4H_9)_2SnS]_3$	65–9(*159*) 63–9(*162*) 64–5(*P87*) 64–6(*129*)	210–4/1(*191*) 260(decomp.)(*31*) 215/1(*109a*)	1.5680(*P47*) 1.5742(n_D^{16}) (*P47*, *P61*, *P22*, *P150*) 1.5768(n_D^{21}) (*109a*)		(*20*, *31*, *44*, *51*, *52*, *60*, *62*, *105b*, *108*, *109a*, *129*, *144d*, *159*, *162*, *164*, *191*, *196*, *198*, *203*, *236*, *P22*, *P38d*, *P47*, *P61*, *P76f*, *P87*, *P95b*, *P103*, *P104*, *P105*, *P106*, *P122*, *P141*, *P144*, *P150*, *P151*)
$[(C_5H_{11})_2SnS]_3$					(*P76h*)
$[(n\text{-}C_6H_{13})_2SnS]_3$					(*P76h*)

$[(\text{cyclo-}C_6H_{11})_2SnS]_3$		(*P76h*)
$[(2,4\text{-}(CH_3)_2\text{—cyclo—}C_6H_9)_2SnS]_3$		(*P76h*)
$[(n\text{-}C_8H_{17})_2SnS]_3$	25–30(*162*) 70–1(*P87*)	(*162*, *164*, *P76f*, *P87*, *P141*, *P144*)
$[(C_{12}H_{25})_2SnS]_3$		(*P122*, *P141*, *P144*)
$[(C_6H_5CH_2)_2SnS]_3$		(*P76h*)
$[(o\text{-}CH_3\text{—}C_6H_4CH_2)_2SnS]_3$		(*P76h*)
$\{[(CH_3)_2CH\text{—}C_6H_4\text{—}CH_2]_2SnS\}_3$	200–5	(*55*)
$[(p\text{-}tert\text{-}C_4H_9\text{—}C_6H_4CH_2)_2SnS]_3$		(*P76h*)
$[(C_6H_5)_2SnS]_3$	179–80.5(*113*) 183–4(*159*) 182–4(*162*) 181–4(*163*, *164*) 182–3(*211*) 165–72(*48*) 185(*129*)	(*48*, *68*, *92b*, *113*, *117*, *129*, *152*, *159*, *162–164*, *191*, *194*, *196*, *198*, *203*, *211*, *P122*, *P141*, *P144*)
$[(2,4\text{-}(iso\text{-}C_3H_7)_2\text{—}C_6H_3)_2SnS]_3$		(*P76h*)
$[(C_6H_5\text{—}C_6H_4)_2SnS]_3$	134–5	(*216*)
$[(1\text{-}C_{10}H_7)_2SnS]_3$	215	(*149*)

(*continued*)

APPENDIX 6 (*continued*)

Compounds of the Type $(R_2SnS)_3$

Compound	mp(°C)	bp(°C/mmHg)	n_D^{20}	d_4^{20}	References
$[(p\text{-}ClC_6H_4)_2SnS]_3$	179				(*104*, *136*)
$[(p\text{-}BrC_6H_4)_2SnS]_3$	228–9				(*104*, *136*)
$[(p\text{-}IC_6H_4)_2SnS]_3$	248				(*104*, *136*)
$[(p\text{-}H_3CC_6H_4)_2SnS]_3$	188–9				(*211*)
$[(m\text{-}CH_3C_6H_4)_2SnS]_3$	121.5–2				(*101*)
$[(p\text{-}CH_3OC_6H_4)_2SnS]_3$	95				(*216*)
$[(p\text{-}C_6H_5OC_6H_4)_2SnS]_3$	127				(*216*)

Compound	M.p. (°C)		Ref.
$[(o\text{-}C_6H_5OC_6H_4)_2SnS]_3$	144–8		(*154*)
$\{[C_2H_5O{-}C(O){-}C_6H_4]_2SnS\}_3$	141.5–2.5		(*68*)
$[(NC{-}CH_2CH_2)_2SnS]_3$	173–5		(*166 P158g*,)
$\{[(CH_3)_3SiCH_2]_2SnS\}_3$	74.4–5.5		(*184*)
$[(n\text{-}C_3H_7)(n\text{-}C_6H_{13})SnS]_3$			(*P76h*)
$[(iso\text{-}C_3H_7)(sec\text{-}C_5H_{11})SnS]_3$			(*P76h*)
$[(C_4H_9)(cyclo\text{-}C_6H_{11})SnS]_3$			(*P76h*)
Six-membered Sn_3S_3 ring: $(C_6H_5)(Cl)Sn$; two $(C_6H_5)(C_6H_5S)Sn$ units; ring S atoms	136–40		(*202*)
$\{[C_5H_5Fe(CO)_2]_2SnS\}_3$	268		(*135*)
$[(C_2H_5)_2SnS_{1.2}]_3$			(*211*)
$[(n\text{-}C_4H_9)_2SnS_{1.4}]_3$			(*211*)
$[(n\text{-}C_4H_9)_2SnS_{1.6}]_3$			(*211*)
Six-membered ring: $(H_3C)(CH_3)Sn$, S, S, two $Sn(CH_3)_2$, $N{-}C_2H_5$			(*60a*)
Six-membered ring: $(H_3C)(CH_3)Sn$, $C_2H_5{-}N$, S, two $Sn(CH_3)_2$, $N{-}C_2H_5$			(*60a*)

APPENDIX 7

Compounds of the Type $(RSnS)_2S$

Compound	mp(°C)	bp(°C/mmHg)	n_D^{20}	d_4^{20}	References
$(CH_3Sn)_4S_6$	250(*105*)(dec.) 200(*71*)				(*71, 105, 108, 147, P49*)
$[(C_2H_5Sn)_2S_3]_n$					(*63b*)
$(n\text{-}C_3H_7Sn)_4S_6$	160(dec).				(*105*)
$(n\text{-}C_4H_9Sn)_4S_6$	150(*105*)(dec.) 135–6(*71*)				(*63a, 71, 105, P38c, P76f, P95*)
$[(C_8H_{17}Sn)_4S_6]_n$	200–220(decomp.)				(*P76f*)
$[(C_6H_5Sn)_2S_3]_n$	255(*71*)(dec.) 250(*194*)(dec.)				(*71, 194*)
$o\text{-}CH_3C_6H_4\text{–}Sn(=S)\text{–}S\text{–}Sn(=S)\text{–}C_6H_4CH_3\text{-}o$					(*102, 103, 127*)
$p\text{-}ClC_6H_4\text{–}Sn(=S)\text{–}S\text{–}Sn(=S)\text{–}C_6H_4Cl\text{-}p$	240(*104*)(dec.) 295(*136*)(dec.)				(*104, 136*)
$p\text{-}BrC_6H_4\text{–}Sn(=S)\text{–}S\text{–}Sn(=S)\text{–}C_6H_4Br\text{-}p$					(*104, 136*)
$p\text{-}IC_6H_4\text{–}Sn(=S)\text{–}S\text{–}Sn(=S)\text{–}C_6H_4I\text{-}p$					(*104, 136*)
$[(NCCH_2CH_2Sn)_2S_3]_n$	135(decomp.)				(*P158g*)

APPENDIX 8

Heterocycles with Tin-Sulfur Bonds in the Ring

Compound	mp(°C)	bp(°C/mmHg)	n_D^{20}	d_4^{20}	References
Four-membered Rings					
$(n\text{-}C_4H_9)_2Sn{<}^{S}_{S}{>}C=N-CN$	145(*P158*) 151.5–2.5(*P158c*)				(*211b*, *P158*, *P158c*)
$(n\text{-}C_8H_{17})_2Sn{<}^{S}_{S}{>}C=N-CN$	135				(*211b*, *P158*)
$(C_6H_5-CH_2)_2Sn{<}^{S}_{S}{>}C=N-CN$					(*P158c*)
$(C_6H_5)_2Sn{<}^{S}_{S}{>}C=N-C{\equiv}N$	173–5				(*211b*)
$(C_4H_9)_2Sn{<}^{S}_{S}{>}C=N-SO_2C_6H_4CH_3$					(*P158c*)
$(C_4H_9)_2Sn{<}^{S}_{S}{>}C=C(CN)_2$	253–4.5				(*P158i*)
$(C_4H_9)Sn{<}^{S}_{S}{>}C=CH-NO_2$	122.5–4.5				(*P158i*)

(*continued*)

APPENDIX 8 (*continued*)

Heterocycles with Tin-Sulfur Bonds in the Ring

Compound	mp(°C)	bp(°C/mmHg)	n_D^{20}	d_4^{20}	References
$(C_6H_5)_2Sn\langle S_2 \rangle C{=}C(CN)_2$	143.5–5				(*P158i*)
$(C_6H_5)_2Sn\langle S_2 \rangle C{=}C(COOC_2H_5)_2$					(*P158i*)
$(C_6H_5)_2Sn\langle S_2 \rangle C{=}CH{-}NO_2$	151.5–3				(*P158i*)
$(C_6H_5CH_2)_2Sn\langle S_2 \rangle C{=}C(CN)_2$					(*P158i*)
Five-membered Rings					
$(n\text{-}C_4H_9)_2Sn\langle S{-}CH_2 / O{-}CH_2 \rangle$					(*P14*, *P136*)
$(CH_3)_2Sn\langle S{-}CH_2 / S{-}CH_2 \rangle$	82(*9, 7*) 82–3(*229*) 81.5–2(*67*) 80–1(*4*)	113–5/5(*229*)			(*4, 7, 9, 67, 229*)
$(CH_3)_2Sn\langle S{-}CH_2 / S{-}CH_2 \rangle \leftarrow$ pyridine (N)					(*67*)

$(C_4H_9)_2Sn(S-CH_2-CH_2-S)$	59–60(*69*, *153*, *156*)	(*69*, *153*, *156*, *P34*)
$(C_6H_5)_2Sn(S-CH_2-CH_2-S)$	109–10(*67*) 108(*56*) 108–9(*69*, *153*, *156*)	(*56*, *67*, *69*, *153*, *156*)
$(C_6H_5)_2Sn(S-CH_2-CH_2-S) \leftarrow NC_5H_5$		(*67*)
$(C_6H_5)_2Sn(S-CH_2-CH_2-S) \leftleftarrows$ phenanthroline		(*67*)
$[Mn(CO)_5]_2Sn(S-CH_2-CH_2-S)$	141–2	(*217*)
$[Re(CO)_5]_2Sn(S-CH_2-CH_2-S)$	150–2	(*217*)
$[\pi C_5H_5Fe(CO)_2]_2Sn(S-CH_2-CH_2-S)$	160–5	(*161a*)

(*continued*)

APPENDIX 8 (*continued*)

Heterocycles with Tin-Sulfur Bonds in the Ring

Compound	mp(°C)	bp(°C/mmHg)	n_D^{20}	d_4^{20}	References
$(CH_3)_2Sn\langle S-CH=CH-S \rangle$ (ring)					(*18*)
$(CH_3)_2Sn\langle S-CH=CH-S \rangle \leftarrow$ pyridine (N)					(*18*)
$(CH_3)_2Sn\langle S-CH=CH-S \rangle \leftarrow$ phenanthroline (N, N)					(*18*)
$(C_2H_5)_2Sn\langle S-CH=CH-S \rangle$ (ring)					(*18*)
$(CH_2{=}CH)_2Sn\langle S-CH=CH-S \rangle$ (ring)					(*18*)
$(C_6H_5)_2Sn\langle S-C(COOC_2H_5)=C(COOC_2H_5)-S \rangle$ (ring)					(*P158d*)

Compound	M.p. (°C)	Ref.
$(CH_3)_2Sn(S-C(CN)=C(CN)-S)$		*(18)*
$(CH_3)_2Sn(S-C(CN)=C(CN)-S) \leftarrow$ pyridine		*(18)*
$(CH_3)_2Sn(S-C(CN)=C(CN)-S) \leftarrow$ 1,10-phenanthroline		*(18)*
$(C_2H_5)_2Sn(S-C(CN)=C(CN)-S)$		*(18)*
$(n\text{-}C_4H_9)_2Sn(S-C(CN)=C(CN)-S)$	165–7(*P158d*) 155(decomp.) (*P158a*)	(*P158a*, *P158d*)
$(n\text{-}C_8H_{17})_2Sn(S-C(CN)=C(CN)-S)$	110	(*P158a*)
$(C_6H_5CH_2)_2Sn(S-C(CN)=C(CN)-S)$		(*P158d*)

(*continued*)

APPENDIX 8 (*continued*)

HETEROCYCLES WITH TIN-SULFUR BONDS IN THE RING

Compound	mp(°C)	bp(°C/mmHg)	n_D^{20}	d_4^{20}	References
$(CH_2{=}CH)_2Sn\langle S{-}C(CN){=}C(CN){-}S\rangle$					(*18*)
$(C_6H_5)_2Sn\langle S{-}C(CN){=}C(CN){-}S\rangle$	210-30(*P158d*) > 200(decomp.) (*P158a*)				(*P158a*, *P158d*)
$(C_4H_9)_2Sn\langle S{-}C(NO_2){=}C(NO_2){-}S\rangle$					(*P158d*)
$(CH_3)_2Sn\langle O{-}C({=}O){-}CH_2{-}S\rangle$	200(decomp.) 257–9				(*144*)
$(C_3H_7)_2Sn\langle O{-}C({=}O){-}CH_2{-}S\rangle$					(*P99*)
$(C_4H_9)_2Sn\langle O{-}C({=}O){-}CH_2{-}S\rangle$	185–7(*P99*)				(*P58a*, *P99*, *P123*)
$(i\text{-}C_5H_{11})_2Sn\langle O{-}C({=}O){-}CH_2{-}S\rangle$					(*P99*)
$(C_6H_{13})_2Sn\langle O{-}C({=}O){-}CH_2{-}S\rangle$					(*P99*)

Compound	m.p. (°C)	Ref.
$(C_4H_9)_2Sn[OC(O)-CH_2-CH(C(=O)-O-)(S-)Sn(C_4H_9)_2]_2$ (cyclic: C—O—Sn(C₄H₉)₂—S—CH)		(*P2*)
$(CH_3)_2Sn(S_2C_6H_3CH_3)$	137(*67*) 95–8(*231*)	(*67, 231*)
$(CH_3)Sn(S_2C_6H_3CH_3)$ ← N (pyridine)	106–8	(*67*)
$(CH_3)_2Sn(S_2C_6H_3CH_3)$ ← N,N (phenanthroline)		(*67*)
$(C_6H_5)_2Sn(S_2C_6H_3CH_3)$	139–50(*156*) 151–2(*67*) 155(*153*)	(*67, 153, 156*)
$(C_6H_5)_2Sn(S_2C_6H_3CH_3)$ ← N (pyridine)	105–6	(*67*)

(*continued*)

APPENDIX 8 (*continued*)

Heterocycles with Tin-Sulfur Bonds in the Ring

Compound	mp(°C)	bp(°C/mmHg)	n_D^{20}	d_4^{20}	References
$[(\pi C_5H_5)Fe(CO)_2]_2Sn(S_2C_6H_3CH_3)$	172–3				(*161a*)
$(C_2H_5)_2Sn(S_2C_8H_4N_2)$	220(*180*) 218–20(*P31*)				(*180, P31*)
$(C_4H_9)_2Sn(S_2C_8H_4N_2)$					(*P31*)
Six-membered Rings					
$(CH_3)_2Sn(S-CH_2)_2CH_2$		113–7/3			(*230*)
$(C_4H_9)_2Sn(S-CH_2)_2CH_2$	63–4				(*69, 153, 156*)
$(C_6H_5)_2Sn(S-CH_2)_2CH_2$	103–4				(*69, 153, 156*)

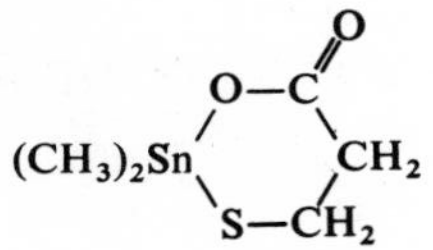

$(CH_3)_2Sn{<}^{O-C(=O)}_{S-CH_2}{>}CH_2$	(*P99*)
$(CH_3)_2Sn{<}^{O-C(=O)}_{S-CH_2}{>}CH-NH_2$	(*P99*)
$(C_2H_5)_2Sn{<}^{O-C(=O)}_{S-CH_2}{>}CH_2$	(*P99*)
$(C_2H_5)_2Sn{<}^{O-C(=O)}_{S-CH_2}{>}CH-NH_2$	(*P99*)
$(n\text{-}C_3H_7)_2Sn{<}^{O-C(=O)}_{S-CH_2}{>}CH_2$	(*P99*)
$(n\text{-}C_3H_7)_2Sn{<}^{O-C(=O)}_{S-CH_2}{>}CH-NH_2$	(*P99*)

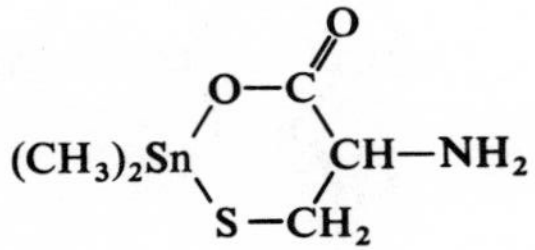

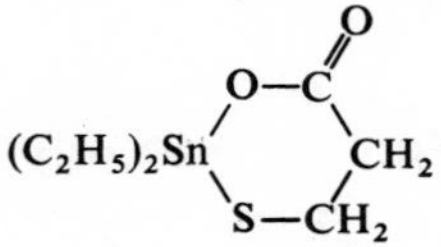

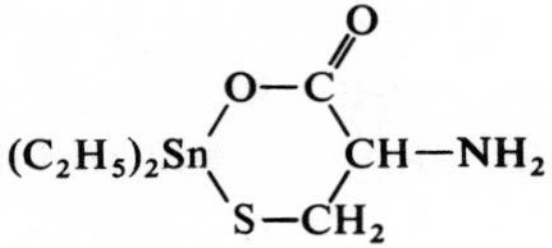

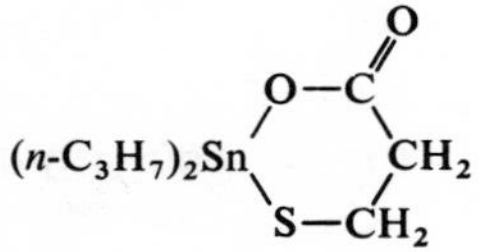

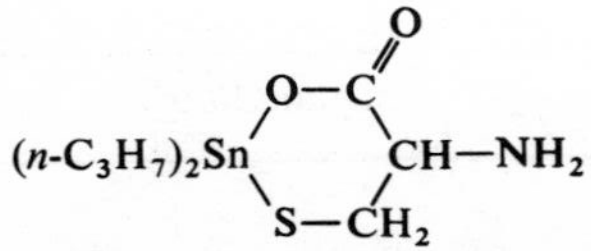

(*continued*)

APPENDIX 8 (*continued*)

HETEROCYCLES WITH TIN-SULFUR BONDS IN THE RING

Compound	mp(°C)	bp(°C/mmHg)	n_D^{20}	d_4^{20}	References
$(i\text{-}C_3H_7)_2Sn\langle O{-}C({=}O){-}CH_2{-}CH_2{-}S\rangle$ (ring: Sn–O–C(=O)–CH₂–CH₂–S–Sn)					(*P99*)
$(i\text{-}C_3H_7)_2Sn\langle O{-}C({=}O){-}CH(NH_2){-}CH_2{-}S\rangle$					(*P99*)
$(i\text{-}C_3H_7)_2Sn\langle O{-}C({=}O){-}CH_2{-}CH(COO(CH_2)_3OH){-}S\rangle$					(*P2*)
$(C_4H_9)_2Sn\langle O{-}C({=}O){-}CH_2{-}CH_2{-}S\rangle$	130(*P39*) 111–30(*P74*) 120–3(*72*)				(*72, 74, 76, P39, P59a, P67, P74, P89a, P92, P99, P147*)
$(C_4H_9)_2Sn\langle O{-}C({=}O){-}CH(NH_2){-}CH_2{-}S\rangle$	205–7(decomp.)				(*P76i, P99*)

$(i\text{-}C_4H_9)_2\overline{Sn\text{—}O\text{—}C(=O)\text{—}CH(NH_2)\text{—}CH_2\text{—}S}$	(*P99*)
$(n\text{-}C_5H_{11})_2\overline{Sn\text{—}O\text{—}C(=O)\text{—}CH_2\text{—}CH_2\text{—}S}$	(*P99*)
$(n\text{-}C_5H_{11})_2\overline{Sn\text{—}O\text{—}C(=O)\text{—}CH(NH_2)\text{—}CH_2\text{—}S}$	(*P99*)
$(i\text{-}C_5H_{11})_2\overline{Sn\text{—}O\text{—}C(=O)\text{—}CH_2\text{—}CH_2\text{—}S}$	(*P99*)
$(i\text{-}C_5H_{11})_2\overline{Sn\text{—}O\text{—}C(=O)\text{—}CH(NH_2)\text{—}CH_2\text{—}S}$	(*P99*)
$(t\text{-}C_5H_{11})_2\overline{Sn\text{—}O\text{—}C(=O)\text{—}CH_2\text{—}CH_2\text{—}S}$	(*P99*)

(*continued*)

APPENDIX 8 (*continued*)

Heterocycles with Tin-Sulfur Bonds in the Ring

Compound	mp(°C)	bp(°C/mmHg)	n_D^{20}	d_4^{20}	References
$(t\text{-}C_5H_{11})_2Sn\langle O{-}C({=}O){-}CH(NH_2){-}CH_2{-}S\rangle$ (ring)					(*P99*)
$(n\text{-}C_6H_{13})_2Sn\langle O{-}C({=}O){-}CH_2{-}CH_2{-}S\rangle$ (ring)					(*P99*)
$(n\text{-}C_6H_{13})_2Sn\langle O{-}C({=}O){-}CH(NH_2){-}CH_2{-}S\rangle$ (ring)					(*P99*)
$(i\text{-}C_6H_{13})_2Sn\langle O{-}C({=}O){-}CH_2{-}CH_2{-}S\rangle$ (ring)					(*P99*)
$(i\text{-}C_6H_{13})_2Sn\langle O{-}C({=}O){-}CH(NH_2){-}CH_2{-}S\rangle$ (ring)					(*P99*)

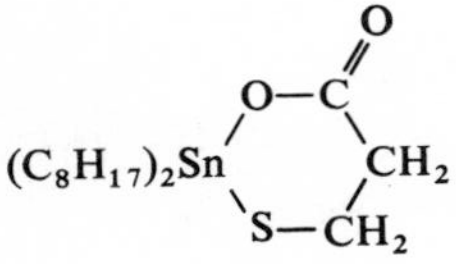 $(C_8H_{17})_2Sn\langle O{-}C({=}O){-}CH_2{-}CH_2{-}S\rangle$	98(*P39*) 90–1(*P73*)	(*P39*, *P67*, *P73*)
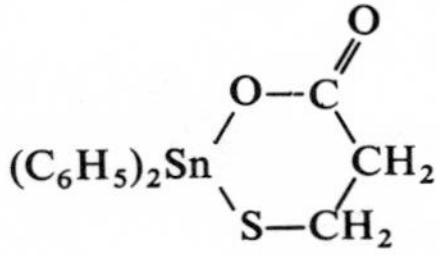 $(C_6H_5)_2Sn\langle O{-}C({=}O){-}CH_2{-}CH_2{-}S\rangle$	220–5(decomp.)	(*P39*, *P140*, *P147*)
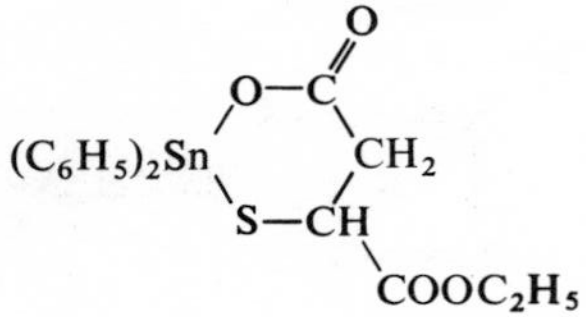 $(C_6H_5)_2Sn\langle O{-}C({=}O){-}CH_2{-}CH(COOC_2H_5){-}S\rangle$		(*P2*)
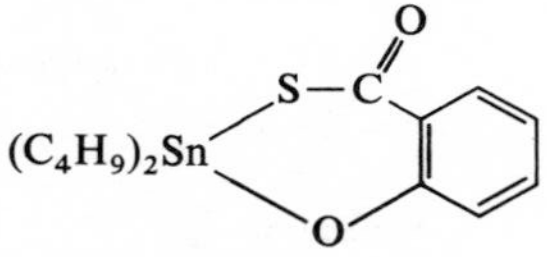 $(C_4H_9)_2Sn\langle S{-}C({=}O){-}C_6H_4{-}O\rangle$		(*P39*)
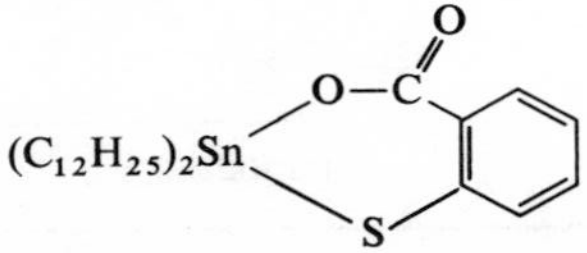 $(C_{12}H_{25})_2Sn\langle O{-}C({=}O){-}C_6H_4{-}S\rangle$		(*P39*, *P140*, *P147*)

(*continued*)

APPENDIX 8 (*continued*)

Heterocycles with Tin-Sulfur Bonds in the Ring

Compound	mp(°C)	bp(°C/mmHg)	n_D^{20}	d_4^{20}	References
$(CH_3)_2Sn$ (–S–C=N–N=C–S–, bridged by S)					(*P154*)
$(n\text{-}C_3H_7)_2Sn$ (–S–C=N–N=C–S–, bridged by S)					(*P154*)
$(n\text{-}C_4H_9)_2Sn$ (–S–C=N–N=C–S–, bridged by S)	120				(*P154*)
$(n\text{-}C_8H_{17})_2Sn$ (–S–C=N–N=C–S–, bridged by S)	52				(*P154*)
$(C_6H_5)_2Sn$ (–S–C=N–N=C–S–, bridged by S)					(*P154*)
$(p\text{-}CH_3C_6H_4)_2Sn$ (–S–C=N–N=C–S–, bridged by S)					(*P154*)

Compound		Ref.
Seven-membered Rings		
$(C_4H_9)_2Sn$ ring: $O-C(=O)-CH_2-CH(SH)-C(=O)-S$		(*P103–P106*)
$(C_4H_9)_2Sn$ ring: $S-C(=O)-CH_2$ / $O-C(=O)-CH(SH)$		(*P103–P106*)
Ten-membered Rings		
$(C_4H_9)_2Sn$ ring: $S-(CH_2)_2-COOCH_2-C(CH_2OOC-(CH_2)_2-SH)(CH_2OOC-C_{11}H_{23})-CH_2-O$		(*P35*)
$(C_4H_9)_2Sn$ ring: $S-(CH_2)_2-COOCH_2-C(CH_2OOC-C_{11}H_{23})(CH_2OOC-C_{11}H_{23})-CH_2-O$		(*P35*)
$(C_4H_9)_2Sn$ ring: $O-CH_2-CH_2-S-Sn(C_4H_9)_2-O-CH_2-CH_2-S$	194/1.5	(*215a*)
$(C_4H_9)_2Sn$ ring: $S-CH_2-C(=O)-O-Sn(C_4H_9)_2-S-CH_2-C(=O)-O$		(*P62*)

(*continued*)

APPENDIX 8 (*continued*)

Heterocycles with Tin-Sulfur Bonds in the Ring

Compound	mp(°C)	bp(°C/mmHg)	n_D^{20}	d_4^{20}	References
$(C_4H_9)_2Sn\langle S{-}CH(CH_3){-}C(O){-}O \,/\, O{-}C(O){-}CH(CH_3){-}S\rangle Sn(C_4H_9)_2$					(*215a*)
Eleven-membered Rings					
$(C_4H_9)_2Sn\langle S{-}CH_2{-}C(=O){-}O{-}CH_2 \,/\, S{-}CH_2{-}C(=O){-}O{-}CH_2\rangle$ (ring closed by $CH_2{-}CH_2$ bond)					(*170, P16, P43, P116*)
$(C_4H_9)_2Sn\langle S{-}CH_2{-}C(=O){-}O{-}CH_2 \,/\, S{-}CH_2{-}C(=O){-}O{-}CH{-}CH_3\rangle$ (ring closed by $CH_2{-}CH$ bond)					(*P16, P43, P116*)
Twelve-membered Rings					
$(C_4H_9)_2Sn\langle S{-}CH_2{-}CH(OH){-}CH_2{-}O \,/\, O{-}CH_2{-}CH(OH){-}CH_2{-}S\rangle Sn(C_4H_9)_2$					(*215a*)

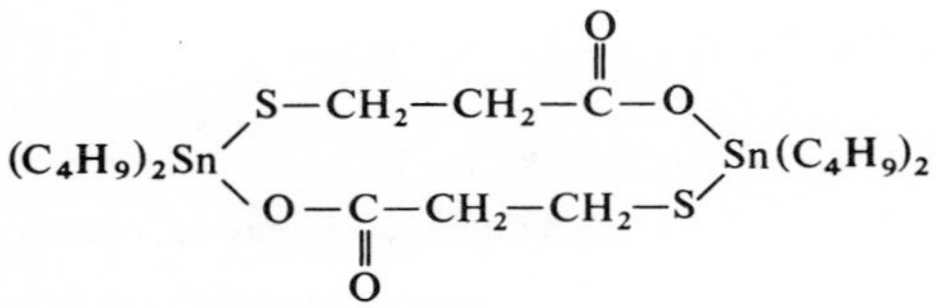

$$(C_4H_9)_2Sn \begin{array}{l} \diagup S-CH_2-CH_2-\overset{O}{\overset{\|}{C}}-O \diagdown \\ \diagdown O-\underset{O}{\underset{\|}{C}}-CH_2-CH_2-S \diagup \end{array} Sn(C_4H_9)_2$$

(*215a*, *P62*)

Thirteen-membered Ring

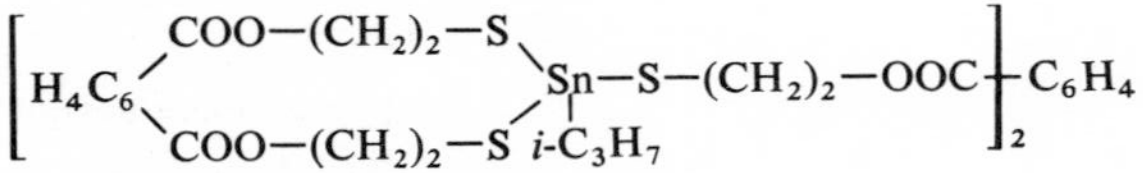

$$\left[H_4C_6 \begin{array}{l} \diagup COO-(CH_2)_2-S \diagdown \\ \diagdown COO-(CH_2)_2-S \diagup \end{array} \underset{i\text{-}C_3H_7}{\underset{|}{Sn}}-S-(CH_2)_2-OOC \right]_2 C_6H_4$$

(*P13*, *P46*, *P134*, *P135*)

Fourteen-membered Rings

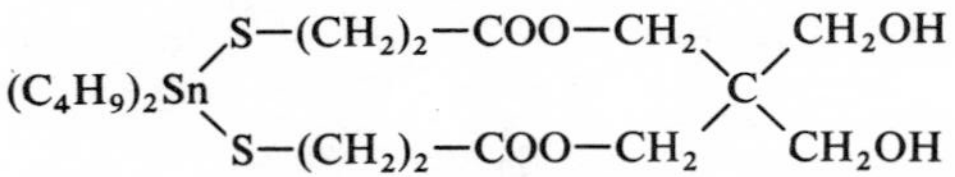

$$(C_4H_9)_2Sn \begin{array}{l} \diagup S-(CH_2)_2-COO-CH_2 \diagdown \\ \diagdown S-(CH_2)_2-COO-CH_2 \diagup \end{array} C \begin{array}{l} \diagup CH_2OH \\ \diagdown CH_2OH \end{array}$$

(*P35*)

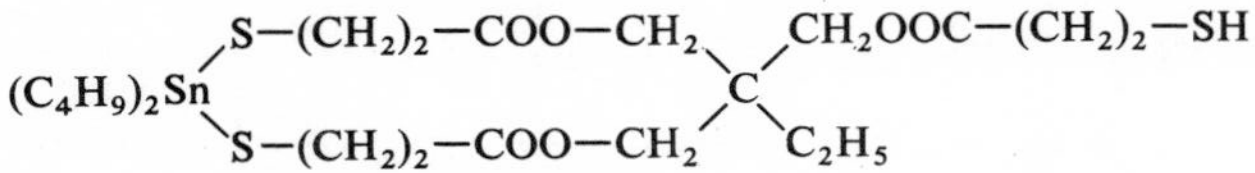

$$(C_4H_9)_2Sn \begin{array}{l} \diagup S-(CH_2)_2-COO-CH_2 \diagdown \\ \diagdown S-(CH_2)_2-COO-CH_2 \diagup \end{array} C \begin{array}{l} \diagup CH_2OOC-(CH_2)_2-SH \\ \diagdown C_2H_5 \end{array}$$

(*P35*)

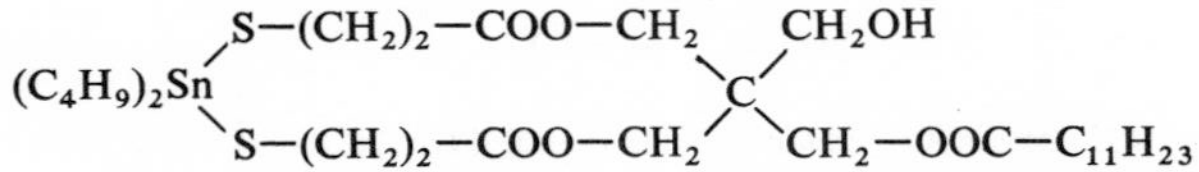

$$(C_4H_9)_2Sn \begin{array}{l} \diagup S-(CH_2)_2-COO-CH_2 \diagdown \\ \diagdown S-(CH_2)_2-COO-CH_2 \diagup \end{array} C \begin{array}{l} \diagup CH_2OH \\ \diagdown CH_2-OOC-C_{11}H_{23} \end{array}$$

(*P35*)

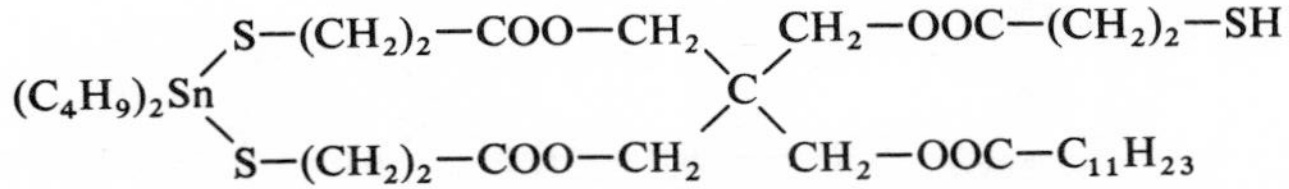

$$(C_4H_9)_2Sn \begin{array}{l} \diagup S-(CH_2)_2-COO-CH_2 \diagdown \\ \diagdown S-(CH_2)_2-COO-CH_2 \diagup \end{array} C \begin{array}{l} \diagup CH_2-OOC-(CH_2)_2-SH \\ \diagdown CH_2-OOC-C_{11}H_{23} \end{array}$$

(*P35*)

$$(C_4H_9)_2Sn \begin{array}{l} \diagup S-(CH_2)_2-COO-CH_2 \diagdown \\ \diagdown S-(CH_2)_2-COO-CH_2 \diagup \end{array} C \begin{array}{l} \diagup CH_2-OOC-C_{11}H_{23} \\ \diagdown CH_2-OOC-C_{11}H_{23} \end{array}$$

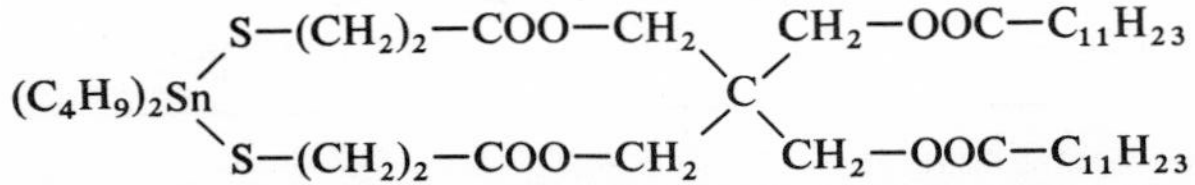

(*P35*)

(*continued*)

APPENDIX 8 (*continued*)

Heterocycles with Tin-Sulfur Bonds in the Ring

Compound	mp(°C)	bp(°C/mmHg)	n_D^{20}	d_4^{20}	References
$(C_4H_9)_2Sn\left[OCH_2(HOCH_2)C(CH_2{-}OOC{-}(CH_2)_2{-}S)_2Sn(C_4H_9)_2\right]_2$					(*P35*)
$(C_4H_9)_2Sn\left[S{-}(CH_2){-}COO{-}CH_2(HOCH_2)C(CH_2{-}OOC{-}(CH_2)_2{-}S)_2Sn(C_4H_9)_2\right]_2$					(*P35*)
$(C_4H_9)_2Sn\left[S{-}(CH_2)_2{-}COO{-}CH_2(C_{11}H_{23}{-}COO{-}CH_2)C(CH_2{-}OOC{-}(CH_2)_2{-}S)_2Sn(C_4H_9)_2\right]_2$					(*P35*)
$(C_4H_9)_2Sn\langle S{-}(CH_2)_3{-}C({=}O){-}O,\ O{-}C({=}O){-}(CH_2)_3{-}S\rangle Sn(C_4H_9)_2$					(*P62*)
Fifteen-membered Rings					
$(n\text{-}C_4H_9)_2Sn\langle S{-}(CH_2)_2{-}O{-}C{=}O,\ S{-}(CH_2)_2{-}O{-}C{=}O\rangle (CH_2)_4$					(*P13*, *P46*, *P134*, *P135*)
Twenty-two-membered Rings					
$(n\text{-}C_4H_9)_2Sn\langle S{-}CH_2{-}COO{-}(CH_2)_2{-}OOC{-}CH_2{-}S,\ S{-}CH_2{-}COO{-}(CH_2)_2{-}OOC{-}CH_2{-}S\rangle Sn(n\text{-}C_4H_9)_2$					(*170*, *P16*, *P43*, *P116*)

Spirans (Carbon as the Central Atom)

$S-CH_2-COO-CH_2 \quad CH_2OOC-CH_2-S$
$(CH_3)_2Sn \quad C \quad Sn(CH_3)_2$ (*P30*)
$S-CH_2-COO-CH_2 \quad CH_2OOC-CH_2-S$

$S-CH_2-COO-CH_2 \quad CH_2-OOC-CH_2-S$
$(C_4H_9)_2Sn \quad C \quad Sn(C_4H_9)_2$ (*P30*)
$S-CH_2-COO-CH_2 \quad CH_2-OOC-CH_2-S$

$S-(CH_2)_2-COO-CH_2 \quad CH_2-OOC-(CH_2)_2-S$
$(C_4H_9)_2Sn \quad C \quad Sn(C_4H_9)_2$ (*P30, P35*)
$S-(CH_2)_2-COO-CH_2 \quad CH_2-OOC-(CH_2)_2-S$

$S-CH_2-COO-CH_2 \quad CH_2-OOC-CH_2-S$
$(C_8H_{17})_2Sn \quad C \quad Sn(C_8H_{17})_2$ (*P30*)
$S-CH_2-COO-CH_2 \quad CH_2-OOC-CH_2-S$

$S-(CH_2)_2-COO-CH_2 \quad CH_2-OOC-(CH_2)_2-S$
$(C_{12}H_{25})_2Sn \quad C \quad Sn(C_{12}H_{25})_2$ (*P30*)
$S-(CH_2)_2-COO-CH_2 \quad CH_2-OOC-(CH_2)_2-S$

$S-(CH_2)_2-COO-CH_2 \quad CH_2-OOC-(CH_2)_2-S$
$(C_6H_5CH_2)_2Sn \quad C \quad Sn(CH_2C_6H_5)_2$ (*P30*)
$S-(CH_2)_2-COO-CH_2 \quad CH_2-OOC-(CH_2)_2-S$

$S-CH_2-COO-CH_2 \quad CH_2-OOC-CH_2-S$
$(C_6H_5)_2Sn \quad C \quad Sn(C_6H_5)_2$ (*P30*)
$S-CH_2-COO-CH_2 \quad CH_2-OOC-CH_2-S$

$S-(CH_2)_2-COO-CH_2 \quad CH_2-OOC-(CH_2)_2-S$
$(C_6H_5)_2Sn \quad C \quad Sn(C_6H_5)_2$ (*P30*)
$S-(CH_2)_2-COO-CH_2 \quad CH_2-OOC-(CH_2)_2-S$

(*continued*)

APPENDIX 8 (*continued*)

Heterocycles with Tin-Sulfur Bonds in the Ring

Compound	mp(°C)	bp(°C/mmHg)	n_D^{20}	d_4^{20}	References
$(C_4H_9)_2Sn$ [$S-(CH_2)_2-COO-CH_2$ / OCH_2] $>C<$ [$CH_2-OOC-(CH_2)_2-S$ / $CH_2-OOC-(CH_2)_2-S$] $Sn(C_4H_9)_2$					(*P35*)
$(C_4H_9)_2Sn$ [$O-CH_2$ / $O-CH_2$] $>C<$ [$CH_2-OOC-(CH_2)_2-S$ / $CH_2-OOC-(CH_2)_2-S$] $Sn(C_4H_9)_2$					(*P35*)
$(CH_2)_4$ [$COO-(CH_2)_2-S-Sn(C_4H_9)$ [$S-(CH_2)_2-OOC$ / $S-(CH_2)_2-OOC$] $(CH_2)_4$ / $COO-(CH_2)_2-S-Sn(C_4H_9)$ [$S-(CH_2)_2-OOC$ / $S-(CH_2)_2-OOC$] $(CH_2)_4$]					(*P13*)
Spirans (Tin as the Central Atom)					
H_2C-S / H_2C-S $>Sn<$ $S-CH_2$ / $S-CH_2$	181–2(*4*) 179–80(*7*) 181(*25*) 182–3(*69*, *153*, *156*)				(*4*, *7*, *25*, *67*, *69*, *153*, *156*)
H_2C-S / H_2C-S $>Sn<$ $S-CH_2$ / $S-CH_2$, $Sn \leftarrow N(C_2H_5)_3$	124–5				(*157*)

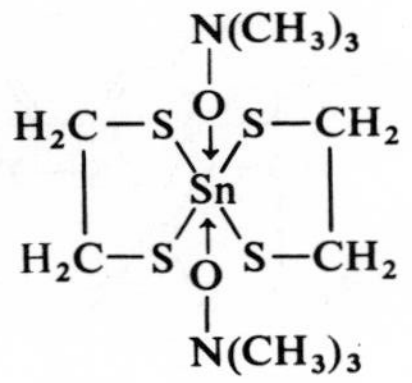 $N(CH_3)_3$ H_2C-S O $S-CH_2$ Sn H_2C-S O $S-CH_2$ $N(CH_3)_3$	135–7	(*157*)
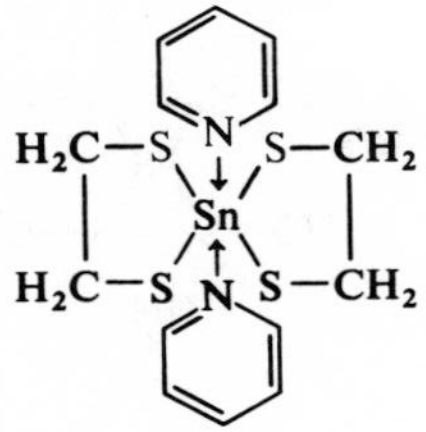 H_2C-S N $S-CH_2$ Sn H_2C-S N $S-CH_2$		(*67, 157*)
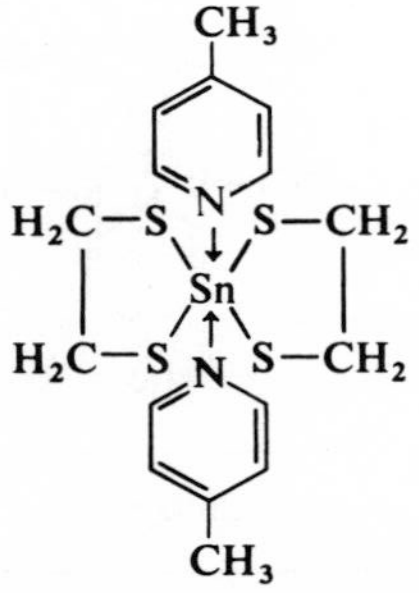 CH_3 H_2C-S N $S-CH_2$ Sn H_2C-S N $S-CH_2$ CH_3	176–8	(*157*)

(*continued*)

APPENDIX 8 (*continued*)

Heterocycles with Tin-Sulfur Bonds in the Ring

Compound	mp(°C)	bp(°C/mmHg)	n_D^{20}	d_4^{20}	References
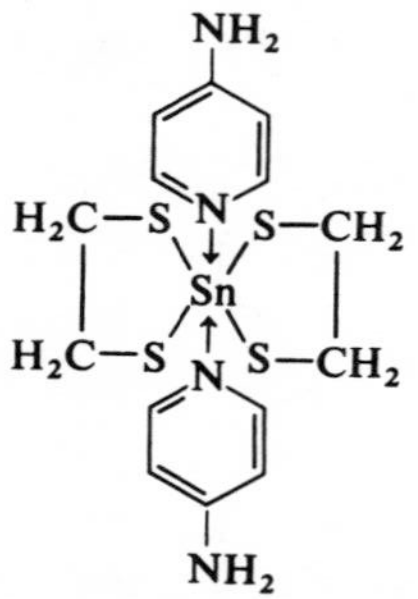 NH_2; N; H_2C-S, $S-CH_2$; Sn; H_2C-S, $S-CH_2$; N; NH_2	232(decomp.)				(*157*)
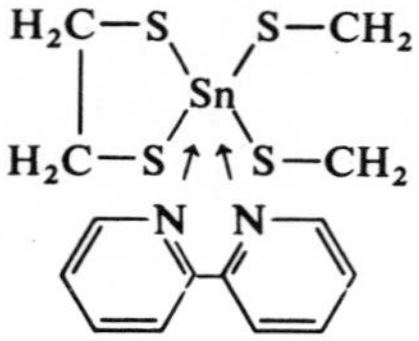 H_2C-S, $S-CH_2$; Sn; H_2C-S, $S-CH_2$; N N	245–8(decomp.)				(*157*)
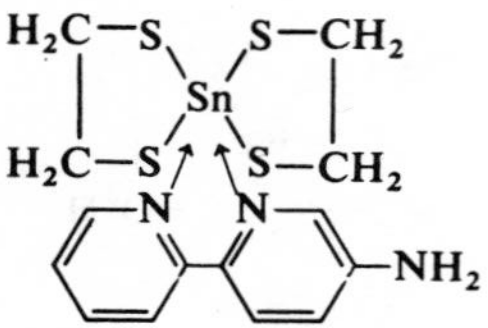 H_2C-S, $S-CH_2$; Sn; H_2C-S, $S-CH_2$; N N; NH_2	171–3				(*157*)

Compound	M.p. (°C)	References
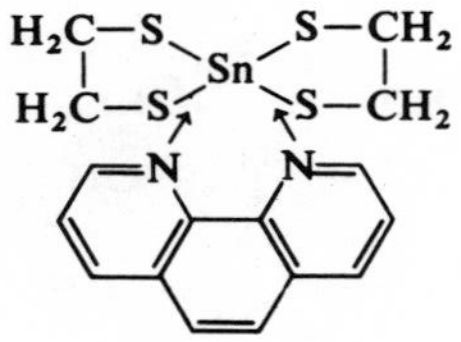 $H_2C-S \diagdown Sn \diagup S-CH_2$ / $H_2C-S \nearrow \nwarrow S-CH_2$ (rings closed H_2C-CH_2), with 1,10-phenanthroline N→Sn coordination	285(decomp.) (*157*)	(*67, 157*)
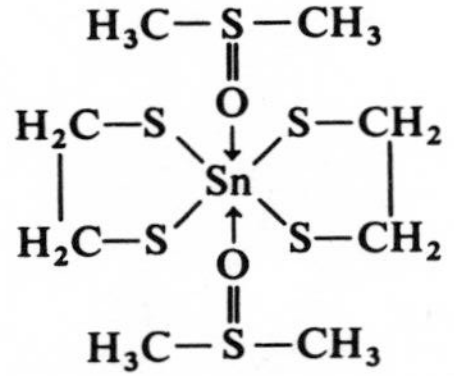 $(H_2C-S)_2Sn(S-CH_2)_2$ with two $O=S(CH_3)_2$ (O→Sn)	140–50	(*157*)
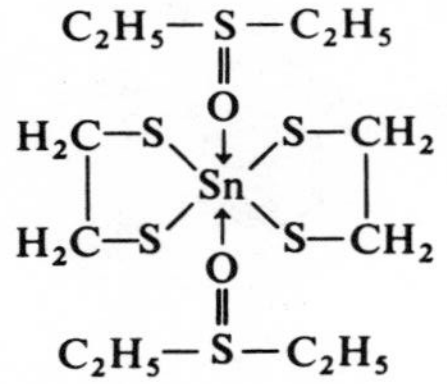 $(H_2C-S)_2Sn(S-CH_2)_2$ with two $O=S(C_2H_5)_2$ (O→Sn)	124–5	(*157*)
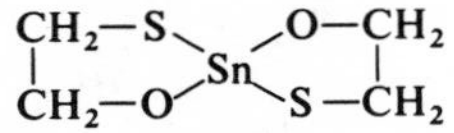 $CH_2-S \diagdown Sn \diagup O-CH_2$ / $CH_2-O \diagup \diagdown S-CH_2$		(*215a*)
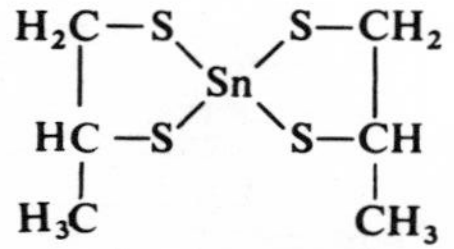 $H_2C-S \diagdown Sn \diagup S-CH_2$ / $H_3C-HC-S \diagup \diagdown S-CH-CH_3$	76–7	(*69, 156*)

(*continued*)

APPENDIX 8 (*continued*)

Heterocycles with Tin-Sulfur Bonds in the Ring

Compound	mp(°C)	bp(°C/mmHg)	n_D^{20}	d_4^{20}	References
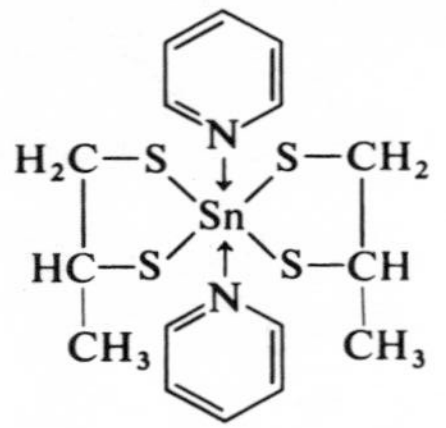 Bis(pyridine)Sn[$SCH_2CH(CH_3)S$]$_2$ (structure: H_2C—S, HC—S, CH_3; S—CH_2, S—CH, CH_3; N→Sn←N)	117–8				(*156*)
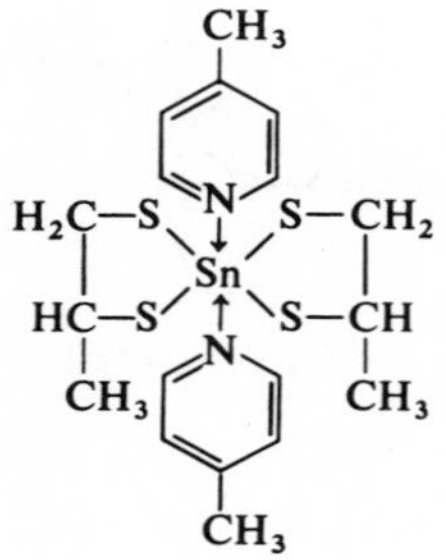 Bis(4-methylpyridine)Sn[$SCH_2CH(CH_3)S$]$_2$ (structure: CH_3; H_2C—S, HC—S, CH_3; S—CH_2, S—CH, CH_3; N→Sn←N; CH_3)	160–2				(*157*)
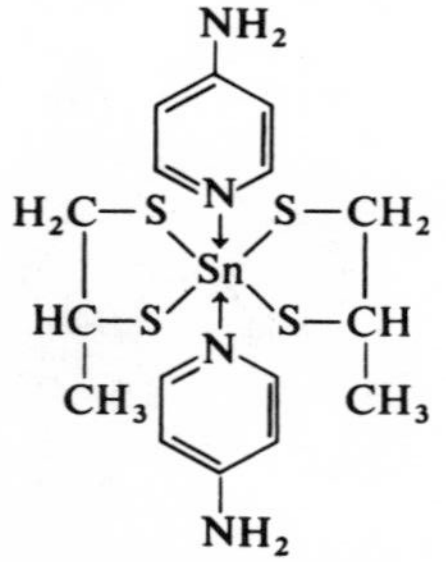 Bis(4-aminopyridine)Sn[$SCH_2CH(CH_3)S$]$_2$ (structure: NH_2; H_2C—S, HC—S, CH_3; S—CH_2, S—CH, CH_3; N→Sn←N; NH_2)	243–5(decomp.)				(*157*)

H_3C CH_3 $HC-S$ $S-CH$ Sn H_2C-S $S-CH_2$ N N	209–10	(*157*)
CH_3 CH_3 $HC-S$ $S-CH$ Sn H_2C-S $S-CH_2$ N N	242–4	(*157*)
CH_3 CH_3 $H_3C-S-CH_3$ O $HC-S$ $S-CH$ Sn H_2C-S $S-CH_2$ O $H_3C-S-CH_3$	109–10	(*157*)
$C_2H_5-S-C_2H_5$ CH_3 CH_3 O $HC-S$ $S-CH$ Sn H_2C-S $S-CH_2$ O $C_2H_5-S-C_2H_5$	79–80	(*157*)

(*continued*)

APPENDIX 8 (*continued*)

Heterocycles with Tin-Sulfur Bonds in the Ring

Compound	mp(°C)	bp(°C/mmHg)	n_D^{20}	d_4^{20}	References
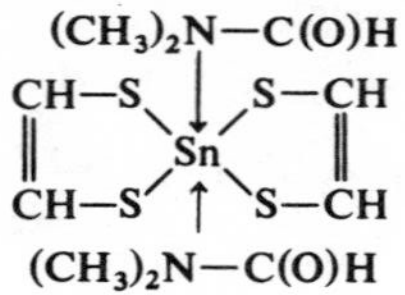 $(CH_3)_2N-C(O)H$ (CH–S, S–CH, Sn, CH–S, S–CH) $(CH_3)_2N-C(O)H$					(*92*)
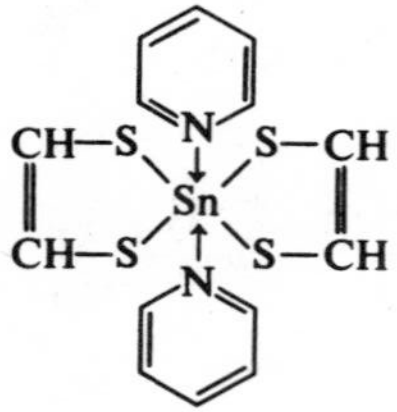 CH–S, S–CH, N, Sn, N, CH–S, S–CH					(*92*)
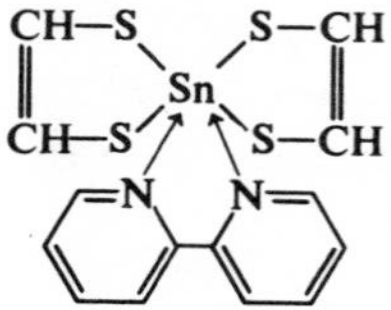 CH–S, S–CH, Sn, CH–S, S–CH, N, N					(*92*)
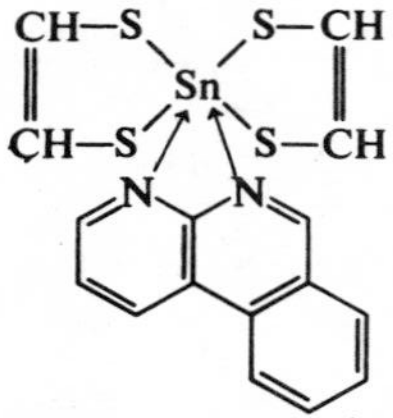 CH–S, S–CH, Sn, CH–S, S–CH, N, N					(*92*)

$Sn(SCH_2CH_2CH_2S)_2$	156–7	(*156*)
$Sn(SCH_2CH_2CH_2S)_2 \leftarrow NC_5H_5$	100–1	(*156*)
$Sn(SCH_2CH_2CH_2S)_2 \leftarrow NC_5H_4CH_3$-4	129–30	(*156*)
$Sn(SCH_2CH_2CH_2S)_2 \cdot$ 2,2′-bipyridyl	190	(*156*)
$Sn(SCH_2CH_2CH_2S)_2 \cdot$ 1,10-phenanthroline	230(decomp.)	(*156*)

(*continued*)

APPENDIX 8 (*continued*)

Heterocycles with Tin-Sulfur Bonds in the Ring

Compound	mp(°C)	bp(°C/mmHg)	n_D^{20}	d_4^{20}	References
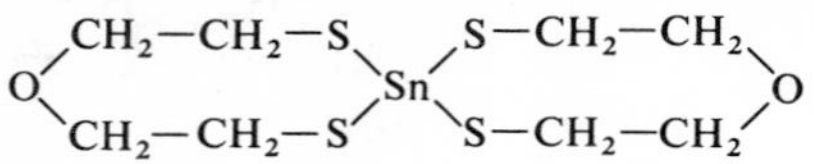 $O(CH_2-CH_2-S)_2Sn(S-CH_2-CH_2)_2O$	124				(*29*)
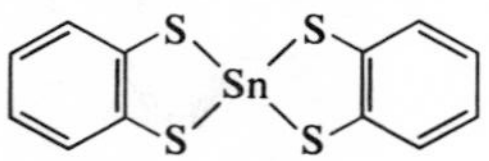 Sn, S, S, S, S	250(decomp.)(*41*)				(*41, 153*)
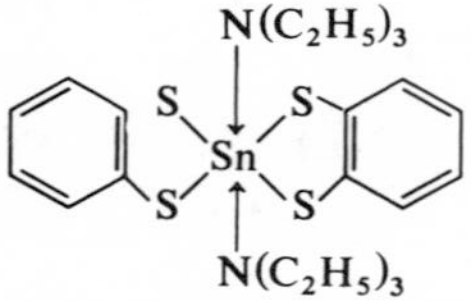 $N(C_2H_5)_3$; S, S, Sn, S, S; $N(C_2H_5)_3$					(*67*)
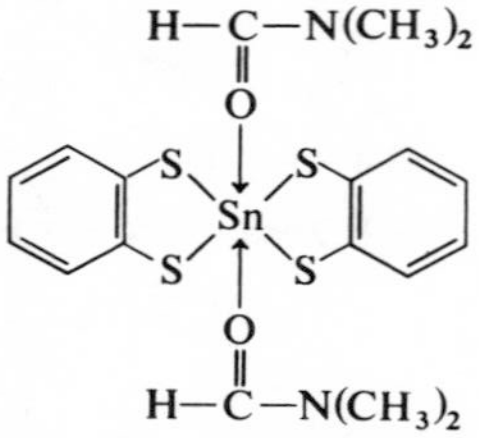 $H-C(=O)-N(CH_3)_2$; O; S, S, Sn, S, S; O; $H-C(=O)-N(CH_3)_2$					(*153, 156*)
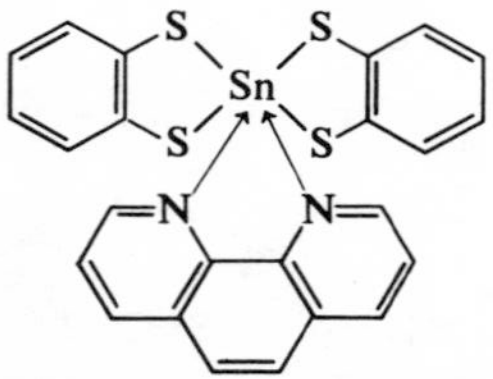 S, S, Sn, S, S; N, N					(*67*)

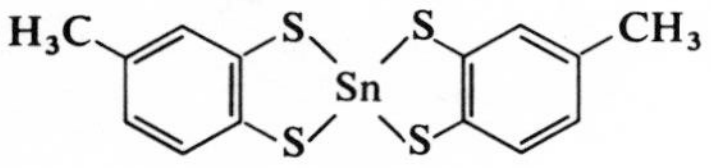	300–5(*153, 156*)	(*67, 70, 153, 156*)
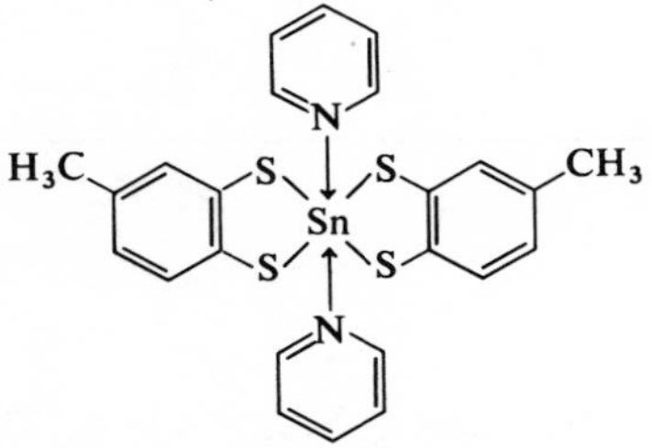		(*67, 153, 156*)
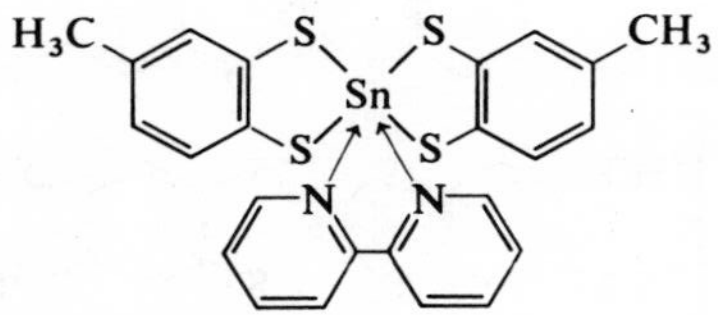	295(decomp.)(*156*)	(*67, 156*)
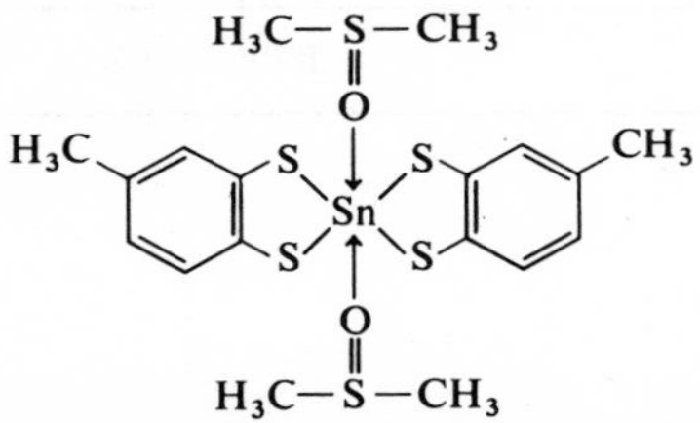		(*67*)

(*continued*)

APPENDIX 8 (*continued*)

Heterocycles with Tin-Sulfur Bonds in the Ring

Compound	mp(°C)	bp(°C/mmHg)	n_D^{20}	d_4^{20}	References
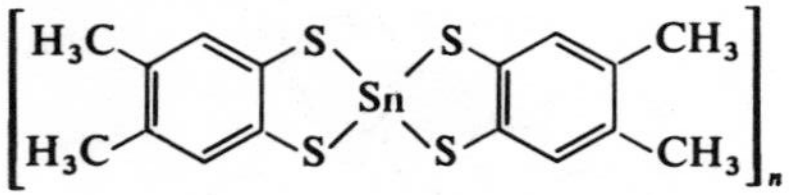 $[H_3C, H_3C, S, S, Sn, S, S, CH_3, CH_3]_n$	360				(*92*)
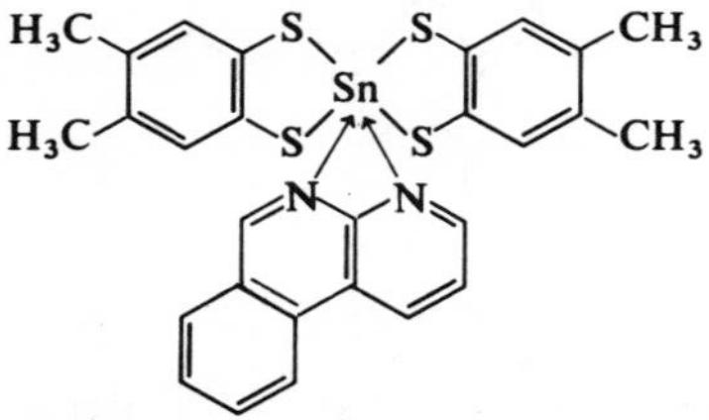 $H_3C, H_3C, S, S, Sn, S, S, CH_3, CH_3, N, N$	285				(*92*)

APPENDIX 9

Polymeric Organotin Sulfides

Compound	mp(°C)	bp(°C/mmHg)	n_D^{20}	d_4^{20}	References
$\left[-\mathrm{Sn}(\mathrm{S-CH_2-CH_2-COOH})(\mathrm{S-CH_2-CH_2-COOH})-\mathrm{CH_2-CH_2}- \right]_n$					(*P36*)
$\left[-\mathrm{Sn}(\mathrm{S-C(O)C_6H_5})(\mathrm{S-C(O)C_6H_5})-\mathrm{CH_2-CH_2-CH_2}- \right]_n$					(*P36*)
$\left[-\mathrm{Sn}(\mathrm{S-C_{12}H_{25}})(\mathrm{OOC-CH_2-COOC_8H_{17}})-\mathrm{CH_2-CH_2-CH_2}- \right]_n$					(*P36*)
$\left[-\mathrm{Sn}(\mathrm{S-C(O)C_6H_5})(\mathrm{OOC-CH_2-COOC_8H_{17}})-(\mathrm{CH_2})_3- \right]_n$					(*P36*)
$\left[-\mathrm{Sn}(\mathrm{S-C(O)C_6H_5})(\mathrm{S-C_{12}H_{25}})-(\mathrm{CH_2})_4- \right]_n$					(*P36*)

(*continued*)

APPENDIX 9 (*continued*)

Polymeric Organotin Sulfides

Compound	mp(°C)	bp(°C/mmHg)	n_D^{20}	d_4^{20}	References
$\left[\text{Sn}(\text{S}-C_{12}H_{25})(\text{S}-CH_2-COOC_8H_{17})-(CH_2)_4-\right]_n$					(*P36*)
$\left[\text{Sn}(\text{S}-CH_2-CH_2-COOH)_2-(CH_2)_5-\right]_n$					(*P36*)
$\left[\text{Sn}(\text{S}-C_{12}H_{25})(\text{S}-CH_2-COOC_8H_{17})-(CH_2)_8-\text{Sn}(\text{S}-C_{12}H_{25})(\text{S}-CH_2COOC_8H_{17})-(CH_2)_2-\right]_n$					(*P36*)
$CH_3-O-\left[\text{Sn}(C_4H_9)_2-O\right]_n-\text{Sn}(C_4H_9)_2-S-C_{12}H_{25}$					(*P50, P140*)
$\left[C_4H_9-\text{Sn}(-)(-O-)-S-CH_2-CH_2-COO\text{-}iso\text{-}C_8H_{17}\right]_n$					(*P93*)

$\left[C_6H_5-\underset{\underset{|}{O}}{\overset{|}{Sn}}-S-CH_2-COO\text{-}iso\text{-}C_8H_{17} \right]_n$		(*P93*)
$\left[-C_4H_9)_2\overset{|}{Sn}-S-(CH_2)_4-S \right]_n$	138	(*P94*)
$\left[(C_4H_9)_2\overset{|}{Sn}-S-C(O)CH_2-S \right]_n$		(*P103–P106*)
$\left[(C_6H_5)_2\overset{|}{Sn}-S-CH_2-C(O)-O \right]_n$	>300	(*56*)
$\left[n-C_4H_9)_2\overset{|}{Sn}-S-CH_2-COO-(CH_2)_2-OOC-CH_2-S \right]_n$		(*170*)
$\left[S-\underset{C_4H_9}{\overset{C_4H_9}{Sn}}-OOC-(CH_2)_4-COO-\underset{C_4H_9}{\overset{C_4H_9}{Sn}} \right]_n$	50–2	(*129*)
$\left[O-\underset{C_4H_9}{\overset{C_4H_9}{Sn}}-S_4-\underset{C_4H_9}{\overset{C_4H_9}{Sn}}-OOC-(CH_2)_2-C(O) \right]_n$		(*129*)
$\left[O-\underset{C_4H_9}{\overset{C_4H_9}{Sn}}-S_4-\underset{C_4H_9}{\overset{C_4H_9}{Sn}}-OOC-(CH_2)_4-C(O) \right]_n$	108–16	(*129*)

(continued)

APPENDIX 9 (*continued*)

POLYMERIC ORGANOTIN SULFIDES

Compound	mp(°C)	bp(°C/mmHg)	n_D^{20}	d_4^{20}	References
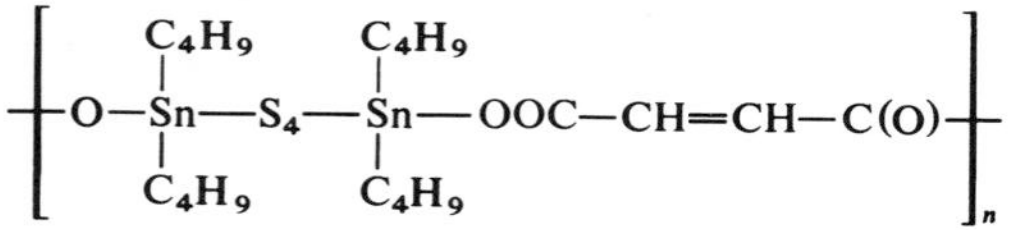 $\left[-O-Sn(C_4H_9)_2-S_4-Sn(C_4H_9)_2-OOC-CH{=}CH-C(O)-\right]_n$					(*129*)
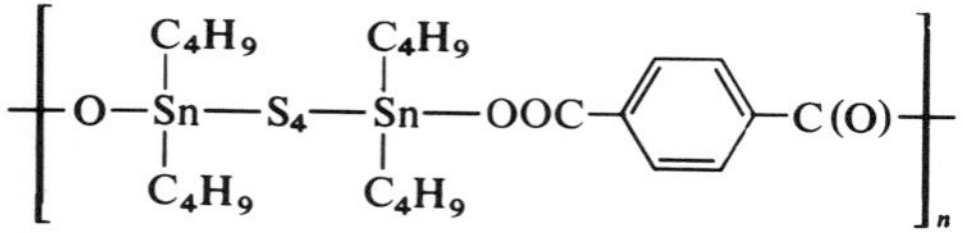 $\left[-O-Sn(C_4H_9)_2-S_4-Sn(C_4H_9)_2-OOC-C_6H_4-C(O)-\right]_n$					(*129*)
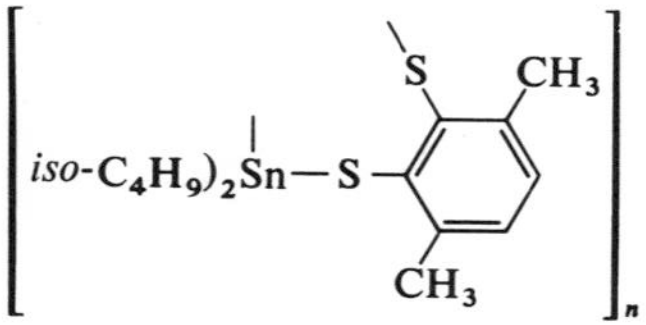 $\left[(iso\text{-}C_4H_9)_2Sn-S-C_6H_2(CH_3)_2-S-\right]_n$					(*P94*)
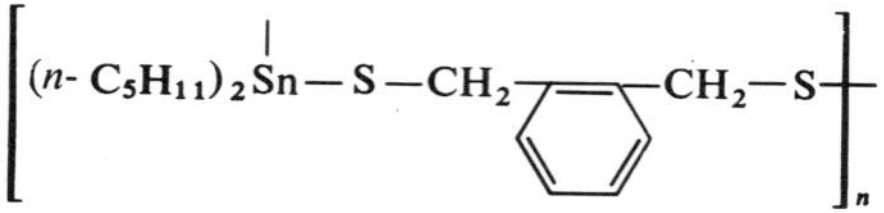 $\left[(n\text{-}C_5H_{11})_2Sn-S-CH_2-C_6H_4-CH_2-S-\right]_n$					(*P72*)
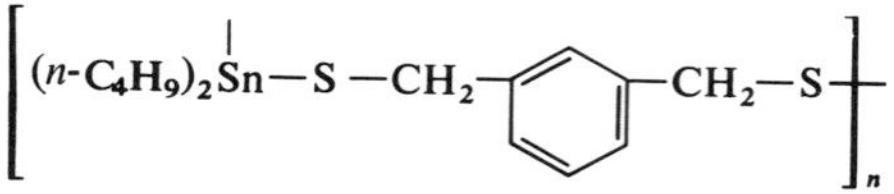 $\left[(n\text{-}C_4H_9)_2Sn-S-CH_2-C_6H_4-CH_2-S-\right]_n$					(*P155*)

$\left[(n\text{-}C_4H_9)_2\overset{|}{Sn}{-}S{-}CH_2{-}C_6H(CH_3)_3{-}CH_2{-}S \right]_n$ (*P155*)

$\left[(C_3H_5)_2\overset{|}{Sn}{-}S{-}CH_2{-}(m\text{-}C_6H_4){-}CH_2{-}S \right]_n$ (*P72*)

$\left[(C_4H_9)_2\overset{|}{Sn}{-}S{-}(CH_2)_2{-}(m\text{-}C_6H_4){-}(CH_2)_2{-}S \right]_n$ (*P72*)

$\left[(n\text{-}C_4H_9)_2\overset{|}{Sn}{-}S{-}CH_2{-}(p\text{-}C_6H_4){-}CH_2{-}S \right]_n$ (*P70, P72, P155*)

$\left[(n\text{-}C_4H_9)_2\overset{|}{Sn}{-}S{-}CH_2{-}C_6H(CH_3)_3{-}CH_2{-}S \right]_n$ (*P155*)

$\left[(n\text{-}C_8H_{17})_2\overset{|}{Sn}{-}S{-}CH_2{-}(p\text{-}C_6H_4){-}CH_2{-}S \right]_n$ (*P70, P72, P155*)

$\left[(C_6H_{11})_2\overset{|}{Sn}{-}S{-}CH_2{-}CH_2{-}(p\text{-}C_6H_4){-}CH_2{-}CH_2{-}S \right]_n$ (*P72*)

(*continued*)

APPENDIX 9 (*continued*)

POLYMERIC ORGANOTIN SULFIDES

Compound	mp(°C)	bp(°C/mmHg)	n_D^{20}	d_4^{20}	References
$[(C_6H_5CH_2)_2Sn{-}S{-}CH_2{-}C_6H_4{-}CH_2{-}S]_n$					(*P72*)
$[S{-}Sn(C_6H_5)_2{-}S{-}CH_2{-}C_6H_4{-}CH_2]_n$					(*P1, P72*)
$[(CH_3{-}C_6H_4{-})_2Sn{-}S{-}(CH_2)_3{-}C_6H_4{-}(CH_2)_3{-}S]_n$					(*P72*)
$C_{12}H_{25}{-}S[Sn{-}S{-}CH_2{-}C_6H_4{-}CH_2{-}S]_nSn{-}S{-}C_{12}H_{25}$					(*P70*)
$[S{-}CH{=}CH{-}S{-}Sn{-}S{-}CH{=}CH{-}S]_n$					(*92*)
$[(C_6H_5)_2Sn{-}S{-}Sn(C_6H_5){-}S{-}Sn(C_6H_5){-}S]$ ring, $-S-$ bridged, $]_n$	250(decomp.)				(*194*)

$\left[-S-Sn(C_6H_5)\langle S-Sn(C_6H_5)(Cl)-S-Sn(C_6H_5)-S\rangle-\right]_n S$ (cyclic Sn_3S_3 ring)	260(decomp.)		*(202)*
$\left[(CH_3)_3SiCH_2)_2Sn-S-\right]_n$	150–65		*(184)*
$[C_4H_9(C_8H_{17}OOCCH_2CH_2S)Sn—O—]_n$			*(P76c)*
$[C_4H_9(C_8H_{17}OSn(C_4H_9)_2OOCCH_2CH_2S)Sn—O—]_n$			*(P76c)*
$[C_4H_9($*iso*-$C_8H_{17}OOCCH_2CH_2S)Sn—S—]_n$			*(P58f)*
Poly(dibutyltin)$_{1.0}$(trimethylolpropanemono-pelargonate-dithioglycolate)$_{1.0}$		1.142(d_4^{25})	*(P38b)*
Poly(dibutyltin)$_{2.0}$(trimethylolpropanemonopelar-gonate-di(thioglycolate))$_{1.0}$(lauryl mercaptan)$_{2.0}$			*(P38b)*
Poly(dibutyltin)$_{2.0}$(pentaerythriolmonohydroxy monopelargonate-di(thioglycolate))$_{1.0}$(lauryl mercaptan)$_{2.0}$		1.130(d_4^{25})	*(P38b)*
Poly(dibutyltin)$_{2.0}$(trimethylolpropane mono-pelargonate-di(thioglycolate))$_{1.0}$(lauryl thiogly-colate)$_{2.0}$		1.068(d_4^{25})	*(P38b)*
Poly(dibutyltin)$_{2.0}$(pentaerythriolmonohydroxy monopelargonate-di(thioglycolate))$_{1.0}$(lauryl thioglycolate)$_{2.0}$		1.140(d_4^{25})	*(P38b)*
Poly(dibutyltin)$_{2.0}$(trimethylolpropanemono-pelargonate-di(mercaptopropionate))$_{1.0}$(lauryl thioglycolate)$_{2.0}$		1.040(d_4^{25})	*(P38b)*
Poly(dibutyltin)$_{2.0}$(pentaerythriolmonohydroxy monopelargonate)$_{1.0}$(lauryl thioglycolate)$_{2.0}$		1.10(d_4^{25})	*(P38b)*

(continued)

APPENDIX 9 (*continued*)

Polymeric Organotin Sulfides

Compound	mp(°C)	bp(°C/mmHg)	n_D^{20}	d_4^{20}	References
Poly(dibutyltin)$_{2.0}$(pentaerythriolmonohydroxy monopelargonate-di(mercaptopropionate))$_{1.0}$ (lauryl thioglycolate)$_{2.0}$(toluene-2,4-diisocyanate)$_{0.6}$					(*P38b*)
Poly(dibutyltin)$_{2.0}$(trimethylolpropanemonopelargonate-di(mercaptopropionate))$_{1.0}$(isooctylthioglycolate)$_{2.0}$				1.10(d_4^{25})	(*P38b*)
Poly(dibutyltin)$_5$(trimethylolpropanemonopelargonate-di(mercaptopropionate))$_4$					(*P38b*)
Poly(dibutyltin)$_4$(trimethylolpropanemonopelargonate-di(mercaptopropionate))$_{3.0}$(isooctylthioglycolate)$_2$				1.181(d_4^{25})	(*P38b*)
Poly(dibutyltin)$_{2.0}$(trimethylolpropanemonopelargonate-di(thioglycolate))$_{1.0}$(lauryl thioglycolate)$_{2.0}$				1.120(d_4^{25})	(*P38b*)
Poly(dibutyltin)$_{2.0}$(trimethylolpropanemonopelargonate-di(thioglycolate))$_{1.0}$(isooctylthioglycolate)$_{2.0}$				1.100(d_4^{25})	(*P38b*)
Poly(dibutyltin)$_{1.0}$(pentaerythrioldipelargonate-dithioglycolate)$_{1.0}$					(*P38b*)
Poly(dibutyltin)$_{4.0}$(pentaerythriol$_{1.0}$(pelargonate)$_{2.0}$-(3-mercaptopropionate)$_{1.5}$)$_{5.0}$				1.140(d_4^{25})	(*P38b*)
Poly(dibutyltin)$_{1.7}$(pentaerythriol(pelargonate)$_{2.0}$-(mercaptopropionate)$_{1.6}$)$_{1.0}$(isooctylthioglycolate)$_{1.7}$				1.090(d_4^{25})	(*P38b*)
Poly(dibutyltin)$_{2.0}$(pentaerythriolmonopelargonate-benzoate dithioglycolate)$_{1.0}$(isooctylthioglycolate)$_{2.0}$				1.178(d_4^{25})	(*P38b*)

Poly(dibutyltin)$_{1.4}$((pentaerythriol)$_{1.0}$(pelargonate)$_{1.75}$-(mercaptopropionate)$_{1.75}$)$_{1.0}$(isooctylthioglycolate)$_{0.8}$	1.15(d_4^{25})	(*P38b*)
Poly(dibutyltin)$_{2.0}$(trimethylolpropanepelargonate-di(thioglycolate))$_{1.0}$(phenoxyethylenethioglycolate)$_{2.0}$	1.257(d_4^{25})	(*P38b*)
Poly(dibutyltin)$_{2.0}$(trimethylolpropane monopelargonate-di(thioglycolate))$_{1.0}$(isooctylthioglycolate)$_{2.0}$		(*P38b*)
Poly(dibutyltin)$_{2.0}$((pentaerythriol)$_{1.0}$(pelargonate)$_{1.8}$-(mercaptopropionate)$_{1.75}$)$_{1.6}$(phenoxyethylthioglycolate)$_{1.2}$	1.179(d_4^{25})	(*P38b*)
Poly(dibutyltin)$_{2.0}$((pentaerythriol)$_{1.0}$(pelargonate)$_{1.75}$-(mercaptopropionate)$_{1.75}$)$_{1.6}$ (*n* dodecylmercaptide)$_{1.2}$	1.130(d_4^{25})	(*P38b*)
Poly(dibutyltin)$_{2.0}$((pentaerythriol)$_{1.0}$(pelargonate)$_{1.75}$-(mercaptopropionate)$_{1.75}$)$_{1.14}$ (*n*-dodecylmercaptide)$_{2.0}$	1.082(d_4^{25})	(*P38b*)
Poly(dibutyltin)$_{2.0}$((pentaerythritol)$_{1.0}$(pelargonate)$_{1.8}$-(mercaptopropionate)$_{1.75}$)$_{1.6}$(isooctylthioglycolate)$_{1.2}$	1.143(d_4^{25})	(*P38b*)
Poly(dibutyltin)$_{2.0}$((pentaerythritol)$_{1.0}$(pelargonate)$_{2.0}$-(mercaptopropionate)$_{1.8}$)$_{1.1}$(phenoxymethylthioglycolate)$_{2.0}$	1.206(d_4^{25})	(*P38b*)

APPENDIX 10

Miscellaneous Organotin Sulfides

Compound	mp(°C)	bp(°C/mmHg)	n_D^{20}	d_4^{20}	References
$CH_3Sn(S)SH$					*(146)*
$(C_2H_5)_3SnSH$					*(227)*
$(CH_3)_3SnSNa$					*(P95b)*
$(CH_3)_2Sn(SNa)_2$					*(P95b)*
$(C_2H_5)_2Sn(SNa)_2$					*(P38f, P95b)*
$(C_4H_9)_2Sn(SNa)_2$	106–10*(P95b)*				*(P34, P38f, P95b)*
$(C_4H_9)_2Sn(S_2Ca)$					*(P95b)*
$(C_8H_{17})_2Sn(SNa)_2$					*(P95b)*
$(C_6H_5)_3Sn$—S(Li)→$Sn(C_6H_5)_3$ / ←S(Li)— (four-membered ring: two Sn bridged by two S, each S bearing Li, with S→Sn coordinate bonds)	>250*(206)*				*(203, 205, 206)*
$(C_6H_5)_3SnSNa$					*(88)*
$(C_6H_5)_2Sn(SLi)_2$					*(203, 207)*
$(C_6H_5)_2Sn(SNa)_2$					*(P95b)*
$C_4H_9Sn(SNa)_3$					*(P95b)*
$(n\text{-}C_4H_9)_3Sn—S—SO_2CH_3$		98–100/0.2			*(P149)*
$(n\text{-}C_4H_9)_2Sn(—S—SO_2CH_3)_2$	106				*(P149)*
$(C_6H_5)_3Sn—S—SO_2CH_3$	>350				*(P149)*
$(C_4H_9)_3Sn—S—P(=O)(OH)—OC_6H_5$					*(P38e)*

$(C_4H_9)_2Sn[-S-\overset{O}{\overset{\Vert}{P}}(OC_2H_5)_2]_2$		197–8	1.5073	(*118a*)
$(C_4H_9)_2Sn[-S-\overset{O}{\overset{\Vert}{P}}(OC_4H_9)_2]_2$		183–4	1.4961	(*118a*)
$(C_6H_{13})_3Sn-S-\overset{O}{\overset{\Vert}{P}}(OC_6H_{13})_2$				(*P38e*)
$(C_6H_5)_3Sn-S-P(OC_6H_5)_2$				(*P38e*)
$(CH_3)_2Sn[-S-\overset{S}{\overset{\Vert}{P}}(C_2H_5)_2]_2$				(*38*)
$(C_2H_5)_3Sn-S-\overset{S}{\overset{\Vert}{P}}(C_2H_5)_2$		190/2		(*37, 38*)
$(n\text{-}C_4H_9)_2Sn[-S-\overset{S}{\overset{\Vert}{P}}(C_2H_5)_2]_2$	40.5(*111*)			(*38, 111*)
$(C_6H_5)_3Sn-S-\overset{S}{\overset{\Vert}{P}}(C_2H_5)_2$	88–9			(*37, 38*)
$(C_6H_5)_2Sn[-S-\overset{S}{\overset{\Vert}{P}}(C_2H_5)_2]_2$	149.5(*111*)			(*38, 111*)
$Cl_2Sn[-S-\overset{S}{\overset{\Vert}{P}}(C_2H_5)_2]_2$	179–80(*37*) 171–2(*111*)			(*37, 38, 111*)

(*continued*)

APPENDIX 10 (*continued*)

Miscellaneous Organotin Sulfides

Compound	mp(°C)	bp(°C/mmHg)	n_D^{20}	d_4^{20}	References
$Br_2Sn[-S-P(=S)(C_2H_5)_2]_2$	175				(*38*)
$(CH_3)_2Sn[-S-P(=S)(C_6H_5)_2]_2$	161–2				(*38*)
$(C_2H_5)_2Sn[-S-P(=S)(C_6H_5)_2]_2$	162–3				(*37, 38*)
$(C_6H_5)_3Sn-S-P(=S)(C_6H_5)_2$	128–30				(*200*)
$Cl_2Sn[-S-P(=S)(C_6H_5)_2]_2$	206–7				(*37, 38*)
$Br_2Sn[-S-P(=S)(C_6H_5)_2]_2$	195				(*37, 38*)
$(CH_3)_3Sn-S-P(=S)(OCH_3)_2$					(*P38e*)
$(CH_3)_3Sn-S-P(=S)(OC_{16}H_{33})_2$					(*P38e*)

$(C_4H_9)_3Sn{-}S{-}P(=S)(-OCH_3)_2$			$1.4965(n_D^{21})$		(*P86*)
$(C_6H_5)_3Sn{-}S{-}P(=S)(-OCH_3)_2$	86				(*110*)
$(C_2H_5)_3Sn{-}S{-}P(=S)(-OC_2H_5)_2$		115–8/0.1			(*110*)
$(n\text{-}C_4H_9)_3Sn{-}S{-}P(=S)(-OC_2H_5)_2$		132–6/0.0002(*110*)			(*110, P71*)
$(C_6H_5)_3Sn{-}S{-}P(=S)(-OC_2H_5)_2$	105(*110*) 124–5(*P86*)				(*110, P86*)
$(C_6H_5)_3Sn{-}S{-}P(=S)(-O\text{-}n\text{-}C_3H_7)_2$	63				(*110*)
$(C_{12}H_{25})_2Sn[-S{-}P(=S)(-O\text{-}i\text{-}C_3H_7)_2]_2$					(*P8, P118*)
$(C_4H_9)_2Sn[-S{-}P(=S)(OC_2H_5)_2]_2$			1.5553	1.3310	(*118a*)
$(C_4H_9)_2Sn[-S{-}P(=S)(-OC_4H_9)_2]_2$					(*P71*)

(continued)

APPENDIX 10 (*continued*)

Miscellaneous Organotin Sulfides

Compound	mp(°C)	bp(°C/mmHg)	n_D^{20}	d_4^{20}	References
$(C_6H_5)_3Sn{-}S{-}P(S)({-}O{-}n{-}C_4H_9)_2$	69.5				(*110*)
$(C_4H_9)_2Sn[{-}S{-}P(S)({-}OC_6H_{13})_2]_2$					(*P71*)
$(C_4H_9)_2Sn[{-}S{-}P(S)({-}OC_8H_{15})_2]_2$					(*P24, P118*)
$(C_8H_{17})_2Sn[{-}S{-}P(S)({-}OC_6H_{11})_2]_2$					(*P71*)
$(n{-}C_4H_9)_2Sn[{-}S{-}P(S)({-}O{-}C_6H_{10}{-}CH_3)_2]_2$ (4-methylcyclohexyl)					(*P8, P118*)
$(C_4H_9)_2Sn[{-}S{-}P(S)({-}O{-}C_6H_4{-}CH_3)_2]_2$ (p-tolyl)					(*P71*)
$(n{-}C_4H_9)_2Sn[{-}S{-}P(S)({-}O{-}C_6H_4{-}C_8H_{17})_2]_2$ (p-octylphenyl)					(*P8, P118*)

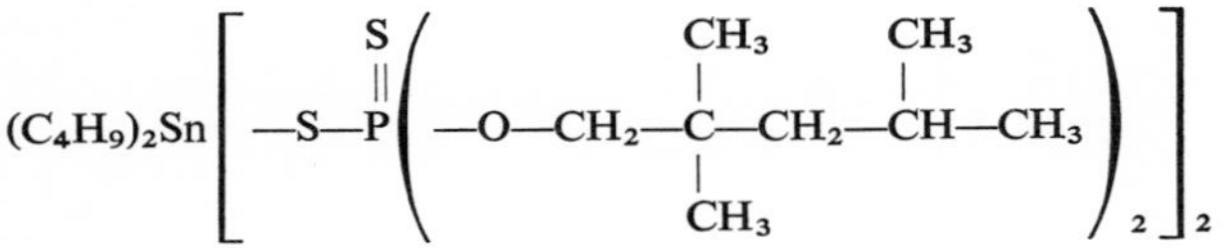

$(C_4H_9)_2Sn\left[-S-\overset{S}{\overset{\Vert}{P}}\left(-O-CH_2-\underset{CH_3}{\overset{CH_3}{C}}-CH_2-\overset{CH_3}{CH}-CH_3\right)_2\right]_2$		(*P71*)
$(C_8H_{17})_2Sn\left[-S-\overset{S}{\overset{\Vert}{P}}\left(-O-CH_2-\underset{CH_3}{\overset{CH_3}{C}}-CH_2-\overset{CH_3}{CH}-CH_3\right)_2\right]_2$		(*P71*)
$(C_8H_{17})_2Sn\left[-S-\overset{S}{\overset{\Vert}{P}}(-O-CH=CH-CH_2-CH_3)_2\right]_2$		(*P71*)
$(C_4H_9)_2Sn\left[-S-\overset{S}{\overset{\Vert}{P}}\left(-O-CH=\overset{C_2H_5}{C}-(CH_2)_3CH_3\right)_2\right]_2$		(*P71*)
$(C_8H_{17})_2Sn\left[-S-\overset{\Vert}{P}\left(-O-CH=\overset{C_2H_5}{C}-(CH_2)_3-CH_3\right)_2\right]_2$		(*P71*)
$(C_4H_9)_2Sn\left[-S-\overset{S}{\overset{\Vert}{P}}\begin{matrix}O-H_2C \\ O-H_2C\end{matrix}C(CH_3)_2\right]_2$		(*P71*)
$(C_8H_{17})_2Sn\left[-S-\overset{S}{\overset{\Vert}{P}}\begin{matrix}O-H_2C \\ O-H_2C\end{matrix}C(CH_3)_2\right]_2$	126–9	(*P71*)

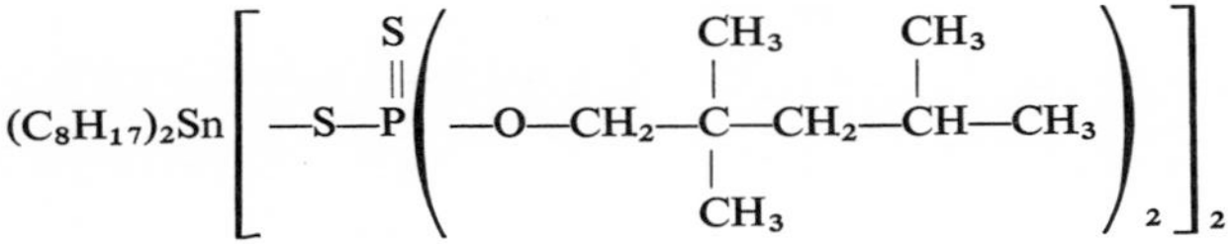

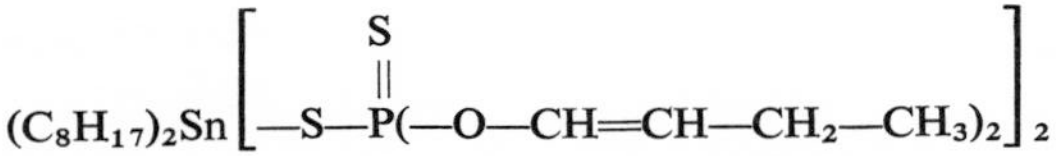

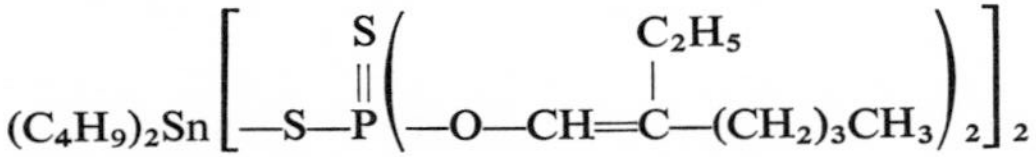

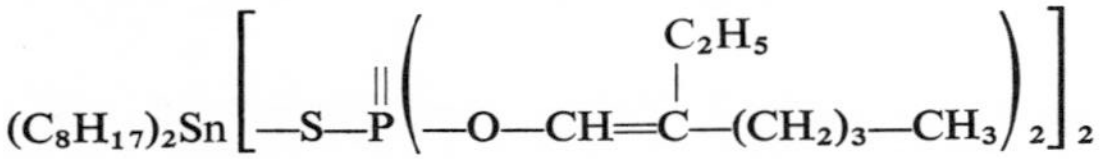

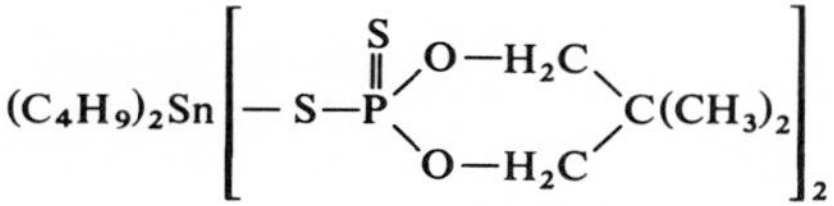

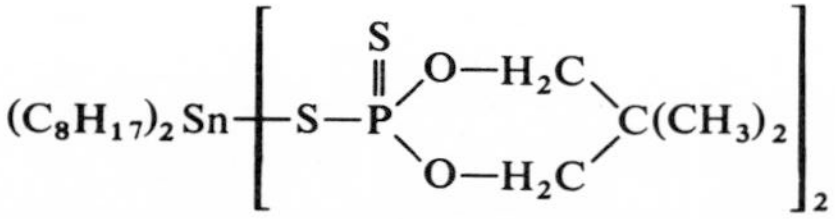

(*continued*)

APPENDIX 10 (*continued*)

MISCELLANEOUS ORGANOTIN SULFIDES

Compound	mp(°C)	bp(°C/mmHg)	n_D^{20}	d_4^{20}	References
$(C_3H_7)_2(C_6H_5)Sn{-}S{-}P(=S)[N(C_{18}H_{37})_2]_2$					(*P38e*)
$(C_6H_5)_3Sn{-}S{-}P(=S)[N(CH_3)_2]_2$					(*P38e*)
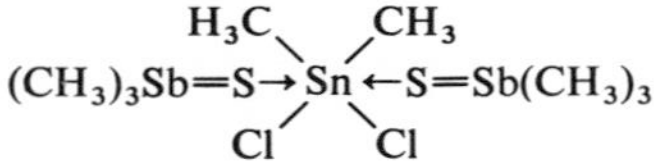 $(CH_3)_3Sb{=}S{\rightarrow}Sn(CH_3)_2Cl_2{\leftarrow}S{=}Sb(CH_3)_3$	147–7.5				(*212, 213*)
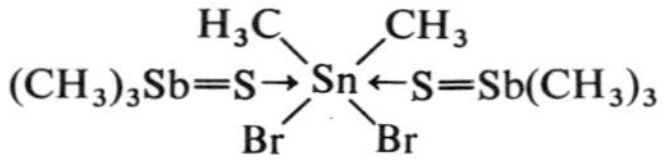 $(CH_3)_3Sb{=}S{\rightarrow}Sn(CH_3)_2Br_2{\leftarrow}S{=}Sb(CH_3)_3$	123–4				(*212*)
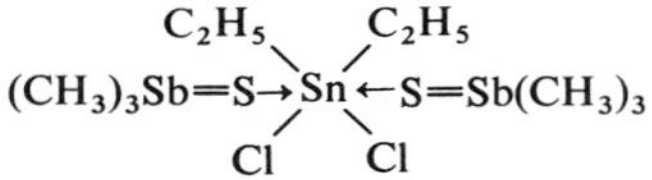 $(CH_3)_3Sb{=}S{\rightarrow}Sn(C_2H_5)_2Cl_2{\leftarrow}S{=}Sb(CH_3)_3$	122–3				(*212*)
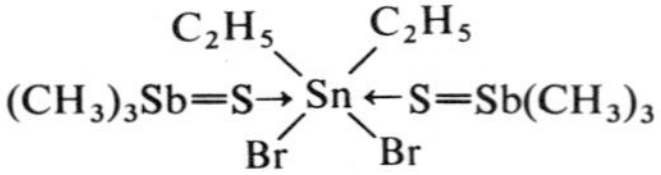 $(CH_3)_3Sb{=}S{\rightarrow}Sn(C_2H_5)_2Br_2{\leftarrow}S{=}Sb(CH_3)_3$	121				(*212*)
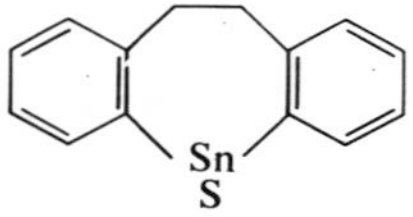 [dibenzo ring structure with Sn=S bridging, $-CH_2CH_2-$] Sn S	274–5				(*112*)

$H_2\left[\begin{array}{l} \quad OOC{-}CH{=}CH{-}COO{-}C_8H_{17} \\ C_4H_9 \diagdown \quad \diagup S{-}C_{12}H_{25} \\ \qquad Sn \\ C_4H_9 \diagup \quad \diagdown S{-}C_{12}H_{25} \\ \quad OOC{-}CH{=}CH{-}COO{-}C_8H_{17} \end{array}\right]$	(*P126, P131*)
$(C_4H_9)_2Sn\left[\begin{array}{l} \quad OOC{-}CH{=}CH{-}COO{-}C_8H_{17} \\ C_4H_9 \diagdown \quad \diagup S{-}C_{12}H_{25} \\ \qquad Sn \\ C_4H_9 \diagup \quad \diagdown S{-}C_{12}H_{25} \\ \quad OOC{-}CH{=}CH{-}COO{-}C_8H_{17} \end{array}\right]$	(*P126, P131*)
$H_2\left[\begin{array}{l} \quad OOC{-}CH{=}CH{-}COO{-}C_8H_{17} \\ C_4H_9 \diagdown \quad \diagup S{-}CH_2{-}COO{-}C_8H_{17} \\ \qquad Sn \\ C_4H_9 \diagup \quad \diagdown S{-}CH{-}COO{-}C_8H_{17} \\ \quad OOC{-}CH{=}CH{-}COO{-}C_8H_{17} \end{array}\right]$	(*P17*)
$(C_4H_9)_2Sn\left[\begin{array}{l} \quad OOC{-}CH{=}CH{-}COO{-}C_8H_{17} \\ C_4H_9 \diagdown \quad \diagup S{-}CH_2{-}COO{-}C_8H_{17} \\ \qquad Sn \\ C_4H_9 \diagup \quad \diagdown S{-}CH_2{-}COO{-}C_8H_{17} \\ \quad OOC{-}CH{=}CH{-}COO{-}C_8H_{17} \end{array}\right]$	(*P17*)
$\left[\begin{array}{l} CH{-}C{-}CH_3 \\ \Vert \quad\ \ \Vert \\ C \quad\ \ S^{\oplus} \\ \vert \ \diagdown S \diagup \\ CH_3 \end{array}\right]_2 SnCl_6$	(*134a*)

(continued)

APPENDIX 10 (*continued*)

Miscellaneous Organotin Sulfides

Compound	mp(°C)	bp(°C/mmHg)	n_D^{20}	d_4^{20}	References
$(CH_3)_3SnSC(C_6H_5)=NC(O)N(CH_3)_2$	124.5–5.1				(*94c*)
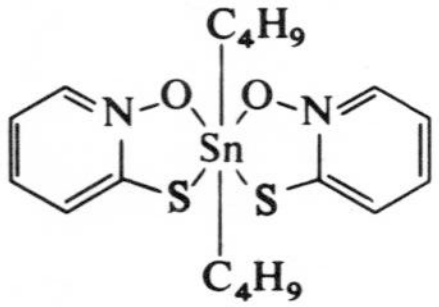 C_4H_9; N–O, O–N; Sn; S, S; C_4H_9	94–6				(*145c*)
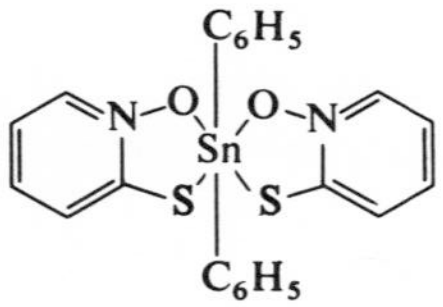 C_6H_5; N–O, O–N; Sn; S, S; C_6H_5	> 210				(*145c*)
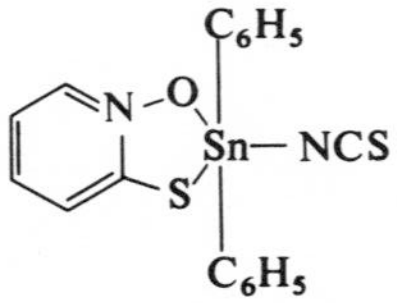 C_6H_5; N–O; Sn—NCS; S; C_6H_5					(*145c*)
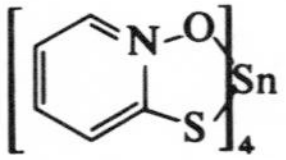 [N–O, S]$_4$Sn	> 242–3				(*145c*)

Compound	M.p. (°C)	Ref.
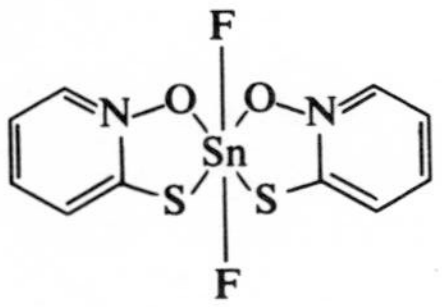 SnF_2 bis(2-pyridinethiolato N-oxide) (structure: F–Sn–F, N–O, S)	282–3	(*145c*)
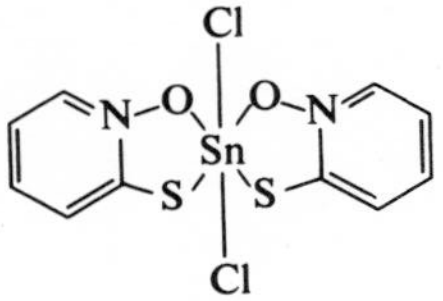 $SnCl_2$ bis(2-pyridinethiolato N-oxide) (structure: Cl–Sn–Cl, N–O, S)	297–8.5	(*145c*)
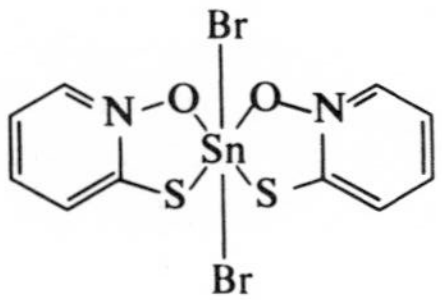 $SnBr_2$ bis(2-pyridinethiolato N-oxide) (structure: Br–Sn–Br, N–O, S)	293–4	(*145c*)
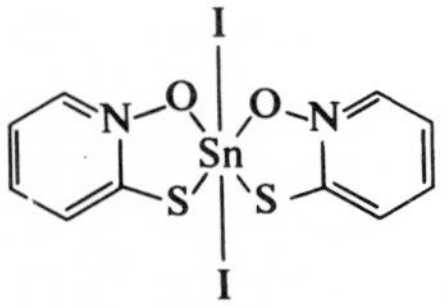 SnI_2 bis(2-pyridinethiolato N-oxide) (structure: I–Sn–I, N–O, S)	281–2	(*145c*)
$[(C_4H_9)_3SnSCH_2O—]_4Si$		(*P158e*)
$[(C_4H_9)_2Sn(SCH_2CH_2O—)_2]_2Si$		(*P158e*)

(*continued*)

APPENDIX 10 (*continued*)

Miscellaneous Organotin Sulfides

Compound	mp(°C)	bp(°C/mmHg)	n_D^{20}	d_4^{20}	References
$C_4H_9Sn[S(CH_2)_2O]_3SiO(CH_2)_2SSn(C_4H_9)[S(CH_2)_2O]_2Si[O(CH_2)_2S]_2Sn(C_4H_9)S(CH_2)_2OSi[O(CH_2)_2S]_3SnC_4H_9$					(*P158e*)
$(C_4H_9)_2Sn[S(CH_2)_2O]_2Si(OC_2H_5)O(CH_2)_2SSn(C_4H_9)_2S(CH_2)_2OSi(OC_2H_5)[O(CH_2)_2S]_2Sn(C_4H_9)_2$					(*P158e*)
$\{(C_4H_9)_2Sn[SCH_2C(O)O(CH_2)_2O]_2\}_2Si$					(*P158e*)
$[(C_4H_9)_2Sn(SCH_2CH_2O)_2]_3[Si(C_4H_9)]_2$					(*P158e*)
$[(C_4H_9)_2Sn(SCH_2CH_2O)_2][Si(C_4H_9)(OC_2H_5)_2]_2$					(*P158e*)
$[(C_4H_9)Sn(SCH_2CH_2O)_3][SiOC_{12}H_{25}]$					(*P158e*)
$[(C_4H_9)Sn(SCH_2CH_2O)_3]_2[Si(C_4H_9)(OC_{18}H_{37})]_3$					(*P158e*)
$[(C_4H_9)Sn(SCH_2CH_2O)_3][Si(C_4H_9)_2(OC_2H_5)]_3$					(*P158e*)
$[(C_4H_9)_3SnSCH_2CH_2O]_3[SiOCH_3]$					(*P158e*)
$[(C_4H_9)_3SnSCH_2CH_2O]_2[Si(C_4H_9)_2]$					(*P158e*)
$[(C_4H_9)_3SnSCH_2CH_2O][Si(C_4H_9)_2(OCH_2CH{=}CH_2)]$					(*P158e*)
$[(CH_3)_2Sn(SCH_2CH_2O)_2][Si(CH_3)(OC_2H_5)]$					(*P158e*)
$[(C_6H_5)_2Sn(SCH_2CH_2O)_2][Si(OC_2H_5)_2]$					(*P158e*)
$[(C_6H_5CH_2)_2Sn(SCH_2CH_2O)_2][Si(CH_2C_6H_5)_2]$					(*P158e*)
$[(o\text{-}CH_3C_6H_4)_2Sn(SCH_2CH_2O)_2][Si(o\text{-}CH_3C_6H_4)(OC_2H_5)]$					(*P158e*)
$[(C_5H_5)_2Sn(SCH_2CH_2O)_2][Si(C_5H_5)(OC_3H_7)]$					(*P158e*)
$\{[CH_3(CH_2)_{16}CH_2]_2Sn(SCH_2CH_2O)_2\}[Si(OC_2H_5)_2]$					(*P158e*)
$\{[CH_3(CH_2)_3CH(C_2H_5)CH_2]_3SnSCH_2CH_2O\}_3[(Si(CH_2CH(C_2H_5)(CH_2)_3CH_3)]$					(*P158e*)

Compound	Ref.
$[(HC{\equiv}C)Sn(SCH_2CH_2O)_3][Si(C{\equiv}CH)_2(OC_2H_5)]_3$	(*P158e*)
$[(C_6H_{11})_2Sn(SCH_2CH_2O)_2][Si(C_6H_{11})_3]_2$	(*P158e*)
$[(C_6H_{11})_2Sn(SCH_2CH_2O)_2][Si(C_6H_{11})(OC_3H_7)]$	(*P158e*)
$[(C_4H_9)_2Sn(SCH_2O)_2][Si(C_4H_9)_2]$	(*P158e*)
$\{(C_4H_9)_2Sn[S(CH_2)_4O]_2\}[Si(OC_2H_5)_2]$	(*P158e*)
$\{(C_4H_9)_2Sn[S(CH_2)_{18}O]_2\}[Si(C_4H_9)(OC_2H_5)]$	(*P158e*)
$[(C_4H_9)_2Sn(SCH(C_{16}H_{33})—CH_2—O)_2][Si(C_4H_9)(OC_2H_5)]$	(*P158e*)
$[(C_4H_9)_2Sn(SCH_3CH{=}CHCH_2O)_2][Si(OC_2H_5)_2]$	(*P158e*)
$[(C_4H_9)_2Sn(SCH_2C(CH_3){=}CHCH_2O)_2][Si(C_4H_9)_2]$	(*P158e*)
$[(C_4H_9)_2Sn(SCH_2C{\equiv}CCH_2O)_2][Si(C_4H_9)(OC_2H_5)]$	(*P158e*)
$[(C_4H_9)_2Sn(SCH_2C{\equiv}CCH_2O)_2][Si(C_4H_9)(OC_3H_7)]$	(*P158e*)
$[(C_4H_9)_2Sn(S—HC<(H_2C—CH_2)(H_2C—CH_2)>CH—O)_2]_2Si$	(*P158e*)
$[(C_4H_9)_2Sn(S—HC<(HC{=}CH)(H_2C—CH_2)>CH—O)_2]_2Si$	(*P158e*)
$[(C_4H_9)_2Sn(S—p\text{-}C_6H_4—O)_2]_2Si$	(*P158e*)

(*continued*)

APPENDIX 10 (*continued*)

Miscellaneous Organotin Sulfides

Compound	mp(°C)	bp(°C/mmHg)	n_D^{20}	d_4^{20}	References
$[(C_4H_9)_2Sn(S{-}C_6H_3(CH_3){-}O)_2]_2Si$					(*P158e*)
$[(C_4H_9)Sn(S{-}C_{10}H_6{-}O)_3][SiOC_2H_5]$					(*P158e*)
$[C_4H_9Sn(SCH_2CH_2O)_3][SiOCH_3]$					(*P158e*)
$[(C_4H_9)_3SnSCH_2CH_2O]_3[SiOC_6H_5]$					(*P158e*)
$[(C_4H_9)_3SnSCH_2CH_2O]_3[SiOCH_2C_6H_5]$					(*P158e*)
$[(C_4H_9)_2Sn(SCH_2CH_2O)_2][Si(C_4H_9)_2(OCH_2C_6H_5)]_2$					(*P158e*)
$[(C_4H_9)_3SnSCH_2CH_2O][Si(C_4H_9)_2(O{-}o{-}C_6H_4CH_3]$					(*P158e*)
$[(C_4H_9)_2Sn(SCH_2CH_2O)_2][Si(C_4H_9)_2(OC_5H_5)]_2$					(*P158e*)
$[(C_4H_9)_3SnSCH_2CH_2O][Si(C_4H_9)_2(OC_5H_5)]$					(*P158e*)
$[(C_4H_9)_3SnSCH_2CH_2O][Si(C_4H_9)(OC_{18}H_{37})_2]$					(*P158e*)
$[(C_4H_9)_2Sn(SCH_2CH_2O)_2][Si(C_4H_9)(OCH_2C{\equiv}CH)]$					(*P158e*)
$[(C_4H_9)_2Sn(SCH_2CH_2O)_2][SiC_4H_9(O{-}HC(CH_2CH_2)_2CH_2)]$					(*P158e*)

Compound	
$[(C_4H_9)_3SnSCH_2CH_2O]_3\left[SiO{-}HC\langle\begin{smallmatrix}HC{=}CH\\H_2C{-}CH_2\end{smallmatrix}\rangle CH_2\right]$	(*P158e*)
$[(C_4H_9)_3SnSCH_2C(O)OCH_2O]_3[SiOC_2H_5]$	(*P158e*)
$\{(C_4H_9)_2Sn[SCH_2C(O)O(CH_2)_4O]_2\}_3[SiOC_2H_5]_2$	(*P158e*)
$\{(C_4H_9)_2Sn[SCH_2C(O)O(CH_2)_{18}O]_2\}_3[SiOC_2H_5]_2$	(*P158e*)
$\left\{(C_4H_9)_2Sn[SCH_2C(O)O\underset{}{\overset{CH_2(CH_2)_{14}CH_3}{\overset{\vert}{C}H}}{-}CH_2{-}O]_2\right\}_3[SiOC_2H_5]_2$	(*P158e*)
$\{(C_4H_9)_2Sn[SCH_2C(O)OCH_2CH{=}CHCH_2O]_2\}_3[SiOC_2H_5]_2$	(*P158e*)
$\left\{(C_4H_9)_2Sn[SCH_2C(O)OCH_2\overset{CH_3}{\overset{\vert}{C}}{=}CHCH_2O]_2\right\}[Si(OC_2H_5)_2]$	(*P158e*)
$[(C_4H_9)_3SnSCH_2C(O)OCH_2C{\equiv}CCH_2O]_2[Si(OC_2H_5)_2]$	(*P158e*)
$\left\{(C_4H_9)Sn[SCH_2C(O)O{-}HC\langle\begin{smallmatrix}H_2C{-}CH_2\\H_2C{-}CH_2\end{smallmatrix}\rangle CH{-}O]_3\right\}_2[Si(OC_{18}H_{37})_2]_3$	(*P158e*)
$\left\{(C_4H_9)Sn[SCH_2C(O)O{-}HC\langle\begin{smallmatrix}HC{=}CH\\H_2C{-}CH_2\end{smallmatrix}\rangle CH{-}O]_3\right\}_2[Si(OC_{12}H_{25})_2]_3$	(*P158e*)
$[(C_4H_9)_3SnSCH_2C(O)O{-}p{-}C_6H_4{-}O][Si(C_4H_9)(OC_2H_5)_2]$	(*P158e*)

(*continued*)

APPENDIX 10 (*continued*)

Miscellaneous Organotin Sulfides

Compound	mp(°C)	bp(°C/mmHg)	n_D^{20}	d_4^{20}	References
$[(C_4H_9)_3SnSCH_2C(O)O{-}C_6H_3(CH_3){-}O][Si(C_4H_9)(OC_2H_5)_2]$					(*P158e*)
$\{(C_4H_9)Sn[SCH_2C(O)O{-}C_{10}H_6{-}O]_3\}[Si(OC_2H_5)_3]_3$					(*P158e*)
$\{(CH_3)Sn[SCH_2C(O)OCH_2CH_2O]_3\}[Si(CH_3)_2(OC_6H_{13})]_3$					(*P158e*)
$\{(C_6H_5)Sn[SCH_2C(O)OCH_2CH_2O]_3\}_2[Si(C_6H_5)_2]_3$					(*P158e*)
$\{(C_6H_5CH_2)_2Sn[SCH_2C(O)OCH_2CH_2O]_2\}[Si(CH_2C_6H_5)(OC_2H_5)]$					(*P158e*)
$\{(o\text{-}CH_3C_6H_4)Sn[SCH_2C(O)OCH_2CH_2O]_3\}_2[Si(OC_2H_5)(o\text{-}C_6H_4CH_3)]_3$					(*P158e*)
$\{(C_5H_5)_2Sn[SCH_2C(O)OCH_2CH_2O]_2\}[Si(OC_2H_5)_2]$					(*P158e*)
$\{(n\text{-}C_{18}H_{37})_2Sn[SCH_2C(O)OCH_2CH_2O]_2\}[Si(n\text{-}C_{18}H_{37})(OC_2H_5)]$					(*P158e*)
$\{[CH_3(CH_2)_3CH(C_2H_5)CH_2]_3SnSCH_2C(O)OCH_2CH_2O\}_3\{Si[CH_2CH(C_2H_5)(CH_2)_3CH_3]\}$					(*P158e*)
$\{(HC{\equiv}C)Sn[SCH_2C(O)OCH_2CH_2O]_3\}[Si(C{\equiv}CH)_2(OC_2H_5)]_3$					(*P158e*)
$[(cyclo\text{-}C_6H_{11})_3SnSCH_2C(O)OCH_2CH_2O]_2[Si(cyclo\text{-}C_6H_{11})(OC_2H_5)]$					(*P158e*)
$[(H_2C\langle HC{=}CH,\ H_2C{-}CH_2\rangle CH)_3SnSCH_2C(O)OCH_2CH_2O]_2[Si(HC\langle HC{=}CH,\ H_2C{-}CH_2\rangle CH_2)(OC_2H_5)]$					(*P158e*)

$[(C_4H_9)_3SnSCH_2CH_2C(O)OCH_2CH_2O][Si(C_4H_9)_2(OC_3H_7)]$	(*P158e*)
$[(C_4H_9)_3SnS(CH_2)_4C(O)OCH_2CH_2O][Si(C_4H_9)_2(OC_2H_5)]$	(*P158e*)
$\{(C_4H_9)_2Sn[S(CH_2)_{18}C(O)OCH_2CH_2O]_2\}[Si(C_4H_9)_3]_2$	(*P158e*)
$\{(C_4H_9)_2Sn[SCH(CH_2(CH_2)_{14}CH_3)CH_2C(O)OCH_2CH_2O]_2\}[Si(C_4H_9)(OC_2H_5)_2]_2$	(*P158e*)
$[(C_4H_9)_3SnSCH_2CH{=}CHCH_2C(O)OCH_2CH_2O]_2[Si(C_4H_9)(OC_2H_5)]$	(*P158e*)
$[(C_4H_9)_3SnSCH_2C(CH_3){=}CHCH_2C(O)OCH_2CH_2O]_2[Si(OC_2H_5)_2]$	(*P158e*)
$[(C_4H_9)_3SnSCH_2C{\equiv}CCH_2C(O)OCH_2CH_2O]_2[Si(C_4H_9)(OC_2H_5)]$	(*P158e*)
$\{(C_4H_9)Sn[SCH_2C{\equiv}CCH_2C(O)OCH_2CH_2O]_3\}_2[Si(C_4H_9)(OC_2H_5)]_3$	(*P158e*)
$\{(C_4H_9)Sn[S{-}cyclo{-}C_6H_{10}C(O)OCH_2CH_2O]_3\}[Si(OC_2H_5)]$	(*P158e*)
$\{(C_4H_9)Sn[SHC(CH{=}CH)(CH_2{-}CH_2)CHC(O)OCH_2CH_2O]_3\}[Si(C_4H_9)]$ (ring: HC=CH / H_2C-CH_2)	(*P158e*)
$\{(C_4H_9)_3Sn[S{-}p{-}C_6H_4{-}C(O)OCH_2CH_2O]_3\}[Si(C_4H_9)]$	(*P158e*)
$\{(C_4H_9)_2Sn[S{-}C_6H_3(CH_3){-}C(O)OCH_2CH_2O]_2\}_3[Si(C_4H_9)]_2$	(*P158e*)
$[(C_4H_9)_3SnS{-}C_{10}H_6{-}C(O)OCH_2CH_2O][Si(C_4H_9)(OC_2H_5)_2]$	(*P158e*)
$[(C_4H_9)_3SnSCH_2C(O)OCH_2CH_2O][Si(C_4H_9)_3]$	(*P158e*)

(*continued*)

APPENDIX 10 (*continued*)

MISCELLANEOUS ORGANOTIN SULFIDES

Compound	mp(°C)	bp(°C/mmHg)	n_D^{20}	d_4^{20}	References
$\{(C_4H_9)Sn[SCH_2C(O)OCH_2CH_2O]_3\}[Si(OC_2H_5)_3]_3$					(*P158e*)
$\{(C_4H_9)Sn[SCH_2C(O)OCH_2CH_2O]_3\}_4[Si]_3$					(*P158e*)
$[(C_4H_9)_3SnSCH_2C(O)OCH_2CH_2O]_4Si$					(*P158e*)
$\{(C_4H_9)_2Sn[SCH_2C(O)OCH_2CH_2O]_2\}_2Si$					(*P158e*)
$[(C_4H_9)_3SnSCH_2C(O)OCH_2CH_2O][Si(C_4H_9)_2]O—o\text{-}C_6H_4—CH_3)]$					(*P158e*)
$[(C_4H_9)_3SnSCH_2C(O)OCH_2CH_2O][Si(C_4H_9)_2(OC_5H_5)]$					(*P158e*)
$[(C_4H_9)_3SnSCH_2C(O)O(CH_2)_{18}O]_3[Si(OC_5H_5)]$					(*P158e*)
$[(C_4H_9)_3SnSCH_2C(O)OCH_2CH_2O]_3\{Si[O(CH_2)_{17}CH_3]\}$					(*P158e*)
$\{(C_4H_9)Sn[SCH_2C(O)OCH_2CH_2O]_3\}_2[Si(C_4H_9)(OC_{18}H_{37})]_3$					(*P158e*)
$\{(C_4H_9)Sn[SCH_2C(O)OCH_2CH_2O]_3\}_2[Si(C_{18}H_{37})(OCH_2C{\equiv}CH)]_3$					(*P158e*)
$\{(C_4H_9)Sn[SCH_2C(O)OCH_2CH_2O]_3\}_2[Si(C_4H_9)(O—cyclo\text{-}C_6H_{11})]_3$					(*P158e*)
$\{(C_4H_9)Sn[SCH_2C(O)OCH_2CH_2O]_3\}_4[Si]_3$					(*P158e*)
$\{(C_4H_9)_2Sn[SCH_2C(O)OCH_2CH_2O]_2\}[Si(C_4H_9)(OC_2H_5)_2]_2$					(*P158e*)
$[(C_4H_9)_3SnSCH_2C(O)OCH_2CH_2O]_3[Si(C_4H_9)]$					(*P158e*)

APPENDIX 11

ORGANOTIN SELENIDES

Compound	mp(°C)	bp(°C/mmHg)	n_D^{20}	d_4^{20}	References
$(CH_3)_3Sn—Se—CH_3$					*(8)*
$(CH_3)_3Sn—Se—C_6H_5$		67–9/0.001	1.6119	1.65	*(3)*
$(C_2H_5)_3Sn—Se—CH_2—C_6H_5$		115–7/0.0001	1.5888		*(138)*
$(C_6H_5)_3Sn—Se—C_6H_5$	87–8*(195)*				*(195, 203)*
$(CH_3)_2Sn(—Se—C_6H_5)_2$					*(15)*
$Sn(—Se\text{-}t\text{-}C_4H_9)_4$	190(decomp.)*(26)*				*(26, 27)*
$Sn(—Se—C_6H_5)_4$	83–83.5				*(26)*
$Sn(—Se—C_6H_4—CH_3)_4$	119				*(26)*
$Sn[—Se—C_6H_4—C(CH_3)_3]_4$	132				*(26)*
$Sn(—Se—C_6H_4—Cl)_4$	201.5				*(26)*
$(CH_3)_3Sn—Se—Sn(CH_3)_3$	–6*(174)*	118/15*(108, 109)* 72/1*(174)* 115–6/0.5*(227)*			*(108, 109, 109a, 174, 188, 203, 227)*
$(C_2H_5)_3Sn—Se—Sn(C_2H_5)_3$		138–40/1.5*(225)*	1.5652*(225)*	1.5710*(225)*	*(36b, 64c, 65, 225, 227a, 227b)*
$(C_4H_9)_3Sn—Se—Sn(C_4H_9)_3$		160/0.0001	1.526		*(139)*
$(C_6H_5)_3Sn—Se—Sn(C_6H_5)_3$	148*(192, 193, 203, 205, 208)*				*(63, 192, 193, 203, 204, 205, 208)*

(continued)

APPENDIX 11 (*continued*)

Organotin Selenides

Compound	mp(°C)	bp(°C/mmHg)	n_D^{20}	d_4^{20}	References
$(C_2H_5)_3Sn—Se—Ge(C_2H_5)_3$		111/0.5	1.5470(*226*)		(*64c*, *226*, *227*)
$(C_6H_5)_3Sn—Se—Ge(C_6H_5)_3$	145(*203*) 133(*205*) 144(*208*) 144–5(*209*)				(*63*, *203–205*, *208*, *209*)
$(C_6H_5)_3Sn—Se—Pb(C_6H_5)_3$	138(*203*, *210*)				(*203*, *204*, *210*)
$(C_6H_5)_2Sn[—Se—Sn(C_6H_5)_3]_2$					(*208*)
H_3C CH_3 Sn Se Se H_3C Sn Sn CH_3 H_3C Se CH_3	120(*108*) 119(*192*, *193*)				(*108*, *109a*, *192*, *193*, *P122*, *P130*)
$[(C_2H_5)_2SnSe]_3$		169–74/18			(*109a*)
C_4H_9 C_4H_9 Sn C_4H_9 Se Se C_4H_9 Sn Sn C_4H_9 Se C_4H_9					(*195*, *203*)
C_6H_5 C_6H_5 Sn Se Se C_6H_5 Sn Sn C_6H_5 C_6H_5 Se C_6H_5	176–7(*208*)				(*203*, *208*)

$[(CH_3Sn)_2Se_3]_n$		(*108*, *218*)
$\left[-Se-Sn(C_2H_5)-Se-Sn(C_2H_5)-Se-\right]_n$		(*218*)
$(C_6H_5)_3Sn-Se-Li$		(*203*, *205*, *208*)
$(C_6H_5)_3Sn-Se-Ag$		(*63*)
$(CH_3)_2Sn[S_2C{=}N(CH_3)_2]_2$ (S, Se chelate: $(CH_3)_2Sn\left[\begin{smallmatrix}S\\Se\end{smallmatrix}C{=}N(CH_3)_2\right]_2$)	195	(*98a*)
$(CH_3)_2ClSn\left(\begin{smallmatrix}S\\Se\end{smallmatrix}C{=}N(CH_3)_2\right)$	126–8	(*98a*)
$[(CH_3)_3Sn]_2Se-Cr(CO)_5$	84(decomp.)	(*210a*)
$[(CH_3)_3Sn]_2Se-Mo(CO)_5$	86(decomp.)	(*210a*)
$[(CH_3)_3Sn]_2Se-W(CO)_5$	107(decomp.)	(*210a*)

APPENDIX 12

ORGANOTIN TELLURIDES

Compound	mp(°C)	bp(°C/mmHg)	n_D^{20}	d_4^{20}	References
$(CH_3)_3Sn—Te—Sn(CH_3)_3$		103/0.1			(*210a*)
$(C_2H_5)_3Sn—Te—Sn(C_2H_5)_3$		144–5/1.5(*225*)	1.5950(*225*, *227c*)	1.668(*227c*)	(*36b*, *64c*, *65*, *225*, *226*, *227a*, *227b*, *227c*, *227g*)
		119–21/1(*226*, *227c*)	1.5972(*226*, *227c*)		
		134–5/1.5(*227c*)	1.5982(*227g*)		
		131–4/1.5(*227g*)			
$(C_6H_5)_3Sn—Te—Sn(C_6H_5)_3$	150(*203*)				(*203–205*, *208*)
	148(*205*)				
	149–50(*208*)				
$(C_2H_5)_3Sn—Te—Si(C_2H_5)_3$		109–12/1(*226*, *227c*)	1.5680(*226*, *227c*)		(*64a*, *64c*, *65*, *226*, *227c*)
$(C_2H_5)_3Sn—Te—Ge(C_2H_5)_3$		126–8/1(*226*, *227c*)	1.5723(*226*, *227c*)		(*64c*, *65*, *226*, *227c*)
$(C_6H_5)_3Sn—Te—Ge(C_6H_5)_3$	145(*203*)				(*203*, *204*, *209*)
	142–6(*209*)				
$(C_6H_5)_3Sn—Te—Pb(C_6H_5)_3$	136(decomp.) (*203*, *208*)				(*203*, *204*, *208*)
$(C_6H_5)_3Sn—Te—Li$					(*203*, *205*, *208*)
$[(C_4H_9)_2SnTe]_3$ (six-membered ring: three $Sn(C_4H_9)_2$ units alternating with three Te)					(*P122*, *P130*)
$[(CH_3)_3Sn]_2Te—Cr(CO)_5$	73(decomp.)				(*210a*)
$[(CH_3)_3Sn]_2Te—W(CO)_5$	71(decomp.)				(*210a*)

REFERENCES

1. E. W. Abel and D. A. Armitage, "Organosulfur Derivatives of Si, Ge, Sn and Pb," in *Advances in Organometallic Chemistry* (F. G. A. Stone and R. West, eds.), Vol. 5, Academic Press, New York, 1967.
2. E. W. Abel, D. A. Armitage, and D. B. Brady, *Trans. Faraday Soc.*, **62**, 3459 (1966).
3. E. W. Abel, D. A. Armitage, and D. B. Brady, *J. Organometal. Chem.*, **5**, 130 (1966).
4. E. W. Abel, D. A. Armitage, and R. P. Bush, *J. Chem. Soc.*, 7098 (1965).
5. E. W. Abel, D. A. Armitage, and S. P. Tyfield, *J. Chem. Soc. A*, 554 (1967).
6. E. W. Abel, A. M. Atkins, B. C. Crosse, and G. V. Hutson, *J. Chem. Soc. A*, 687 (1968).
7. E. W. Abel and D. B. Brady, *J. Chem. Soc.*, 1192 (1965).
8. E. W. Abel and D. B. Brady, *J. Organometal. Chem.*, **11**, 145 (1968).
9. E. W. Abel, D. B. Brady, and B. C. Crosse, *J. Organometal. Chem.*, **5**, 260 (1966).
10. E. W. Abel, R. P. Bush, C. R. Jenkins, and T. Zobel, *Trans. Faraday Soc.*, **60**, 1214 (1964).
11. E. W. Abel and B. C. Crosse, *J. Chem. Soc. A*, 1377 (1966).
12. E. W. Abel and B. C. Crosse, *J. Chem. Soc. A*, 1141 (1966).
13. E. W. Abel and B. C. Crosse, *Organometal. Chem. Rev.*, **2**, 443 (1967).
14. E. W. Abel, B. C. Crosse, and D. B. Brady, *J. Am. Chem. Soc.*, **87**, 4397 (1965).
15. E. W. Abel, B. C. Crosse, and G. V. Hutson, *Chem. Ind. London*, 238 (1966).
16. E. W. Abel, B. C. Crosse, and G. V. Hutson, *J. Chem. Soc. A*, 2014 (1967).
17. E. W. Abel, B. C. Crosse, and T. J. Leedham, unpublished.
17a. E. W. Abel, J. P. Crow, and J. N. Wingfield, *Chem. Commun.*, 967 (1969).
17b. E. W. Abel and S. M. Illingworth, *J. Chem. Soc.*, A1094 (1969).
18. E. W. Abel and C. R. Jenkins, *J. Chem. Soc. A*, 1344 (1967).
19. E. W. Abel and C. R. Jenkins, unpublished.
20. A. Y. Aleksandrov, N. N. Delyagin, K. P. Mitrofanov, L. S. Polak, and K. S. Spinel, *Zhur. Eksp. Teor. Fiz.*, **43**, 1242 (1962); through *CA*, **58**, 7554 (1963).
21. A. Y. Aleksandrov, N. N. Delyagin, K. P. Mitrofanov, L. S. Polak, and K. S. Spinel, *Soviet Phys. JETP*, **16**, 879 (1963).
22. H. H. Anderson, *J. Org. Chem.*, **19**, 1766 (1954).
23. H. H. Anderson, *J. Am. Chem. Soc.*, **79**, 4913 (1957).
24. H. H. Anderson and J. A. Vasta, *J. Org. Chem.*, **19**, 1300 (1954).
24a. K. N. Anisimov, B. V. Lokshin, N. E. Kolobova, and V. V. Skripkin, *Izvest. Akad. Nauk S.S.S.R., Ser. Khim.*, 1024 (1968); through *CA*, **69**, 47717 (1968).
24b. A. Apsitis and E. Jansons, *Latv. PSR Zinat. Akad. Vestis, Kim. Ser.*, 400 (1968); through *CA*, **70**, 43679 (1969).
25. H. J. Backer and W. Drenth, *Rec. trav. chim.*, **70**, 559 (1951).
26. H. J. Backer and J. B. G. Hurenkamp, *Rec. trav. chim.*, **61**, 803 (1942).
27. H. J. Backer and H. A. Klasens, *Rec. trav. chim.*, **61**, 500 (1942).
28. H. J. Backer and J. Kramer, *Rec. trav. chim.*, **52**, 916 (1933).
29. H. J. Backer and J. Kramer, *Rec. trav. chim.*, **53**, 1101 (1934).
30. H. J. Backer and F. Stienstra, *Rec. trav. chim.*, **54**, 607 (1935).
30a. P. Bamberg, B. Ekstroem, and B. Sjoeberg, *Acta Chem. Scand.*, **22**, 367 (1968).
31. G. A. Baum and W. J. Considine, *J. Polymer Sci.*, **B1**, 517 (1963); through *CA*, **59**, 14117 (1963).
32. D. Blake, G. E. Coates, and J. M. Tate, *J. Chem. Soc.*, 618 (1961).
33. A. J. Bloodworth and A. G. Davies, *Proc. Chem. Soc. London*, 264 (1963).

34. A. J. Bloodworth and A. G. Davies, *Proc. Chem. Soc. London*, 315 (1963).
35. A. J. Bloodworth and A. G. Davies, *J. Chem. Soc.*, 5238 (1965).
36. A. J. Bloodworth, A. G. Davies, and S. C. Vasishtha, *J. Chem. Soc. C*, 1309 (1967).
36a. A. J. Bloodworth, A. G. Davies, and S. C. Vasishtha, *J. Chem. Soc. C*, 2640 (1968).
36b. M. N. Bochkarev, L. P. Sanina, and N. S. Vyazankin, *Zhur. Obshchei. Khim.*, **39**, 135 (1969); through *CA*, **70**, 96876 (1969).
37. F. Bonati, S. Cenini, and R. Ugo, *Inst. Lombardo Rend. Sci.*, **A99**, 825 (1965).
38. F. Bonati, S. Cenini, and R. Ugo, *J. Organometal. Chem.*, **9**, 395 (1967).
39. F. Bonati and R. Ugo, *J. Organometal. Chem.*, **10**, 257 (1967).
39a. F. Bonati, G. Minghetti, and S. Cenini, *Inorg. Chim. Acta*, **2**, 375 (1968).
39b. V. A. Bork and P. I. Selivokhin, *Plasticheskie Massy*, 56 (1968); through *CA*, **69**, 1006 (1968).
40. R. W. Bost and P. Borgstrom, *J. Am. Chem. Soc.*, **51**, 1922 (1929).
40a. D. C. Bradley and M. H. Gitlitz, *J. Chem. Soc.*, A1152 (1969).
40b. C. A. Brighton, *Plast. Polym.*, **36**, 549 (1968); through *CA*, **70**, 97679 (1969).
41. H. P. Brown and J. A. Austin, *J. Am. Chem. Soc.*, **62**, 673 (1940).
42. M. P. Brown, R. Okawara, and E. G. Rochow, *Spectrochim. Acta*, **16**, 595 (1960).
43. A. B. Bruker, L. D. Balashova, and L. Z. Soborovskii, *Zhur. Obshchei. Khim.*, **36**, 75 (1966); through *CA*, **64**, 14211*g* (1966).
44. J. J. Burke and P. C. Lauterbur, *J. Am. Chem. Soc.*, **83**, 326 (1961).
44a. V. Chromy and L. Srp, *Chem. Listy.*, **61**, 1509 (1967); through *CA*, **69**, 1503 (1968).
44b. N. A. D. Carey and H. C. Clark, *Can. J. Chem.*, **46**, 643 (1968).
44c. H. G. Carr, *Soc. Plast. Eng. J.* **25**, 72 (1969); through *CA*, **72**, 41901 (1970).
45. M. P. Claesson, *Bull. Soc. Chim. France*, **25**, 183 (1876).
46. G. E. Coates, M. L. H. Greene, and K. Wade, *Organometallic Compounds*, 3rd Ed. Vol. I, 1967.
47. W. J. Considine, J. J. Ventura, A. J. Gibbons, and A. Ross, *Can. J. Chem.*, **41**, 1239 (1963); through *CA*, **58**, 13975 *f* (1963).
48. H. M. J. C. Creemers, Doct. Diss., Utrecht Univ., 1967.
49. H. M. J. C. Creemers and J. G. Noltes, *Rec. trav. chim.*, **84**, 1589 (1965).
50. H. M. J. C. Creemers, F. Verbeek, and J. G. Noltes, *J. Organometal. Chem.*, **8**, 469 (1967).
51. R. A. Cummins, *Australian J. Chem.*, **16**, 985 (1963); through *CA*, **60**, 5304 (1964).
52. R. A. Cummins and P. Dunn, *Australia Commonwealth Dept. Supply Defence Standards Lab.*, **266**, 106 (1963); through *CA*, **60**, 11503 (1964).
53. C. W. N. Cumper, A. Melnikoff, and A. I. Vogel, *J. Chem. Soc. A*, 242 (1966).
54. C. W. N. Cumper, A. Melnikoff, and A. I. Vogel, *J. Chem. Soc. A*, 246 (1966).
54a. S. B. Damle and W. J. Considine, *J. Organometal. Chem.*, **19**, 207 (1969).
55. O. Danek, *Coll. Czech. Chem. Commun.*, **26**, 2035 (1961); through *CA*, **56**, 4788 (1962).
56. W. E. Davidson, K. Hills, and M. C. Henry, *J. Organometal. Chem.*, **3**, 285 (1965).
57. A. G. Davies, *Trans. N.Y. Acad. Sci.*, **26**, 923 (1964).
57a. A. G. Davies, *Chemistry in Britain*, **4**, 403 (1968); through *CA*, **69**, 105697 (1968).
58. A. G. Davies and P. G. Harrison, *J. Chem. Soc. C*, 298 (1967).
59. A. G. Davies and P. G. Harrison, *J. Chem. Soc. C*, 1313 (1967).
60. A. G. Davies and P. G. Harrison, *J. Organometal. Chem.*, **8**, P19 (1967).
60a. A. G. Davies, P. G. Harrison, J. D. Kennedy, T. N. Mitchell, and R. J. Puddephatt, *J. Chem. Soc.*, A1136 (1969).
60b. A. G. Davies and J. D. Kennedy, *J. Chem. Soc.*, *C*, 759 (1970).
61. A. G. Davies, T. N. Mitchell, and W. R. Symes, *J. Chem. Soc. C*, 1311 (1966).
62. A. G. Davies and G. J. D. Peddle, *Chem. Commun.*, 96 (1965).

63. R. E. Dessy, W. Kitching, and T. Chivers, *J. Am. Chem. Soc.*, **88**, 453 (1966).
63a. C. Dörfelt, A. Janeck, D. Kobelt, E. F. Paulus, and H. Scherer, *J. Organometal. Chem.*, **14**, P22 (1968).
63b. M. Devaud, *Rev. Chim. Miner.*, **4**, 921 (1967); through *CA*, **69**, 87122 (1968).
63c. M. Donadille, M. A. Delmas, and J. C. Maire, *J. Organometal. Chem.*, **15**, 224 (1968).
64. M. Dub, *Organometallic Compounds*, 2nd Edn., Vol. II, 1967.
64a. A. N. Egorochkin, S. Y. Khorshev, N. S. Vyazankin, M. N. Bochkarev, O. A. Kruglaya, and G. S. Semchikova, *Zhur. Obshchei. Khim.*, **37**, 2308 (1967); through *CA*, **68**, 8823 (1968).
64b. A. N. Egorochkin, S. Y. Khorshev, N. S. Vyazankin, E. N. Gladyshev, V. T. Bychkov. and O. A. Kruglaya, *Zhur. Obshchei. Khim.*, **38**, 276 (1968).
64c. A. N. Egorochkin, N. S. Vyazankin, M. N. Bochkarev, and S. Y. Khorshev, *Zhur. Obshchei. Khim.*, **37**, 1165 (1967); through *CA*, **68**, 2805 (1968).
65. A. N. Egorochkin, N. S. Vyazankin, G. A. Razuvaev, O. A. Kruglaya, and M. N. Bochkarev, *Dokl. Akad. Nauk S.S.S.R.*, **170**, 333 (1966); through *CA*, **66**, 4780 (1967).
66. L. M. Epstein, *U.S. Atomic Energy Comm.*, WERL-2989-1 (1965); through *CA*, **63**, 12501 (1965).
67. L. M. Epstein and D. K. Straub, *Inorg. Chem.*, **4**, 1551 (1965).
68. I. T. Eskin, A. N. Nesmeyanov, and K. A. Kocheshkov, *J. Gen. Chem. U.S.S.R.*, **8**, 35 (1938); through *CA*, **32**, 5386 (1938).
69. A. Finch, R. C. Poller, and D. Steele, *Trans. Faraday Soc.*, **61**, 2628 (1965).
70. F. H. Finck, J. A. Turner, and D. A. Payne, *J. Am. Chem. Soc.*, **88**, 1571 (1966).
70a. B. W. Fitzsimmons, *Chem. Commun.*, 1485 (1968).
71. J. A. Forstner and E. L. Muetterties, *Inorg. Chem.*, **5**, 552 (1966).
72. A. H. Frye and R. W. Horst, *Int. J. Appl. Radiation Isotopes*, **15**, 169 (1964); through *CA*, **61**, 4189 (1964).
73. A. H. Frye, R. W. Horst, and M. A. Paliobagis, *Am. Chem. Soc., Div. Polymer Chem., Preprints*, **4**, 260 (1963); through *CA*, **62**, 698 (1965).
74. A. H. Frye, R. W. Horst, and M. A. Paliobagis, *J. Polymer Sci.*, **A2**, 1765 (1964); through *CA*, **60**, 16056 (1964).
75. A. H. Frye, R. W. Horst, and M. A. Paliobagis, *J. Polymer Sci.*, **A2**, 1785 (1964); through *CA*, **60**, 16056 (1964).
76. A. H. Frye, R. W. Horst, and M. A. Paliobagis, *J. Polymer Sci.*, **A2**, 1801 (1964); through *CA*, **60**, 16056 (1964).
77. R. Garzuly, *Organometalle* (1927).
78. H. Geissler and H. Kriegsmann, *J. Organometal. Chem.*, **11**, 85 (1968).
79. T. A. George, K. Jones, and M. F. Lappert, *J. Chem. Soc.*, 2157 (1965).
79a. T. A. George and M. F. Lappert, *J. Organometal. Chem.* **14**, 327 (1968).
79b. T. A. George and M. F. Lappert, *J. Chem. Soc.*, A992 (1969).
80. R. Geyer and H. J. Seidlitz, *Z. Chem.*, **7**, 114 (1967); through *CA*, **66**, 10383 (1967).
81. M. Gielen and N. Sprecher, *Organometal. Chem. Rev.*, **1**, 455 (1966).
82. K. Gingold, E. G. Rochow, D. Seyferth, A. C. Smith, and R. West, *J. Am. Chem. Soc.*, **74**, 6306 (1952).
83. F. Gliniecki, Dissertation, Univ. Marburg, 1964.
83a. I. P. Goldstein, E. N. Guryanova, N. N. Zemlyanskii, O. P. Syutkina, E. M. Panov, and K. A. Kocheshkov, *Izvest. Akad. Nauk S.S.S.R., Ser. Khim.*, 2201 (1967); through *CA*, **68**, 2442 (1968).
83b. I. P. Goldstein, E. N. Guryanova, N. N. Zemlyanskii, O. P. Syutkina, E. M. Panov, and K. A. Kocheshkov, *Dokl. Akad. Nauk S.S.S.R.*, **175**, 836 (1967); through *CA*, **68**, 7476 (1968).

83c. V. I. Goldanskii, V. V. Khrapov, and R. A. Stukan, *Organometal. Chem. Rev. A*, **4**, 225 (1969).

83d. V. I. Goldanskii, E. F. Makarov, R. A. Stukan, T. N. Sumakarova, V. A. Trukhtanov, and V. V. Khrapov, *Dokl. Phys. Chem.*, **156**, 474 (1964).

83e. R. Gould, *Stabilization of Polymers and Stabilizer Processes*, Advances in Chemistry, No. 85, Amer. Chem. Soc., Washington, D.C., 1968.

84. T. Harada, *Bull. Chem. Soc. Japan*, **17**, 281 (1942); through *CA*, **41**, 4444 (1947).

85. T. Harada, *Bull. Chem. Soc. Japan*, **17**, 283 (1942); through *CA*, **41**, 4444 (1947).

86. T. Harada, *Rep. Sci. Res. Inst. Japan*, **24**, 177 (1948); through *CA*, **45**, 2356 (1951).

87. T. Harada and T. Okubo, *Sci. Papers Inst. Phys. Chem. Res. Tokyo*, **42**, 59 (1947); through *CA*, **43**, 7900 (1949).

87a. P. G. Harrison, *Organometal. Chem. Rev.*, **A4**, 379 (1969).

88. M. C. Henry and W. E. Davidson, *Can. J. Chem.*, **41**, 1276 (1963); through *CA*, **58**, 13975 (1963).

89. R. H. Herber, H. A. Stöckler, and W. T. Reichle, *J. Chem. Phys.*, **42**, 2447 (1965); through *CA*, **62**, 12628 (1965).

89a. S. Hirotoshi, *Enka Biniiru To Porima*, **10**, 23 (1970); through *CA*, **72**, 79496 (1970).

90. H. E. Hirschland and C. K. Banks, *Adv. Chem. Ser.*, **23**, 204 (1959); through *CA*, **54**, 4347 (1960).

90a. M. G. Hogben, R. S. Gay, A. J. Oliver, J. A. J. Thompson, and W. A. G. Graham, *J. Am. Chem. Soc.*, **91**, 291 (1969).

90b. S. Homrowski, *Rocz. Panstw. Zakl. Hig.*, **19**, 329 (1968); through *CA*, **69**, 85100 (1968).

90c. M. Honda, M. Komura, Y. Kawasaki, T. Tanaka, and R. Okawara, *J. Inorg. Nucl. Chem.*, **30**, 3231 (1968).

91. M. Honda, Y. Kawasaki, and T. Tanaka, *Tetrahedron Letters*, 3313 (1967).

91a. K. A. Hooton, *Preparative Inorg. Reactions*, **4**, 85 (1968).

92. E. Hoyer, W. Dietzsch, H. Müller, A. Zschunke, and W. Schroth, *Inorg. Nucl. Chem. Letters*, **3**, 457 (1967).

92a. H. Huber and J. Wimmer, *Kunststoffe*, **58**, 786 (1968); through *CA*, **70**, 48074 (1969).

92b. B. K. Hunter and L. W. Reeves, *Can. J. Chem.*, **46**, 1399 (1968).

92c. Y. Ishii and K. Itoh, *Asahi Garasu Kogyo Gijutsu Shoreikai Kenkyu Hokoku*, **14**, 39 (1968); through *CA*, 72, 43812 (1970).

93. R. K. Ingham, S. D. Rosenberg, and H. Gilman, *Chem. Rev.*, **60**, 459 (1960).

93a. K. Itoh, Y. Fukumoto, and Y. Ishii, *Tetrahedron Letters.*, 3199 (1968).

93b. K. Itoh, Y. Kato, and Y. Ishii, *J. Org. Chem.*, **34**, 459 (1969).

94. K. Itoh, I. K. Lee, I. Matsuda, S. Sakai, and Y. Ishii, *Tetrahedron Letters*, 2667 (1967).

94a. K. Itoh, K. Matsuzaki, and Y. Ishii, *J. Chem. Soc. C*, 2709 (1968).

94b. K. Itoh, I. Matsuda, and Y. Ishii, *Tetrahedron Letters*, 2675 (1969).

94c. K. Itoh, I. Matsuda, T. Katsuura, and Y. Ishii, *J. Organometal. Chem.*, **19**, 347 (1969).

95. K. Itoh, S. Sakai, and Y. Shii, *Yuki Gosei Kagaku Kyokai Shi*, **24**, 729 (1966); through *CA*, **65**, 16998 (1966).

96. W. Jasching, *Kunststoffe*, **52**, 458 (1962); through *CA*, **57**, 15349 (1962).

96a. C. R. Jenkins, *J. Organometal. Chem.*, **15**, 441 (1968).

96b. A. D. Jenkins, M. F. Lappert, and R. C. Srivastava, *J. Organometal. Chem.*, **23**, 165 (1970).

97. K. Jones and M. F. Lappert, *Proc. Chem. Soc. London*, 358 (1962).

98. K. Jones and M. F. Lappert, *Organometal. Chem. Rev.*, **1**, 67 (1966).

98a. T. Kamitani and T. Tanaka, *Inorg. Nucl. Chem. Letters*, **6**, 91 (1970).

99. H. A. Klasens and H. J. Backer, *Rec. trav. chim.*, **58**, 941 (1939).

99a. O. R. Klimmer, *Arzneim. Forsch.*, **19**, 934 (1969); through *CA*, **71**, 79257 (1969).

99b. P. Klimsch and P. Kuehnert, *Plaste Kaut.*, **16**, 242 (1969); through *CA*, **70**, 107023 (1969).

100. K. A. Kocheshkov, *Ber.*, **66**, 1661 (1933).

100a. K. A. Kocheshkov, N. N. Zemlyanskii, and N. J. Sheverdina, in *Methods of Elemento-organic Chemistry, Germanium, Tin and Lead* (A. N. Nesmejanov and K. A. Kocheskov, eds.) (1968).

101. K. A. Kocheshkov and M. M. Nad, *J. Gen. Chem., U.S.S.R.*, **4**, 1434 (1934); through *CA*, **29**, 3660 (1935).

102. K. A. Kocheshkov and M. M. Nad, *Ber.*, **67**, 717 (1934).

103. K. A. Kocheshkov and M. M. Nad, *Zhur. Obshchei. Khim.*, **5**, 1158 (1935); through *CA*, **30**, 1036 (1936).

104. K. A. Kocheshkov and A. N. Nesmeyanov, *Ber.*, **64**, 628 (1931).

104a. D. A. Kochkin and I. N. Azerbaev, "Olovo-i Svinets-Organicheskii Monomery i Polimery" (1968); through *CA*, **70**, 20211 (1969).

105. M. Komura and R. Okawara, *Inorg. Nucl. Chem. Letters*, **2**, 93 (1966).

105a. E. Kostiner, M. L. N. Reddy, D. S. Urch, and A. G. Massey, *J. Organometal. Chem.*, **15**, 383 (1968).

105b. V. Kotkhekar and V. S. Shpinel, *Zhur. Strukt. Khim.*, **10**, 37 (1969); through *CA*, **70**, 110326 (1969).

106. C. A. Kraus and W. V. Sessions, *J. Am. Chem. Soc.*, **47**, 2361 (1925).

107. E. Krause and A. von Grosse, *Die Chemie der metallorganischen Verbindungen*, 1937.

107a. D. N. Kravtsov, E. M. Rokhlina, and A. N. Nesmeyanov, *Izvest. Akad. Nauk S.S.S.R., Ser. Khim.*, 1035 (1968).

108. H. Kriegsmann and H. Hoffmann, *Z. Chem.*, **3**, 268 (1963).

109. H. Kriegsmann, H. Hoffmann, and H. Geissler, *Z. Anorg. Allgem. Chem.*, **341**, 24 (1965).

109a. H. Kriegsmann, H. Hoffmann, and H. Geissler, *Z. Anorg. Allgem. Chem.*, **359**, 58 (1968).

110. H. Kubo, *Agr. Biol. Chem. Tokyo*, **29**, 43 (1965); through *CA*, **63**, 7032 (1965).

110a. W. Kuchen and H. Hertel, *Angew. Chem.*, **81**, 127 (1969).

111. W. Kuchen, A. Judat, and J. Metten, *Chem. Ber.*, **98**, 3981 (1965).

112. H. G. Kuivila and O. F. Beumel, *J. Am. Chem. Soc.*, **80**, 3250 (1958).

113. H. G. Kuivila and E. R. Jakusik, *J. Org. Chem.*, **26**, 1430 (1961).

114. P. Kulmitz, *Jahresber.*, 375 (1860).

115. P. Kulmitz, *J. prakt. Chem.*, **80**, 60 (1860).

116. E. J. Kupchik and P. J. Calabretta, *Inorg. Chem.*, **3**, 905 (1964).

117. E. J. Kupchik and P. J. Calabretta, *Inorg. Chem.*, **4**, 973 (1965).

117a. E. J. Kupchik and E. F. McInerney, *J. Organometal. Chem.*, **11**, 291 (1968).

117b. E. J. Kupchik and C. T. Theisen, *J. Organometal. Chem.*, **11**, 627 (1968).

118. B. R. Laliberte, W. Davidson, and M. C. Henry, *J. Organometal. Chem.*, **5**, 526 (1966).

118a. G. A. Lapitskii, L. S. Granenkina, P. S. Khokhlov, and N. K. Bliznyuk, *Zhur. Obshchei. Khim.*, **38**, 2787 (1968); through *CA*, **70**, 78100 (1969).

119. M. F. Lappert and B. Prokai, "Insertion Reactions of Compounds of Metals and Metalloids Involving Unsaturated Substrates" in *Advances in Organometallic Chemistry*, **5** (1967).

120. A. J. Leusink, Diss., Utrecht Univ., 1966.

121. A. J. Leusink, H. A. Budding, and J. G. Noltes, *Rec. trav. chim.*, **85**, 151 (1966).

122. D. H. Lohmann, *J. Organometal. Chem.*, **4**, 382 (1965).

123. D. H. Lorenz and E. I. Becker, *J. Org. Chem.*, **28**, 1707 (1963).

123a. C. R. Lucas and M. E. Peach, *Inorg. Nucl. Chem. Letters*, **5**, 73 (1969).

124. J. G. A. Luijten and S. Pezarro, *Brit. Plast.*, **30**, 183 (1957); through *CA*, **51**, 10947 (1957).

125. J. G. A. Luijten and G. J. M. van der Kerk, *A Survey of the Chemistry and Applications of Organotin Compounds*, Tin Research Institute, London, 1952.

126. J. G. A. Luijten and G. J. M. van der Kerk, *Investigations in the Field of Organotin Chemistry*, Tin Research Institute, London, 1955.

126a. J. G. A. Luijten, *Organometal. Chem. Rev. B*, **5**, 687 (1969).

127. S. Matsuda and S. Kikkawa, *Yuki Gosei Kagaku Kyokai Shi*, **24**, 281 (1966); through *CA*, **64**, 17630 (1966).

128. R. C. Mehrotra, V. D. Gupta, and D. Sukhani, *J. Inorg. Nucl. Chem.*, **29**, 1577 (1967).

128a. R. C. Mehrotra, V. D. Gupta, and D. Sukhani, *Inorg. Chim. Acta Rev.*, **2**, 111 (1968).

128b. R. C. Mehrotra, V. D. Gupta, and D. Sukhani, *Indian J. Chem.*, **7**, 708 (1969); through *CA*, **71**, 81488 (1969).

129. S. Migdal, D. Gertner, and A. Zilka, *Can. J. Chem.*, **45**, 2987 (1967).

130. J. Miskowiec, *Polimery*, **7**, 255 (1962); through *CA*, **58**, 11528 (1963).

131. K. Moedritzer, *Organometal. Chem. Rev.*, **1**, 179 (1966).

132. K. Moedritzer and J. R. Van Wazer, *Inorg. Chem.*, **3**, 943 (1964).

133. H. Muecke, *Monatsber. Deut. Akad. Wiss. Berlin*, **3**, 668 (1961); through *CA*, **57**, 15350 (1962).

134. M. M. Nad and K. A. Kocheshkov, *Zhur. Obshchei. Khim.*, **8**, 42 (1938); through *CA*, **32**, 5387 (1938).

134a. M. Nakatani, Y. Takahashi, and A. Ouchi, *J. Inorg. Nucl. Chem.*, **31**, 3330 (1969).

135. A. N. Nesmeyanov, K. N. Anisimov, N. E. Kolobova, and V. V. Skripkin, *Izvest. Akad. Nauk S.S.S.R., Ser. Khim.*, 1292 (1966).

135a. A. N. Nesmeyanov, V. I. Goldanskii, V. V. Khrapov, V. Y. Rochev, D. N. Kravtsov, V. M. Pachavskaya, and E. M. Rokhlina, *Dokl. Akad. Nauk S.S.S.R.*, **181**, 921 (1968); through *CA*, **69**, 105684 (1968).

136. A. N. Nesmeyanov and K. A. Kocheshkov, *J. Gen. Chem. U.S.S.R.*, **1**, 219 (1931); through *CA*, **26**, 2182 (1932).

136a. A. N. Nesmeyanov, N. E. Kolobova, M. Y. Zakharova, B. V. Lokshin, and K. N. Anisimov, *Izvest. Akad. Nauk S.S.S.R., Ser. Khim.*, **529** (1969); through *CA*, **71**, 61523 (1969).

137. W. P. Neumann, *Die Organische Chemie des Zinns*, 1967.

138. W. P. Neumann and E. Heymann, *Liebigs Ann. Chem.*, **683**, 11 (1965).

138a. W. P. Neumann, H. Lind, and G. Alester, *Chem. Ber.*, **101**, 2845 (1968).

139. W. P. Neumann, B. Schneider, and R. Sommer, *Liebigs Ann. Chem.*, **692**, 1 (1966).

139a. D. C. Nguen, V. S. Fajnberg, J. I. Baukov, and I. F. Lucenko, *Zhur. Obshchei. Khim.*, **38**, 191 (1968).

140. J. G. Noltes, *Rec. trav. chim.*, **84**, 799 (1965).

141. J. G. Noltes and M. J. Janssen, *Rec. trav. chim.*, **82**, 1055 (1963).

142. J. G. Noltes and M. J. Janssen, *J. Organometal. Chem.*, **1**, 346 (1964).

143. J. G. Noltes and G. J. M. van der Kerk, *Chem. Ind. London*, 294 (1959); through *CA*, **53**, 21757 (1959).

144. M. O'Hara, R. Okawara, and Y. Nakamura, *Bull. Chem. Soc. Japan*, **38**, 1379 (1965); through *CA*, **63**, 18137 (1965).

144a. Y. Oki, *Yuki Gosei Kagaku Kyokai Shi*, **26**, 688 (1968); through *CA*, **69** 78001 (1968).

144b. A. J. Oliver and W. A. G. Graham, *J. Organometal. Chem.*, **19**, 17 (1969).

144c. J. Otera, T. Kadowaki, and R. Okawara, *J. Organometal. Chem.*, **19**, 213 (1969).

144d. R. V. Parish and R. H. Platt, *J. Chem. Soc. A*, 2145 (1969).

145. M. Pang and E. I. Becker, *J. Org. Chem.*, **29**, 1948 (1964).
145a. M. E. Peach, *Can. J. Chem.*, **46**, 211 (1968).
145b. M. E. Peach, *Can. J. Chem.*, **46**, 2699 (1968); through *CA*, **69**, 83092 (1968).
145c. D. Petridis, F. P. Mullins, and C. Curran, *Inorg. Chem.*, **9**, 1270 (1970).
146. P. Pfeiffer and R. Lehnhardt, *Ber.*, **36**, 1054 (1903).
147. P. Pfeiffer and R. Lehnhardt, *Ber.*, **36**, 3027 (1903).
148. P. Pfeiffer, R. Lehnhardt, H. Luftensteiner, R. Prade, K. Schnurmann, and P. Truskier, *Z. anorg. Chem.*, **68**, 102 (1910).
149. E. I. Pikina, T. V. Talalaeva, and K. A. Kocheshkov, *Zhur. Obshchei. Khim.*, **8**, 1844 (1938); through *CA*, **33**, 5839 (1939).
150. J. Pollak, *Monatsh. Chem.*, **34**, 1673 (1913).
151. F. H. Pollard, G. Nickless, and D. J. Cooke, *J. Chromatog.*, **17**, 472 (1965).
152. R. C. Poller, *J. Inorg. Nucl. Chem.*, **24**, 593 (1962).
153. R. C. Poller, *Proc. Chem. Soc. London*, 312 (1963).
154. R. C. Poller, *J. Chem. Soc.*, 706 (1963).
155. R. C. Poller, *J. Organometal. Chem.*, **3**, 321 (1965).
156. R. C. Poller and J. A. Spillman, *J. Chem. Soc. A*, 958 (1966).
157. R. C. Poller and J. A. Spillman, *J. Chem. Soc. A*, 1024 (1966).
158. R. C. Poller and J. A. Spillman, *J. Organometal. Chem.*, **6**, 668 (1966).
159. R. C. Poller and J. A. Spillman, *J. Organometal. Chem.*, **7**, 259 (1967).
160. S. V. Ponomarev and I. F. Lutsenko, *Zhur. Obshchei. Khim.*, **34**, 3450 (1964); through *CA*, **62**, 2787 (1965).
161. V. Potschka, Dissertation, Univ. Marburg, 1964.
161a. P. Powell, *Inorg. Chem.*, **7**, 2458 (1968).
161b. G. A. Razuvaev, O. S. Dyachkovskaya, and V. I. Fionov, *Dokl. Akad. Nauk S.S.S.R.*, **177**, 1113 (1967); through *CA*, **68**, 6678 (1968).
161c. G. A. Razuvaev and N. S. Vyazankin, *Pure Appl. Chem.*, **19**, 353 (1969).
162. W. T. Reichle, *J. Org. Chem.*, **26**, 4634 (1961).
163. W. T. Reichle, *J. Polymer Sci.*, **49**, 521 (1961); through *CA*, **55**, 21013 (1961).
164. W. T. Reichle, *Inorg. Chem.*, **1**, 650 (1962).
165. W. T. Reichle, *Inorg. Chem.*, **5**, 87 (1966).
166. G. H. Reifenberg and W. J. Considine, *J. Organometal. Chem.*, **10**, 279 (1967).
167. A. Rieche, A. Grimm, and H. Mücke, *Kunststoffe*, **52**, 265 (1962).
168. A. Rieche, A. Grimm, and H. Mücke, *Kunststoffe*, **52**, 398 (1962).
168a. H. F. Reiff, B. R. La Liberte, W. E. Davidson, and M. C. Henry, *J. Organometal. Chem.*, **15**, 247 (1968).
169. S. A. Riethmayer, *Kunststoff-Rundschau*, **10**, 277 (1963); through *CA*, **60**, 5706 (1964).
170. S. A. Riethmayer, *Kunststoff-Rundschau*, **10**, 345 (1963); through *CA*, **60**, 5706 (1964).
170a. J. K. Ruff and R. B. King, *Inorg. Chem.*, **8**, 180 (1969).
171. I. Ruidisch, Habilitationsschrift, Univ. Marburg, 1965.
172. I. Ruidisch, H. Schmidbaur, and H. Schumann, "Organoelement Halides of Ge, Sn and Pb" in *Halogen Chemistry* (V. Gutmann, ed.), Vol. II, Academic Press, New York, 1967.
173. I. Ruidisch and M. Schmidt, *Chem. Ber.*, **96**, 1424 (1963).
174. I. Ruidisch and M. Schmidt, *J. Organometal. Chem.*, **1**, 160 (1963).
175. N. A. Rybakova, N. K. Taikova, and E. N. Zilberman, *Trudy Khim. Khim. Tekhnol.*, **2**, 183 (1959); through *CA*, **54**, 10838 (1960).
175a. S. Sakai, Y. Kobayashi, and Y. Ishii, *Chem Commun.*, 235 (1970).
175b. A. Sasaki, *Japan Plast. Age*, **7**, 33 (1969); through *CA*, **71**, 71259 (1969).

175c. A. Sasaki, M. Motoyoshi, Y. Hiramatsu, and H. Hosaka, *Plast. Age* (*Osaka*), 91 (1968); through *CA*, **69**, 78000 (1968).
176. G. S. Sasin, *J. Org. Chem.*, **18**, 1142 (1953).
177. G. S. Sasin and R. Sasin, *J. Org. Chem.*, **20**, 387 (1955).
178. G. S. Sasin, A. L. Borror, and R. Sasin, *J. Org. Chem.*, **23**, 1366 (1958).
179. R. Sasin and G. S. Sasin, *J. Org. Chem.*, **20**, 770 (1955).
180. K. Sasse, R. Wegler, G. Unterstenhöfer, and F. Grewe, *Angew. Chem.*, **72**, 973 (1960).
181. R. Sayre, *J. Chem. Eng. Data*, **6**, 560 (1961); through *CA*, **56**, 8154 (1962).
181a. U. Schöllkopf and N. Rieber, *Angew. Chem.*, **79**, 906 (1967).
181b. H. Schumann, O. Stelzer, and W. Gick, *Angew. Chem.*, **81**, 256 (1969).
182. D. Seyferth, *Naturwissenschaften*, **44**, 34 (1957).
183. D. Seyferth, *J. Am. Chem. Soc.*, **79**, 2133 (1957).
184. D. Seyferth, *J. Am. Chem. Soc.*, **79**, 5881 (1957).
184a. D. Seyferth, *Organometal. Chem. Rev. B*, **4**, 242 (1968).
185. D. Seyferth and R. B. King, *Annual Survey of Organometallic Chemistry*, Vol. I, Elsevier, Amsterdam, 1965.
186. D. Seyferth and R. B. King, *Annual Survey of Organometallic Chemistry*, Vol. II, Elsevier, Amsterdam, 1966.
187. D. Seyferth and R. B. King, *Annual Survey of Organometallic Chemistry*, Vol. III, Elsevier, Amsterdam, 1967.
188. H. Schmidbaur and I. Ruidisch, *Inorg. Chem.*, **3**, 599 (1964).
189. M. Schmidt, *Österr. Chemiker-Ztg.*, **64**, 236 (1963).
190. M. Schmidt, *Pure Appl. Chem.*, **13**, 15 (1966).
191. M. Schmidt, H. J. Dersin, and H. Schumann, *Chem. Ber.*, **95**, 1428 (1962).
191a. M. Schmidt and J. F. Jaggard, *J. Organometal. Chem.*, **17**, 283 (1969).
192. M. Schmidt and H. Ruf, *Angew. Chem.*, **73**, 64 (1961).
193. M. Schmidt and H. Ruf, *Chem. Ber.*, **96**, 784 (1963).
194. M. Schmidt and H. Schumann, *Chem. Ber.*, **96**, 462 (1963).
195. M. Schmidt and H. Schumann, *Chem. Ber.*, **96**, 780 (1963).
196. M. Schmidt and H. Schumann, *Z. Anorg. Allgem. Chem.*, **325**, 130 (1963).
196a. M. Schmidt, H. Schumann, F. Gliniecki, and J. F. Jaggard, *J. Organometal. Chem.*, **17**, 277 (1969).
197. H. Schumann, Habilitationsschrift, Univ. Würzburg, 1967.
198. H. Schumann, *Z. Anorg. Allgem. Chem.*, **354**, 192 (1967).
198a. H. Schumann, *Angew. Chem.*, **81**, 970 (1969).
199. H. Schumann and P. Jutzi, *Chem. Ber.*, **101**, 24 (1968).
200. H. Schumann, P. Jutzi, A. Roth, P. Schwabe, and E. Schauer, *J. Organometal. Chem.*, **10**, 71 (1967).
201. H. Schumann, P. Jutzi, and M. Schmidt, *Angew. Chem.*, **77**, 812 (1965).
202. H. Schumann and M. Schmidt, *Chem. Ber.*, **96**, 3017 (1963).
203. H. Schumann and M. Schmidt, *Angew. Chem.*, **77**, 1049 (1965).
204. H. Schumann and M. Schmidt, *J. Organometal. Chem.*, **3**, 485 (1965).
204a. H. Schumann and I. Schumann-Ruidisch, *J. Organometal. Chem.*, **18**, 355, (1969).
205. H. Schumann, K. F. Thom, and M. Schmidt, *Angew. Chem.*, **75**, 138 (1963).
206. H. Schumann, K. F. Thom, and M. Schmidt, *J. Organometal. Chem.*, **1**, 167 (1963).
207. H. Schumann, K. F. Thom, and M. Schmidt, *J. Organometal. Chem.*, **2**, 97 (1964).
208. H. Schumann, K. F. Thom, and M. Schmidt, *J. Organometal. Chem.*, **2**, 361 (1964).
209. H. Schumann, K. F. Thom, and M. Schmidt, *J. Organometal. Chem.*, **4**, 22 (1965).
210. H. Schumann, K. F. Thom, and M. Schmidt, *J. Organometal. Chem.*, **4**, 28 (1965).
210a. H. Schumann and R. Weis, *Angew. Chem.*, **82**, 256 (1970).
211. W. T. Schwartz and H. W. Post, *J. Organometal. Chem.*, **2**, 425 (1964).

211a. H. Seidler, H. Woggon, M. Haertig, and W. J. Uhde, *Nahrung*, **13**, 257 (1969); through *CA*, **72**, 2251 (1970).

211b. R. Seltzer, *J. Org. Chem.*, **33**, 3896 (1968).

212. M. Shindo, Y. Matsumura, and R. Okawara, *J. Organometal. Chem.*, **11**, 299 (1968).

213. M. Shindo and R. Okawara, *Inorg. Nucl. Chem. Letters*, **3**, 75 (1967).

213a. M. F. Shostakovskii, R. G. Mirskov, and V. M. Vlasov, *Khim. Atsetilena*, 171 (1968); from *Ref. Zhur. Khim.* Abstr. No. 22Zh415; through *CA*, **71**, 81487 (1969).

213b. T. N. Srivastava, S. N. Bhattacharya, S. K. Tandon, J. Dasgupta, B. J. Jaffri, and O. P. Srivastava, *Indian J. Microbiol.*, **8**, 65 (1968); through *CA*, **71**, 28109 (1969).

214. T. N. Srivastava and S. K. Tandon, *Indian J. Appl. Chem.*, **26**, 171 (1963); through *CA*, **60**, 15900 (1964).

214a. C. H. Stapfer, *J. Paint Technol.*, **41**, 309 (1969); through *CA*, **71**, 69585 (1969).

214b. C. H. Stapfer, *Inorg. Chem.*, **9**, 421 (1970).

214c. C. H. Stapfer and R. D. Dworkin, *Inorg. Chem.*, **9**, 421 (1970).

214d. C. H. Stapfer, K. L. Leung, and R. H. Herber, *Inorg. Chem.*, **9**, 970 (1970).

214e. A. Sturis and J. Bankovskis, *Latv. PSR Zinat. Akad. Vestis, Kim. Ser.*, 751 (1968); through *CA*, **70**, 61747 (1969).

215. D. Sukhani, V. D. Gupta, and R. C. Mehrotra, *J. Organometal. Chem.*, **7**, 85 (1967).

215a. D. Sukhani, V. D. Gupta, and R. C. Mehrotra, *Aust. J. Chem.*, **21**, 1175 (1968); through *CA*, **69**, 77386 (1968).

215b. I. G. Sulimov and M. D. Stadnichuk, *Khim. Prakt. Primen. Kremniorg. Soedin., Tr. Sovesch.*, 57 (1966); through *CA*, **72**, 54422 (1970).

216. T. V. Talalaeva, N. A. Zaitseva, and K. A. Kocheshkov, *Zhur. Obshchei. Khim.*, **16**, 901 (1946); through *CA*, **41**, 2015 (1947).

216a. T. Tanaka, *Organometal. Chem. Rev.*, **5**, 1 (1970).

216b. T. Tanaka and T. Abe, *Inorg. Nucl. Chem. Letters*, **4**, 569 (1968).

217. A. J. Thompson and W. A. G. Graham, *Inorg. Chem.*, **6**, 1365 (1967).

218. A. Tschakirian and P. Bevillard, *Bull. Soc. Chim. France*, 1300 (1950); through *CA*, **45**, 6110 (1951).

219. E. V. Van den Berghe, D. F. Van de Vondel, and G. P. Van der Kelen, *Inorg. Chim. Acta*, **1**, 97 (1967).

220. R. F. Van der Heide, *Z. Lebensm. Untersuch. Forsch.*, **124**, 198 (1964).

221. R. F. Van der Heide, *Z. Lebensm. Untersuch. Forsch.*, **124**, 348 (1964); through *CA*, **61**, 7708 (1964).

221a. D. F. Van de Vondel, E. V. Van den Berghe, and G. P. Van der Kelen, *J. Organometal. Chem.*, **23**, 105 (1970).

222. G. J. M. Van der Kerk and J. G. A. Luijten, *J. Appl. Chem.*, **7**, 369 (1957); through *CA*, **52**, 291 (1958).

222a. G. J. M. Van der Kerk and J. G. A. Luijten, *Arzneim. Forsch.*, **19**, 932 (1969); through *CA*, **71**, 61437 (1969).

222b. G. J. M. Van der Kerk, J. G. A. Luijten, J. G. Noltes, and H. M. J. C. Creemers, *Chimia*, **23**, 313 (1969).

223. G. J. M. Van der Kerk, J. G. A. Luijten, J. C. Van Egmond, and J. G. Noltes, *Chimia*, **16**, 36 (1962).

223a. D. L. Venezky, *Encycl. Polymer. Sci. Technol.*, **7**, 664 (1967); through *CA*, **69**, 1854 (1968).

224. H. Verity-Smith, *The Development of Organotin Stabilizers*, Tin Research Institute, 1959.

225. N. S. Vyazankin, M. N. Bochkarev, and L. P. Sanina, *Zhur. Obshchei. Khim.*, **36**, 166 (1966); through *CA*, **64**, 14212 (1966).

226. N. S. Vyazankin, M. N. Bochkarev, and L. P. Sanina, *Zhur. Obshchei. Khim.*, **36**, 1154 (1966).
227. N. S. Vyazankin, M. N. Bochkarev, and L. P. Sanina, *Zhur. Obshchei. Khim.*, **36**, 1961 (1966).
227a. N. S. Vyazankin, M. N. Bochkarev, and L. P. Sanina, *Zhur. Obshchei. Khim.*, **37**, 1545 (1967); through *CA*, **68**, 2900 (1968).
227b. N. S. Vyazankin, M. N. Bochkarev, and L. P. Sanina, *Zhur. Obshchei. Khim.*, **38**, 414 (1968).
227c. N. S. Vyazankin, M. N. Bochkarev, and L. P. Sanina, *Zhur. Obshchei. Khim.*, **37**, 1037 (1967); through *CA*, **68**, 1262 (1968).
227d. N. S. Vyazankin, M. N. Bochkarev, L. P. Sanina, A. N. Egorochkin, and S. Y. Khorshev, *Zhur. Obshchei. Khim.*, **37**, 2576 (1967); through *CA*, **68**, 8430 (1968).
227e. N. S. Vyazankin and O. A. Kruglaya, *Uspekhi Khim.*, **35**, 1388 (1966).
227f. N. S. Vyazankin, G. A. Razuvaev, and O. A. Kruglaya, *Organometal. Chem. Rev. A*, **3**, 323 (1968).
227g. N. S. Vyazankin, L. P. Sanina, G. S. Kalinina, and M. N. Bochkarev, *Zhur. Obshchei. Khim.*, **38**, 1800 (1968).
227h. M. Wada and R. Okawara, *Kagaku No Ryoiki*, **20**, 19 (1966); through *CA*, **68**, 9259 (1968).
227i. J. L. Wardell and D. W. Grant, *J. Organometal. Chem.*, **20**, 91 (1969).
227k. L. B. Weisfeld, G. A. Thacker, and L. Giamundo, *Advan. Chem. Ser.*, **85**, 38 (1968); through *CA*, **70**, 38428 (1969).
228. M. Wieber, Habilitationsschrift, Univ. Marburg, 1965.
229. M. Wieber and M. Schmidt, *Z. Naturforsch.*, **18b**, 846 (1963).
230. M. Wieber and M. Schmidt, *J. Organometal. Chem.*, **1**, 336 (1964).
231. M. Wieber and M. Schmidt, *J. Organometal. Chem.*, **2**, 129 (1964).
232. P. Woodward, L. F. Dahl, E. W. Abel, and B. C. Crosse, *J. Am. Chem. Soc.*, **87**, 5251 (1965).
232a. H. Woggon and W. J. Uhde, *Plaste. Kaut.*, **16**, 88 (1969); through *CA*, **70**, 76530; (1968).
233. H. Wuyts and A. Vangindertaelen, *Bull. Soc. Chim. Belges*, **30**, 323 (1921).
234. K. Yasuda and R. Okawara, *Inorg. Nucl. Chem. Letters*, **3**, 135 (1967).
235. R. X. Zhuo, H. S. Xu, X. Y. Cheng, C. L. Fan, Y. L. Mo, and Z. F. Yu, *Acta Chim. Sinica*, **32**, 196 (1966).
236. J. J. Zuckerman, *J. Inorg. Nucl. Chem.*, **29**, 2191 (1967).

P1. Belgian Pat. 632,271; through *CA*, **60**, 15909 (1964).
P2. Belgian Pat. 658,003; through *CA*, **64**, 3802 (1966).
P3. British Pat. 719,421; through *CA*, **50**, 397 (1956).
P4. British Pat. 719,733; through *CA*, **49**, 5512 (1955).
P5. British Pat. 723,296.
P6. British Pat. 728,953; through *CA*, **49**, 13693 (1955).
P7. British Pat. 728,954; through *CA*, **49**, 13693 (1955).
P8. British Pat. 737,392; through *CA*, **50**, 9010 (1956).
P9. British Pat. 737,508; through *CA*, **50**, 6507 (1956).
P10. British Pat. 740,392.
P11. British Pat. 740,397; through *CA*, **50**, 13987 (1956).
P12. British Pat. 742,975; through *CA*, **51**, 2033 (1957).
P13. British Pat. 743,304; through *CA*, **50**, 16828 (1956).
P14. British Pat. 743,313; through *CA*, **50**, 16829 (1956).
P15. British Pat. 747,239.

P16. British Pat. 748,228; through *CA*, **51**, 8128 (1957).
P17. British Pat. 749,722; through *CA*, **51**, 2859 (1957).
P18. British Pat. 750,106; through *CA*, **51**, 1247 (1957).
P19. British Pat. 759,382; through *CA*, **51**, 13918 (1957).
P20. British Pat. 781,452; through *CA*, **52**, 3864 (1958).
P21. British Pat. 781,905; through *CA*, **52**, 2049 (1958).
P22. British Pat. 792,309; through *CA*, **52**, 17805 (1958).
P23. British Pat. 800,168; through *CA*, **53**, 7023 (1959).
P24. British Pat. 806,535; through *CA*, **53**, 13496 (1959).
P25. British Pat. 841,151; through *CA*, **54**, 26007 (1960).
P26. British Pat. 855,214; through *CA*, **55**, 10958 (1961).
P27. British Pat. 866,484; through *CA*, **55**, 21672 (1961).
P28. British Pat. 892,137; through *CA*, **58**, 3557 (1963).
P29. British Pat. 902,560; through *CA*, **58**, 2798 (1963).
P30. British Pat. 903,068; through *CA*, **57**, 15152 (1962).
P31. British Pat. 919,248; through *CA*, **60**, 1775 (1964).
P32. British Pat. 1,018,805; through *CA*, **64**, 11251 (1966).
P33. British Pat. 1,020,291; through *CA*, **64**, 15926 (1966).
P34. British Pat. 1,020,612; through *CA*, **64**, 14219 (1966).
P35. British Pat. 1,027,781; through *CA*, **64**, 19910 (1966).
P36. British Pat. 1,047,949; through *CA*, **66**, 1874 (1967).
P37. British Pat. 1,061,747; through *CA*, **67**, 1173 (1967).
P38. British Pat. 1,069,165; through *CA*, **67**, 3111 (1967).
P38a. British Pat. 1,089,243; through *CA*, **68**, 7572 (1968).
P38b. British Pat. 1,129,725; through *CA*, **70**, 12306 (1969).
P38c. British Pat. 1,138,786; through *CA*, **70**, 48273 (1969).
P38d. British Pat. 1,151,927; through *CA*, **71**, 13755 (1969).
P38e. British Pat. 1,163,738; through *CA*, **72**, 2550 (1970).
P38f. British Pat. 1,173,466; through *CA*, **72**, 43846 (1970).
P39. German Pat. 1,020,331*a*; through *CA*, **53**, 19880 (1959); German Pat. 1,020,331*b*; through *CA*, **53**, 19880 (1959); German Pat. 1,020,331*c*; through *CA*, **53**, 19880 (1959).
P40. German Pat. 1,020,332; through *CA*, **53**, 19881 (1959).
P41. German Pat. 1,020,333; through *CA*, **53**, 19880 (1959).
P42. German Pat. 1,020,334; through *CA*, **55**, 383 (1961).
P43. German Pat. 1,020,335; through *CA*, **53**, 19889 (1959).
P44. German Pat. 1,020,336; through *CA*, **53**, 19975 (1959).
P45. German Pat. 1,020,337; through *CA*, **55**, 383 (1961).
P46. German Pat. 1,020,338; through *CA*, **53**, 19881 (1959).
P47. German Pat. 1,046,053.
P48. German Pat. 1,073,496; through *CA*, **55**, 10319 (1961).
P49. German Pat. 1,078,772; through *CA*, **55**, 13927 (1961).
P50. German Pat. 1,080,555; through *CA*, **54**, 20336 (1960).
P51. German Pat. 1,088,709; through *CA*, **55**, 20511 (1961).
P52. German Pat. 1,150,814; through *CA*, **64**, 8403 (1966).
P53. German Pat. 1,167,836.
P54. German Pat. 1,178,853; through *CA*, **62**, 7796 (1965).
P55. German Pat. 1,227,658; through *CA*, **66**, 4416 (1967).
P56. German Pat. 1,232,736; through *CA*, **66**, 8095 (1967).
P57. German Pat. 1,232,739; through *CA*, **66**, 8096 (1967).
P58. German Pat. 1,234,722; through *CA*, **67**, 2107 (1967).

P58a. German Pat. 1,276,643; through *CA*, **70**, 29060 (1969).
P58b. German Pat. 1,801,274; through *CA*, **71**, 13811 (1969).
P58c. German Pat. 1,801,275; through *CA*, **71**, 13812 (1969).
P58d. German Pat. 1,801,276; through *CA*, **71**, 39176 (1969).
P58e. German Pat. 1,801,277; through *CA*, **71**, 13831 (1969).
P58f. German Pat. 1,815,168; through *CA*, **71**, 113741 (1969).
P58g. German Pat. 1,919,927; through *CA*, **72**, 43874 (1970).
P59. French Pat. 71,532; through *CA*, **57**, 11235 (1962).
P59a. French Pat. 93,958; through *CA*, **72**, 44546 (1970).
P60. French Pat. 1,055,906.
P61. French Pat. 1,111,320.
P62. French Pat. 1,319,129; through *CA*, **59**, 8789 (1963).
P63. French Pat. 1,320,051; through *CA*, **59**, 4122 (1963).
P64. French Pat. 1,320,343; through *CA*, **59**, 4121 (1963).
P65. French Pat. 1,320,473; through *CA*, **59**, 8788 (1963).
P66. French Pat. 1,339,457; through *CA*, **60**, 3007 (1964).
P67. French Pat. 1,352,692; through *CA*, **60**, 15910 (1964).
P68. French Pat. 1,355,999; through *CA*, **61**, 3148 (1964).
P69. French Pat. 1,359,490; through *CA*, **62**, 4170 (1965).
P70. French Pat. 1,360,741; through *CA*, **61**, 14858 (1964).
P71. French Pat. 1,365,375; through *CA*, **61**, 14711 (1964).
P72. French Pat. 1,369,815; through *CA*, **62**, 586 (1965).
P73. French Pat. 1,374,539; through *CA*, **62**, 6637 (1965).
P74. French Pat. 1,386,988; through *CA*, **63**, 10135 (1965).
P75. French Pat. 1,393,677; through *CA*, **63**, 4474 (1965).
P76. French Pat. 1,453,490; through *CA*, **66**, 9019 (1967).
P76a. French Pat. 1,477,892; through *CA*, **68**, 4830 (1968).
P76b. French Pat. 1,505,426; through *CA*, **69**, 97580 (1968).
P76c. French Pat. 1,527,274; through *CA*, **70**, 115337 (1969).
P76d. French Pat. 1,531,398; through *CA*, **71**, 4173 (1969).
P76e. French Pat. 1,533,524; through *CA*, **72**, 55660 (1970).
P76f. French Pat. 1,537,462; through *CA*, **71**, 13830 (1969).
P76g. French Pat. 1,540,230; through *CA*, **71**, 4175 (1969).
P76h. French Pat. 1,546,216; through *CA*, **71**, 23521 (1969).
P76i. French Pat. 1,566,449; through *CA*, **71**, 113447 (1969).
P77. Japanese Pat. 8337 (60); through *CA*, **57**, 6146 (1962).
P78. Japanese Pat. 18387 (60); through *CA*, **56**, 4794 (1962).
P79. Japanese Pat. 8178 (62); through *CA*, **59**, 11729 (1963).
P80. Japanese Pat. 8478 (62); through *CA*, **59**, 11729 (1963).
P81. Japanese Pat. 8479 (62); through *CA*, **59**, 11729 (1963).
P82. Japanese Pat. 9223 (62); through *CA*, **59**, 5197 (1963).
P83. Japanese Pat. 22145 (63); through *CA*, **60**, 2952 (1964).
P84. Japanese Pat. 1811 (64); through *CA*, **61**, 2000 (1964).
P85. Japanese Pat. 22069 (65); through *CA*, **64**, 3603 (1966).
P86. Japanese Pat. 4576 (66).
P87. Japanese Pat. 19333 (66); through *CA*, **66**, 3639 (1967).
P88. Japanese Pat. 19416 (66); through *CA*, **66**, 4416 (1967).
P89. Japanese Pat. 865 (67); through *CA*, **67**, 3174 (1967).
P89a. Japanese Pat. 19177 (67); through *CA*, **68**, 7649 (1968).
P89b. Japanese Pat. 24047 (67); through *CA*, **68**, 8493 (1968).
P89c. Japanese Pat. 18764 (68); through *CA*, **70**, 68524 (1969).

P89d. Japanese Pat. 19534 (68); through *CA*, **70**, 58028 (1969).
P89e. Japanese Pat. 69,02506; through *CA*, **71**, 82526 (1969).
P90. Dutch Pat. 109,491; through *CA*, **62**, 9173 (1965).
P91. Dutch Pat. 113,311; through *CA*, **66**, 2759 (1967).
P92. Dutch Pat. 6,504,150; through *CA*, **64**, 8406 (1966).
P93. Dutch Pat. 6,600,846; through *CA*, **66**, 364 (1967).
P94. Dutch Pat. 6,603,742; through *CA*, **66**, 8117 (1967).
P94a. Dutch Pat. 6,604,827.
P95. Dutch Pat. 6,606,681; through *CA*, **66**, 8095 (1967).
P95a. Dutch Pat. 6,700,013; through *CA*, **67**, 10259 (1967).
P95b. Dutch Pat. 6,703,505; through *CA*, **67**, 11019 (1967).
P96. Polish Pat. 47,960; through *CA*, **61**, 14710 (1964).
P97. Polish Pat. 48,178; through *CA*, **62**, 1688 (1965).
P98. Polish Pat. 49,815; through *CA*, **64**, 16082 (1966).
P99. U.S. Pat. 2,636,891; through *CA*, **48**, 3397 (1954).
P100. U.S. Pat. 2,641,588; through *CA*, **48**, 5207 (1954).
P101. U.S. Pat. 2,641,596; through *CA*, **48**, 5208 (1954).
P102. U.S. Pat. 2,648,650; through *CA*, **48**, 10056 (1954).
P103. U.S. Pat. 2,702,775; through *CA*, **49**, 7816 (1955).
P104. U.S. Pat. 2,702,776; through *CA*, **49**, 7816 (1955).
P105. U.S. Pat. 2,702,777; through *CA*, **49**, 7816 (1955).
P106. U.S. Pat. 2,702,778; through *CA*, **49**, 7816 (1955).
P107. U.S. Pat. 2,704,756; through *CA*, **50**, 2687 (1956).
P108. U.S. Pat. 2,713,580; through *CA*, **50**, 5762 (1956).
P109. U.S. Pat. 2,713,585; through *CA*, **50**, 5725 (1956).
P110. U.S. Pat. 2,726,227; through *CA*, **50**, 6095 (1956).
P111. U.S. Pat. 2,726,254; through *CA*, **50**, 6095 (1956).
P112. U.S. Pat. 2,731,440; through *CA*, **50**, 11714 (1956).
P113. U.S. Pat. 2,731,441; through *CA*, **51**, 1656 (1957).
P114. U.S. Pat. 2,731,482; through *CA*, **51**, 13918 (1957).
P115. U.S. Pat. 2,731,484; through *CA*, **50**, 11715 (1956).
P116. U.S. Pat. 2,752,325; through *CA*, **51**, 1264 (1957).
P117. U.S. Pat. 2,759,906; through *CA*, **51**, 3663 (1957).
P118. U.S. Pat. 2,786,812; through *CA*, **51**, 10892 (1957).
P119. U.S. Pat. 2,786,813; through *CA*, **51**, 10892 (1957).
P120. U.S. Pat. 2,786,814; through *CA*, **51**, 10892 (1957).
P121. U.S. Pat. 2,789,102; through *CA*, **51**, 10939 (1957).
P122. U.S. Pat. 2,789,103; through *CA*, **51**, 10939 (1957).
P123. U.S. Pat. 2,789,963; through *CA*, **51**, 17235 (1957).
P124. U.S. Pat. 2,801,258; through *CA*, **51**, 18007 (1957).
P125. U.S. Pat. 2,809,956; through *CA*, **52**, 3863 (1958).
P126. U.S. Pat. 2,830,067; through *CA*, **53**, 222 (1959).
P127. U.S. Pat. 2,832,750; through *CA*, **52**, 16200 (1958).
P128. U.S. Pat. 2,832,751; through *CA*, **53**, 1847 (1959).
P129. U.S. Pat. 2,832,752; through *CA*, **53**, 1848 (1959).
P130. U.S. Pat. 2,858,325; through *CA*, **53**, 3758 (1959).
P131. U.S. Pat. 2,868,819; through *CA*, **53**, 13056 (1959).
P132. U.S. Pat. 2,870,119; through *CA*, **53**, 10844 (1959).
P133. U.S. Pat. 2,870,182; through *CA*, **53**, 13057 (1959).
P134. U.S. Pat. 2,872,468; through *CA*, **53**, 7100 (1959).
P135. U.S. Pat. 2,883,363; through *CA*, **53**, 14586 (1959).

P136. U.S. Pat. 2,885,415; through *CA*, **53**, 18972 (1959).
P137. U.S. Pat. 2,891,922; through *CA*, **53**, 19880 (1959).
P138. U.S. Pat. 2,904,569; through *CA*, **54**, 2175 (1960).
P139. U.S. Pat. 2,904,570; through *CA*, **54**, 2175 (1960).
P140. U.S. Pat. 2,914,506; through *CA*, **54**, 20336 (1960).
P141. U.S. Pat. 2,918,456; through *CA*, **54**, 11534 (1960).
P142. U.S. Pat. 2,934,548; through *CA*, **54**, 19486 (1960).
P143. U.S. Pat. 2,954,363; through *CA*, **55**, 6931 (1961).
P144. U.S. Pat. 2,967,167; through *CA*, **55**, 10944 (1961).
P145. U.S. Pat. 2,998,441; through *CA*, **56**, 6000 (1962).
P146. U.S. Pat. 3,015,644; through *CA*, **57**, 2425 (1962).
P147. U.S. Pat. 3,027,350; through *CA*, **57**, 2429 (1962).
P148. U.S. Pat. 3,029,267; through *CA*, **57**, 8617 (1962).
P149. U.S. Pat. 3,095,434; through *CA*, **59**, 14023 (1963).
P150. U.S. Pat. 3,108,126; through *CA*, **52**, 17805 (1958).
P151. U.S. Pat. 3,132,070; through *CA*, **61**, 6311 (1964).
P152. U.S. Pat. 3,147,285; through *CA*, **62**, 11973 (1965).
P153. U.S. Pat. 3,183,238; through *CA*, **63**, 1816 (1965).
P154. U.S. Pat. 3,201,408; through *CA*, **63**, 13315 (1965).
P155. U.S. Pat. 3,208,969; through *CA*, **63**, 18736 (1965).
P156. U.S. Pat. 3,234,032; through *CA*, **64**, 12966 (1966).
P157. U.S. Pat. 3,243,403; through *CA*, **64**, 19949 (1966).
P158. U.S. Pat. 3,316,284; through *CA*, **67**, 6089 (1967).
P158a. U.S. Pat. 3,317,573; through *CA*, **68**, 3877 (1968).
P158b. U.S. Pat. 3,328,441; through *CA*, **68**, 4829 (1968).
P158c. U.S. Pat. 3,365,478; through *CA*, **68**, 9268 (1968).
P158d. U.S. Pat. 3,397,217; through *CA*, **69**, 7254 (1968).
P158e. U.S. Pat. 3,395,164; through *CA*, **69**, 68052 (1968).
P158f. U.S. Pat. 3,410,797; through *CA*, **70**, 21592 (1969).
P158g. U.S. Pat. 3,412,120; through *CA*, **70**, 96951 (1969).
P158h. U.S. Pat. 3,417,117; through *CA*, **70**, 78149 (1969).
P158i. U.S. Pat. 3,429,905; through *CA*, **70**, 115333 (1969).
P158k. U.S. Pat. 3,450, 668; through *CA*, **71**, 39847 (1969).
P158l. U.S. Pat. 3,458,472; through *CA*, **71**, 102625 (1969).
P159. DDR Pat. 63,490; through *CA*, **70**, 68521 (1969).
P160. S. African. Pat. 67,07058; through *CA*, **70**, 96958 (1969).

7. ORGANOTIN COMPOUNDS WITH Sn–N BONDS

K. JONES

Department of Chemistry, University of Manchester Institute of Science and Technology, Manchester, England

M. F. LAPPERT

School of Molecular Sciences, University of Sussex Brighton, England

I. Introduction

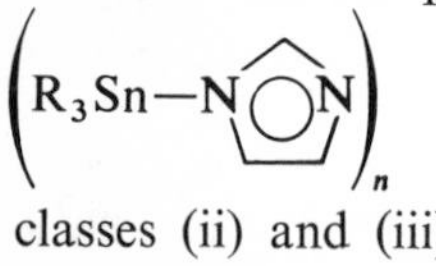

This chapter reviews the chemistry of organotin-nitrogen compounds up to July 1968 [with further references (see Sec. VI), added in proof, up to June, 1970]. These compounds may be divided into three classes: (i) compounds having an N → Sn coordinate link [which are regarded as outside the range of this chapter, but see Ref. (*133*)]; (ii) aminostannanes, $\gtrless$Sn—N$<$, and their *N*- mono- or *N,N*-disubstituted derivatives, in which tin and nitrogen are formally four- and three-coordinate, respectively; and (iii) certain tin pseudohalides (see Sec. V) and related derivatives, in which tin and nitrogen are formally four- and two-coordinate, respectively (these have $\gtrless$Sn—N=C=O, $\gtrless$Sn—N=C=S, $\gtrless$Sn—N=$\overset{+}{N}$=$\overset{-}{N}$, $\gtrless$Sn—N=C=C$<$, $\gtrless$Sn—N(CN)$_2$ (*82*) $\gtrless$Sn—N=C=C(CN)$_2$ (*82*) $\gtrless$Sn—N=C$<$, or $\gtrless$Sn—N=PX$_3$ skeletal structures). The term "formal" is used in (ii) and (iii) in connection with description of coordination numbers, to describe empirical formulas, in which autocomplexation is ignored; for example, compounds such as $(R_3Sn—N\bigcirc N)_n$ and $(R_3SnN_3)_n$ are regarded as within the scope of classes (ii) and (iii), respectively. Indeed, five-coordination for tin, though not usual for Sn—N compounds of classes (i) and (ii), is not particularly rare. Earlier reviews are available both for compounds of type (ii) (*43*, *76*, *113*, *117*) and type (iii) (*97*, *175*, *180*). All compounds belonging to classes (ii) and (iii) are listed in the summarizing Appendices; the chemistry of class (ii) is described in detail in Secs. (II)–(IV); and for class (iii), because of the slower pace of advance since the appearance of earlier reviews, more summararily in Sec. V. Further discussion in this section is concerned with class (ii).

Compounds of type $\geqslant Sn{-}N\begin{smallmatrix} R \\ R' \end{smallmatrix}$ (where R and R′ may be the same or different, and each is H, alkyl, or aryl groups) may be regarded as the parent species for class (ii). They are named variously as aminostannanes, stannylamines, stannazanes, or tin amides. They are known (see Appendices) with primary (although not definitively), secondary, or tertiary nitrogen; and may be acyclic or cyclic. The latter are at present only known as six-membered $(SnN)_3$ or four-membered (SnNGeN) rings. More complex derivatives of the parents include *inter alia* hydrazino $(\geqslant Sn{-}N({-}){-}N\leqslant)$, ureido- $(\geqslant Sn{-}N({-}){-}C({=}O){-}N\leqslant)$, acetamido- $(\geqslant Sn{-}N({-}){-}C({=}O){-}CH_3)$, guanidino- $(\geqslant Sn{-}N({-}){-}C({=}NR){-}N\leqslant)$, and carbamato- $(\geqslant Sn{-}NR{-}C({=}O){-}OR')$ stannanes. They may have one or more Sn—N bond.

It has been reported that $Me_2Sn(NMe_2)_2$ may give rise to mild explosions, either when chloroform solutions are examined in a sealed nmr tube, or at the conclusion of distillation through a long packed column when the still temperature is around 200°C (*135*). The present authors, while working extensively with Sn—N compounds since 1960, are unable to confirm this observation. We interpret the explosions as being caused by chemical reactions; it is now known (*42, 54*) (see Sec. III.A) that Me_3SnNMe_2 and chloroform interact with liberation of dimethylamine.

The subject has developed very rapidly. In 1961, $Sn(NEt_2)_4$ was first reported (*181*), and the preparation of organotin-amides followed in the next year (*2, 72, 164, 187*). At the time of writing, more than 200 compounds have been characterized and their chemistry has been described in upwards of 150 publications. Although the topic is of recent interest, some of the main lines of development have clearly emerged and have relevance far beyond the realms of Sn—N chemistry, as was predicted at an early stage (*58, 72–75*).

The reactivity of Sn—N compounds has been compared with that of the Grignard reagents (*73, 76*) and may be attributed to a combination of a weak and highly polar $Sn^{\delta+}{-}N^{\delta-}$ bond. In general, reactions are with polar substrates (see Sec. III). The considerable scope of Sn—N compounds in synthesis is already being exploited. Chemically, the most widely studied have been the (dialkylamino)trialkylstannanes, and especially Me_3SnNMe_2. Several important classes of reactions were first established with aminostannanes and have since proved to be of far wider scope. For instance, the addition of phenyl isocyanate (or similar unsaturated substrates) to aminostannanes, e.g. Eq. (33), an example of an "insertion" reaction (*72*):

$$\begin{matrix} M{-}N\!\!< \\ \curvearrowright \\ B{=}A \end{matrix} \longrightarrow M{-}B{-}A{-}N\!\!<$$

has been extended to amido-, as well as related (e.g., Cl, OR, SR, Ar), derivatives of many elements (*96*).

Sn—N compounds may be regarded as stannylamines. This lays stress on the comparative aspect with nitrogen derivatives of C, Si, Ge, and Pb(IV). It was predicted (*58*, *74–76*) that metal-nitrogen bond strengths decrease in the series SiN > CN > GeN > SnN > PbN; and that SnN < SnP, unlike SiN > SiP (i.e., Sn(IV) a class *b* and Si a class *a* acceptor). Furthermore, it was anticipated (*74*, *76*) that the basicity of the nitrogen decreased from CN > PbN > SnN > GeN > SiN. Predictions were also made as to stereochemistry (*74*, *76*). For example, whereas trisilylamine, and therefore probably tris-(trimethylsilyl)amine, has D_{3h} skeletal symmetry (*56*), in accord with $2sp^2$(N)–$3sp^3$(Si) σ-orbitals and d_π-p_πSi—N interaction, $(Me_3Sn)_3N$ was expected to be more nearly like a *tert*-alkylamine (C_{3v} skeletal symmetry) with nitrogen in a pyramidal approximately sp^3-situation (*64*). Some data are now available (see Sec. IV) and all these generalizations appear to be broadly valid.

II. Sn—N Bond-Making

That simple amido-derivatives of tin(IV) were not synthesized before 1961 is almost certainly due to the fact that the reaction of an amine with a tin halide leads to the formation of a stable complex, which upon heating dissociates rather than dehydrohalogenates. In a comparative study of the behavior of group IV halides towards amines, Si, and Ge were found to react [Eq. (1)] according to (i), whereas Sn and Pb according to (ii) (*6*). How-

$$\begin{aligned} &{\gt}M{-}X + 2\,R_2NH \xrightarrow{\text{(i)}} {\gt}M{-}NR_2 + R_2NH{\cdot}HX \\ &{\gt}M{-}X + 2\,R_2NH \xrightarrow{\text{(ii)}} {\gt}M{-}X{\cdot}2\,R_2NH \end{aligned} \qquad (1)$$

ever, metathesis is likely to be general if the Sn—N product has unusually high stability, as in the formation (*92*) of triphenylstannylpurines from Ph_3SnCl, benzylamine, and the purine. Alternatively, if a decomposition path is made available, the equilibrium will also be shifted. Examples are found (*89*) in the formation of $(R_3Sn)_2S$ from R_3SnX, NH_3, and CS_2, or in $(Ph_2SnS)_3$ from Ph_2SnCl_2 or $(Ph_2SnO)_n$ with NH_3 and CS_2; compounds having $\geq Sn-N<$ and $\geq Sn-S-C(:S)N<$ (*88*) skeletons are likely intermediates. By contrast, tin(IV) halides in liquid ammonia are ammonolyžed (*12*, *13*, *159*):

$$SnX_4 + 2\,NH_3 \rightarrow SnX_4{\cdot}2\,NH_3 \xrightarrow{-NH_4X} SnX_3{\cdot}NH_2 \xrightarrow[-NH_4X]{2\,NH_3} SnX(NH_2)_3 \quad (2)$$

A. Transmetallation Reactions

The reaction of lithium salts of secondary amines with organotin halides affords a route to aminostannanes (*2*, *14*, *72*, *74*, *92*, *126*, *143–145*, *157*, *158*, *165*, *181*, *187*, *189*):

$$R_{4-n}SnX_n + n\,LiNR'R'' \longrightarrow R_{4-n}Sn(NR'R'')_n + n\,LiX \quad (3)$$

Solutions of lithium salts of secondary amines may be prepared just prior to adding the tin halide (*74*, *189*):

$$RLi + R'R''NH \longrightarrow LiNR'R'' + RH \qquad (R = Bu, Ph) \quad (4)$$

Monolithium salts of primary amines were likewise obtained as shown in Eq. (5) (*74*), but their subsequent reaction with trialkylstannyl halide did not invariably afford the secondary tin amine, Eq. (6) (*74*):

$$n\text{-}BuLi + R'NH_2 \longrightarrow LiNHR' + n\text{-}BuH \quad (5)$$

$$R_3SnX + LiNHR' \longrightarrow R_3SnNHR' + LiX \quad (6)$$

Where R = Me and R′ = aryl, the *N*-aryl(trimethylstannyl)amines were isolated. However, when R = Me and R′ = Me or Et, the products were *N*-methyl- and *N*-ethyl-hexamethyldistannazanes. Presumably, *N*-alkyl-(trimethylstannyl)amines were formed initially, and then underwent further reaction:

$$2\,R_3SnNHR' \longrightarrow (R_3Sn)_2NR' + R'NH_2 \quad (7)$$

or less likely:

$$R_3SnNHR' + R_3SnX \longrightarrow (R_3Sn)_2NR' + HX \quad (8)$$

The reaction of lithium nitride (*81*) (which may be regarded as the trilithium salt of ammonia) with trimethylstannyl chloride gave tris(trimethylstannyl)-amine (*100*, *182*):

$$3\,Me_3SnCl + Li_3N \longrightarrow (Me_3Sn)_3N + 3\,LiCl \quad (9)$$

Likewise sodium or lithium amide reacted with organotin halides (*165*), alkoxides (*146*), or oxides:

$$R_3SnX + NaNH_2 \longrightarrow (R_3Sn)_3N \quad (10)$$

$$(R = Ph, Me; X = Hal, OR, or OSnR_3)$$

$$R_3SnCl + MNH_2 \longrightarrow (R_3Sn)_3N + MCl \quad (11)$$

$$(R = Me, Et, Pr; M = Li, Na)$$

A similar situation has been reported in Sn—P chemistry; tris(trimethylstannyl)phosphine was the only product isolated from the reaction of trimethylstannyl bromide and sodium phosphide (*26*). Presumably, (trimethylstannyl)phosphine is formed first, but undergoes ready condensation:

$$3\ Me_3SnBr + 3\ NaPH_2 \longrightarrow (Me_3Sn)_3P + 2\ PH_3 + 3\ NaBr \quad (12)$$

The use of sodium or potassium salts of secondary amines has also been claimed (*14*, *187*), and especially for the preparation of mixed SiNSn compounds, Eq. (13) (*148*), and for heterocyclic *N*-derivatives, Eq. (14) (*77*, *78*, *109*, *110*, *114*, *115*):

$$R_3SnCl + (Me_3Si)_2NNa \longrightarrow R_3SnN(SiMe_3)_2 + NaCl \quad (13)$$

$$R_3SnX + MN\lt \longrightarrow R_3SnN\lt + MX \quad (14)$$

Amidomagnesium bromide has been used, Eq. (15) (*164*). Transmetallation between silicon and tin, as in Eq. (16), has been reported, but is not general [see Eqs. (17) (*2*) and (18) (*59*)]. The greater stability of the complex in Eq. (17) was attributed to steric hindrance to the approach of the Si and Br atoms as required for the intramolecular elimination of Me_3SiBr:

$$R_3SnCl + \gt NMgX \longrightarrow R_3SnN\lt + MgXCl \quad (15)$$

$$Me_3SnBr + Me_3SiNHEt \rightarrow 1{:}1\ complex \xrightarrow{heat} Me_3SiBr + Me_3SnNHEt \quad (16)$$

$$Me_3SnBr + Me_3SiNEt_2 \longrightarrow 1{:}1\ complex\ (stable) \quad (17)$$

$$Me_3SnNMe_2 + Me_3SiCl \longrightarrow Me_3SnCl + Me_3SiNMe_2 \quad (18)$$

Nitraminostannanes have been obtained, by use of silver salts (*188*):

$$R_3SnX + R'N(NO_2)Ag \longrightarrow R_3Sn{-}NR'(NO_2) + AgX \quad (19)$$

A further example of the use of *N*-lithio salts for generating Sn—N bonds is shown in Eq. (20) (*144*). Several reactions related to Eq. (20) have been exploited for synthesis of other mixed group IV compounds containing tertiary nitrogen (Si_2NSn, SiGeNSn, CGeNSn, and CSiNSn) (*143–145*, *149*,

150, 152, 154, 157, 158). There was no clear evidence for the formation of Sn—N bonds when Ph_2SnCl_2 and $\mathrm{Ph(Li)N{-}N(Li)Ph}$ were mixed (*57*).

Also of interest is the use of an *Sn*-lithio salt, as in Eq. (21) (*126*):

$$\mathrm{Me_2SnCl_2 + (Et_3SiNLi)_2GeMe_2 \longrightarrow Et_3SiN\begin{smallmatrix}\diagup GeMe_2 \diagdown \\ \diagdown SnMe_2 \diagup\end{smallmatrix}NSiEt_3} \quad (20)$$

$$\mathrm{Et_3SnLi + Me_2C(CN){-}I \longrightarrow Et_3SnN{=}C{=}CMe_2} \quad (21)$$

B. Transamination Reactions

An exchange reaction occurred when a secondary amine was heated with an aminostannane derived from another secondary amine (*72, 74*):

$$\mathrm{R_3SnNR'_2 + R''_2NH \longrightarrow R_3SnNR''_2 + R'_2NH} \quad (22)$$

The order of displacement was found to be in accordance with boiling points. Thus, in any such exchange reaction the more volatile amine was released, e.g., $\mathrm{Bu^n_2N > Et_2N > Me_2N}$ (*74*). The advantages of this route as a preparative method are that mild conditions are required and the reactions are quantitative. This is particularly useful in experiments using isotopically labeled amines (*71*). Further examples of the scope of this method are shown in Eqs. (23) and (24) (*147*), (25) (*50*), (26) (*108*), and (27) (*83*):

$$\mathrm{Me_2Sn(NEt_2)_2 + \begin{matrix}HNRCH_2\diagdown \\ HNRCH_2\diagup\end{matrix}(CH_2)_n \longrightarrow Me_2Sn\begin{matrix}\diagup RN{-}CH_2 \diagdown \\ \diagdown RN{-}CH_2 \diagup\end{matrix}(CH_2)_n + 2\,Et_2NH} \quad (23)$$

$R = \mathrm{Me, Me_3Si}$; $n = 0, 1, 2$

$$\mathrm{Me_2Sn(NEt_2)_2 + C_6H_4(NHSiMe_3)_2 \longrightarrow C_6H_4(NSiMe_3)_2SnMe_2 + 2\,Et_2NH} \quad (24)$$

$$\mathrm{3\,Me_3SnNMe_2 + PhNHNH_2 \longrightarrow (Me_3Sn)_2NNPhSnMe_3 + 3\,Me_2NH} \quad (25)$$

$$\mathrm{MeSn(NMe_2)_3 + (Me_3Si)_2NH \longrightarrow MeSn(NMe_2)_2N(SiMe_3)_2 + Me_2NH} \quad (26)$$

$$\mathrm{Me_3SnNMe_2 + \begin{matrix}H_2C\\ |\\ H_2C\end{matrix}\!>\!NH \longrightarrow Me_3Sn{-}N\!\triangleleft + Me_2NH} \quad (27)$$

Aminostannanes derived from secondary amines also underwent transamination with primary amines, as shown in Eqs. (28) and (29) and thus, in contrast to Eq. (22), there is steric, rather than volatility control (*72, 74*):

$$R_3SnNR'_2 + R''NH_2 \longrightarrow R_3SnNHR'' + R'_2NH \tag{28}$$

$$2\,R_3SnNR'_2 + R''NH_2 \longrightarrow (R_3Sn)_2NR'' + 2\,R'_2NH \tag{29}$$

Where R = Me, R′ = Me or Et, and R″ = aryl, the *N*-aryl(trialkylstannyl)-amine was obtained, Eq. (28). Where R = Me, R′ = Me or Et, and R″ = Me or Et, only the distannazane, Eq. (29), was isolated, even in the presence of excess amine. Thus, condensation must occur very rapidly (*74*):

$$2\,R_3SnNHR'' \longrightarrow (R_3Sn)_2NR'' + R''NH_2 \tag{30}$$

It appears that steric effects play a further role in determining the facility of Eq. (30). Thus, *N-tert*-butyl(tri-*n*-butylstannyl)amine, Eq. (28), was the only tin-bearing product from n-Bu_3SnNMe_2/*tert*-$BuNH_2$ (*74*).

The formation of distannazanes, from secondary stannyl amines, e.g., Eq. (30), was extended to provide a synthesis of cyclostannazanes (*72, 74*):

$$n\,Me_2Sn(NMe_2)_2 + n\,RNH_2 \longrightarrow (Me_2SnNR)_n + 2n\,Me_2NH \tag{31}$$

Molecular weight determination (cryoscopic in benzene) indicated a cyclic trimer for R = Et.

Similarly, transamination accompanied by condensation occurred, when ammonia was passed through a light petroleum solution of (dimethylamino)-trimethylstannane (*72, 74*):

$$3\,Me_3SnNMe_2 + NH_3 \longrightarrow (Me_3Sn)_3N + 3\,Me_2NH \tag{32}$$

Transamination has been examined by ^{1}H-nmr spectroscopy (see Sec. IV) (*137*) for the following pairs of reactants (i) $Me_2Sn(NEt_2)_2/Et_2NH$, (ii) $Me_2Sn(NEt_2)_2/Me_2NH$, (iii) $Me_2Sn(NMe_2)_2/Me_2NH$, (iv) $Me_2Sn(NMe_2)_2/Me_2NH$, and (v) $Me_3SnNHPh/PhNH_2$. In (i) or (ii), transamination was too slow to be measured under ambient conditions of temperature and pressure; in (iii), exchange was faster than transamination, with the latter frozen at −40°C, but the former process still rapid at −68°C; for (iv), both exchange and transamination at 40°C was faster than 250 sec^{-1}; and for (v), both exchange and transamination were slower than 120 sec^{-1} under ambient conditions.

C. Addition Reactions

This section relates mainly to insertion reactions of the type

$$\geq Sn{-}X + {-}N{=}Y \longrightarrow \geq Sn{-}\underset{|}{N}{-}Y{-}X,$$

where —N=Y is a nitrogen-containing unsaturated substrate (a 1,2-dipole).

The reaction of (dimethylamino)trimethylstannane with phenyl isocyanate, Eq. (33), was the first such reaction to be described (*72*). Others, Eqs. (34–36), were later reported (*58, 73*). Further examples of insertion into an Sn—N bond are with (i) tetracyanoethylene to afford $Me_3Sn{-}N{=}C(NMe_2){-}C(CN){=}C(CN)_2$ (*61*), and (ii) p-$Me_2N \cdot C_6F_4 \cdot CN$ to give $Me_3SnN{=}C(NMe_2){-}C_6F_4 \cdot NMe_2$-$p$ (*99*); PhCH=CHCN did not react (*99*). With respect to insertion of RNCO, $Bu_3Sn{-}\overset{|}{N}R$ is more readily cleaved than $Bu_3Sn{-}\overset{|}{N}Ar$ (*24*), and $Bu_3SnNRCONArSnBu_3$ is more reactive as an addendum than $Bu_3SnNRCOOSnBu_3$. The generality of Eq. (37) was recognized (*72*) and has been confirmed for M = B, Si, Ge, P, As, Sb, S, Ti, Zr, Hf, Nb, Ta, and Cr; a review is available (*96*):

$$Me_3SnNMe_2 + PhNCO \longrightarrow Me_3Sn{-}N(Ph){-}C({=}O){-}NMe_2 \tag{33}$$

$$Me_3SnNMe_2 + PhCN \longrightarrow Me_3Sn{-}N{=}C(Ph){-}NMe_2 \tag{34}$$

$$Me_3SnNMe_2 + PhNSO \longrightarrow Me_3Sn{-}N(Ph){-}S({=}O){-}NMe_2 \tag{35}$$

$$Me_3SnNEt_2 + p\text{-}CH_3C_6H_4N{=}C{=}NC_6H_4CH_3\text{-}p \longrightarrow Me_3Sn{-}N(C_6H_4CH_3\text{-}p){-}C(NEt_2){=}NC_6H_4CH_3\text{-}p \tag{36}$$

$$\begin{matrix} M{-}N \\ + \\ A{=}B \end{matrix} \longrightarrow \begin{matrix} M \quad N \\ | \quad\; | \\ A{-}B \end{matrix} \tag{37}$$

The essential characteristic of the reagent A=B appears to be that it should be susceptible to attack by nucleophiles.

A second type of insertion reaction relates to Sn—P cleavage (*155*):

$$Ph_3SnPPh_2 + PhNCX \xrightarrow{X=O \text{ or } S} Ph_3Sn{-}N(Ph){-}C({=}X){-}PPh_2$$

A third type of system is that involving fission of an Sn—O bond. Typical reagents have been Bu_3SnOMe or $(Bu_3Sn)_2O$. The topic is reviewed more extensively in Chap. 4. It will be seen that some of these reactions are of interest for providing syntheses of various organic nitrogen compounds not otherwise readily accessible. Others of these reactions have implications in connection with the mechanism of some processes catalyzed by organotin species. For the purpose of this chapter, however, it should be noted that substituted organotin carbamates have similarly been prepared, Eq. (39), by addition of organotin alkoxides to phenyl isocyanate (*16, 20, 21, 24, 119*), and R_3SnOR', $(R_3Sn)_2O$, and $R_2Sn(OR')_2$ additions analogous to Eqs. (33)–(36) have been described, Eq. (41) (*17–23, 25, 52, 53, 129*). It has further (*20, 21, 24*) been found that more than 1 mole of the 1,2-dipole may be inserted, for example as in Eq. (40).

The Sn—O bond is less reactive than Sn—N. Bu_3SnOPh is less reactive than Bu_3SnOMe [which is also less reactive than $(Bu_3Sn)_2O$] towards carbodiimides (*25*). The 1,2-dipoles decrease in reactivity from carbodiimides > PhNCO > PhNCS (which gives (*25*) an Sn—S and not an Sn—N compound). Nitriles are not effective 1,2-dipoles unless there are strong electron-withdrawing groups, as in Cl_3CCN (*53*). Phenyl isocyanate dimer behaves similarly to the monomer (*23*):

$$R_3SnOR' + PhNCO \longrightarrow R_3Sn{-}\underset{\underset{Ph}{|}}{N}{-}\underset{\underset{O}{\|}}{C}{-}OR' \qquad (39)$$

$$R_3Sn{-}\underset{\underset{Ph}{|}}{N}{-}\underset{\underset{O}{\|}}{C}{-}OR' + R''NCO \longrightarrow R_3Sn\underset{\underset{R''}{|}}{N}{-}\underset{\underset{O}{\|}}{C}{-}\underset{\underset{Ph}{|}}{N}{-}\underset{\underset{O}{\|}}{C}{-}OR' \qquad (40)$$

$$R_3SnOR' + {-}N{=}Y \longrightarrow R_3Sn{-}\underset{|}{N}{-}Y{-}OR' \qquad (41)$$

Isocyanates have also been reported to insert into Sn—H bonds, Eq. (42) (*130, 131*). The reactions follows a polar mechanism, with a postulated transition state $\left[\begin{array}{c} PhN{\cdots}C{\cdots}O \\ \vdots \\ H \\ \vdots \\ SnR_3 \end{array}\right]$ (*102, 103*):

$$R_3SnH + PhNCO \longrightarrow R_3Sn{-}\underset{\underset{Ph}{|}}{N}{-}\underset{\underset{O}{\|}}{C}{-}H \qquad (42)$$

Related Sn—H/—N=Y reactions (43)–(47) have been described (*118, 120, 122, 127, 128*). Reaction (47) represents a rare example of a 1 : 4-insertion (*127*):

$$R_3SnH + R'N{=}NR'' \longrightarrow R_3Sn{-}\underset{\underset{R'}{|}}{N}{-}\underset{\underset{R''}{|}}{N}H \qquad (43)$$

$$\mathrm{Et_3SnH + EtOC(=O)N{=}NC(=O)OEt \longrightarrow Et_3Sn{-}N(C(=O)OEt){-}N(C(=O)OEt){-}SnEt_3} \tag{44}$$

$$\mathrm{R_3SnH + PhCH{=}NAr \longrightarrow R_3SnN(Ar)CH_2Ph} \quad (\mathrm{Ar} = \mathrm{Ph},\ p\text{-tolyl}) \tag{45}$$

$$\mathrm{Et_3SnH + C_6H_{11}{-}N{=}C{=}N{-}C_6H_{11} \longrightarrow Et_3Sn{-}N(C_6H_{11}){-}CH{=}N{-}C_6H_{11}} \tag{46}$$

$$\mathrm{R_3SnH + R'CH{=}C(CN)_2 \longrightarrow R_3SnN{=}C{=}C(CN){-}CH_2R'} \tag{47}$$

Related to Eq. (43) are observations that the rate of homolysis of $PhN{=}NSO_2Ph$ is increased tenfold in presence of R_3SnH (*121*), and that R_3SnH catalyzes the decomposition of symmetrical dicyano-azoalkanes (*126*).

For the isocyanate insertions, Eqs. (33), (39), (40), (42), it has been assumed that Sn—N rather than Sn—O products are formed, i.e., that reactions follow pathway (48*a*) rather than (48*b*). The evidence is not yet conclusive (*21*):

$$\mathrm{R_3SnX + R'NCO} \begin{cases} \xrightarrow{(a)} \mathrm{R_3Sn{-}N(R'){-}C(=O){-}X} \\ \xrightarrow{(b)} \mathrm{R_3Sn{-}O{-}C(=NR'){-}X} \end{cases} \tag{48}$$

Finally, a rather different type of addition reaction, is the 1,1-insertion of a nitrene (*156*):

$$\mathrm{Ph_3SnPPh_2 + 2\,PhN_3 \longrightarrow Ph_3Sn{-}N(Ph){-}P(=NPh)Ph_2 + 2\,N_2} \tag{49}$$

$$\mathrm{(Ph_3Sn)_2PPh + 3\,PhN_3 \longrightarrow Ph_3Sn{-}N(Ph){-}P(=NPh)(Ph){-}N(Ph){-}SnPh_3 + 3\,N_2} \tag{50}$$

$$\mathrm{(Me_3Sn)_2PPh + 3\,PhN_3 \longrightarrow Me_3Sn{-}N(Ph){-}SnMe_3 + PhP(NPh)_2 + 3\,N_2} \tag{51}$$

D. Metathetical Reactions from Sn—O Compounds

Aminostannanes are, in general, readily susceptible to hydrolysis. In some cases, however, it has been possible to prepare, Eq. (52), compounds containing

Sn—N bonds, from an organotin oxide and an amine (*77*, *110*), particularly a cyclic amine (*48*, *77*, *78*, *92*, *109–111*), RNHC(:O)X (X = H, Me, or OMe) (*55*, *129*), or $RN(NO_2)H$ (*188*):

$$(R_3Sn)_2O + 2\,HN\!< \longrightarrow 2\,R_3Sn{-}N\!< + H_2O$$

In many of these reactions it has also been shown that several water-stabilizing effects are operative. From infrared spectral data and other evidence, it is likely that the tin atom in these aminostannaes is 5-coordinate, e.g., (*109*). This partial coordination-saturation of the tin, together with the low solubility in the reaction medium, probably accounts for the hydrolytic stability. A further feature of reactions (52) is that the nitrogen reagent HN< has low pK_a [e.g., is an imidazole (*70*), purine (*92*), nitramine (*188*), urea, or secondary amide (*55*, *129*)].

E. Miscellaneous Methods

(Trimethylstannyl)amine was postulated to be formed in the reaction of trimethylstannane with liquid ammonia, though it was not isolated (*84*). However, from trimethylstannylsodium and bromobenzene in liquid ammonia, a moisture-sensitive organotin compound appeared essentially to be tris(trimethylstannyl)amine (*27*):

$$Me_3SnNa + PhBr + NH_3 \longrightarrow (Me_3Sn)_3N + Me_3SnPh \qquad (53)$$

Aminodimethylstannylsodium was prepared during the conductometric titration of a solution of sodium in liquid ammonia with dimethylstannane (*80*):

$$Me_2SnNa_2 + NH_3 \longrightarrow Me_2Sn(NH_2)Na + NaH \qquad (54)$$

Other reactions in which Sn—N compounds may feature as intermediates (though they were not isolated) are represented by Eq. (55) (*79*, *85*, *90*) and Eq. (56) (*91*):

$$>\!Sn{-}H + H{-}NRR' \longrightarrow H_2 + \;>\!Sn{-}Sn\!< \qquad (55)$$

$$Ph_4Sn + Br{-}N\!\!\begin{array}{l}C(=O){-}CH_2\\ |\\ C(=O){-}CH_2\end{array} \xrightarrow[(2)\,H_2O]{(1)\,\text{heat}} (Ph_2SnO)_n + H{-}N\!\!\begin{array}{l}C(=O){-}CH_2\\ |\\ C(=O){-}CH_2\end{array} \qquad (56)$$

Cleavage of a Sn—C bond has been effected by use of the (protic nitrogen compounds) nitramines (*188*):

$$R_4Sn + H{-}NR'(NO_2) \longrightarrow R_3Sn{-}NR'(NO_2) + RH \qquad (57)$$

Reactions leading to Sn—N compounds include metathetical exchange reactions and reverse insertions of the type described in Sec. III.F. These are illustrated by Eq. (58) (*53*) and Eq. (59) (*19*, *20*, *22*, *24*); the question of mechanism of Eq. (59) is discussed in Chap. 4:

$$Bu_2SnCl_2 + Bu_2Sn\left(\begin{array}{c} O \\ \| \\ N-COMe \\ | \\ Et \end{array}\right)_2 \longrightarrow 2\ Bu_2Sn\begin{array}{c} Cl \quad\ O \\ |\quad\ \ \| \\ -N-COMe \\ | \\ Et \end{array} \qquad (58)$$

$$2\ Bu_3Sn-\underset{Ph}{\underset{|}{N}}-\underset{O}{\underset{\|}{C}}-OSnBu_3 \longrightarrow Bu_3Sn-\underset{Ph}{\underset{|}{N}}-\underset{O}{\underset{\|}{O}}-\underset{Ph}{\underset{|}{N}}-SnBu_3 + CO_2 + (Bu_3Sn)_2O \qquad (59)$$

Reactions of organotin pseudohalides which give rise to other Sn—N compounds are (60) (*161*), (61) (*112*), and (62) (*106*). Reaction (61) is a 1,3-cycloaddition. Another interesting example is represented by the general reaction (63), in which A=B represents a variety of 1,3-dipolarophiles, including R—N=C=NR, RNCS, RNCO, CS_2, PhCN, CH_2=CHCN, and MeOOCC≡CCOOMe (*95*):

$$Me_3SnNC + Fe(CO)_5 \longrightarrow Me_3Sn-N=C-Fe(CO)_4 + CO \qquad (60)$$

$$Bu_3SnN_3 + EtOOCC\equiv CCOOEt \longrightarrow Bu_3Sn-N\begin{array}{c} N=N \\ \diagdown\ \ \ | \\ C=C \\ |\quad\ | \\ EtO_2C\quad CO_2Et \end{array} \qquad (61)$$

$$R_3SnN_3 + PX_3 \longrightarrow R_3Sn-N=PX_3 + N_2 \qquad (62)$$

$$(Me_3Sn)_2CN_2 + A=B \longrightarrow Me_3SnC\begin{array}{c} A-B \\ \\ N \end{array}NSnMe_3 \qquad (63)$$

III. Sn—N Bond-Breaking

The chemistry of the simpler derivatives, such as (dimethylamino)trimethylstannane, Me_3SnNMe_2, has been explored in considerable detail. The principal reactions may be divided into six categories. These six types are valid for metal amides generally, although the tin amides are undoubtedly among the most reactive.

The reactions (Sec. III.A) with protic species, Eq. (64) (*73*), appear to require that HA should have a pK_a value of less than ~25, although kinetic acidity (cf., HA=C_6F_5H) is probably of greater significance than thermodynamic acidity. Among the more interesting examples of Eq. (64) are those where HA = C_5H_6, $CHCl_3$, C_6F_5H, RC≡CH, NH_3, Ph_2PH, AsH_3, and SbH_3:

$$\geqslant Sn-N\lessgtr + HA \longrightarrow \geqslant Sn-A + HN\lessgtr \qquad (64)$$

The interaction of amido- and hydrido-derivatives of metals and metalloids results in the synthesis of compounds having metal-metal bonds (*31*, *32*). Equation (65) provides an example (Sec. III.B) relevant to tin chemistry. The metal hydride LM—H need not necessarily be acidic [such as π-$C_5H_5Mo(CO)_3H$, π-$C_5H_5W(CO)_3H$, $(OC)_5MnH$, or $(F_3P)_4RhH$] but may also be neutral or basic [e.g., *trans*-$(Ph_3P)_2PtClH$, Ph_3GeH, or Me_3SnH]:

$$\geqslant Sn-N\lessgtr + LMH \longrightarrow \geqslant Sn-ML + HN\lessgtr \qquad (65)$$

The metathetical exchange reaction (Sec. III.C) provides a means of aminating the moiety "Q" (*59*), as shown in Eq. (66). The groups U being replaced include F, Cl, Br, H, OR, OCOR, R and examples of U—Q are F—C_6F_5, F—CF=CFCl, F—BF_2, Cl—$SiMe_3$, Br—$Mn(CO)_5$, H—BH_2, MeO—COMe, MeCOO—COMe, and Bu—BBu_2.

$$\geqslant Sn-N\lessgtr + U-Q \longrightarrow \geqslant Sn-U + \geqslant N-Q \qquad (66)$$

It is not inevitable that the Sn—N bond be cleaved. Thus, the next reaction class (Sec. III.D) is concerned with amidotin compounds as bases.

When a compound has both replaceable H and halogen (X), a combination of reactions with (i) hydrides and (ii) the metathetical exchange reaction becomes feasible—namely dehydrohalogenation (Sec. III.E) (*33*), as illustrated in Eq. (67). Compounds H$\leftarrow$B$\rightarrow$Cl which are easily dehydrochlorinated by Me_3SnNMe_2 include $(Ph_3P)_3IrHCl_2$ and BuCl:

$$\geqslant Sn-N\lessgtr + H\leftarrow B\rightarrow Cl \longrightarrow \geqslant SnCl \cdot HN\lessgtr + B \qquad (67)$$

Finally, a much studied system, Eq. (68), is the insertion reaction (Sec. III.F) (*49*, *72*, *96*). In general, Y is a 1,2-dipole, such as O=CO, S=CS, S=CO, O=SO, RN=CO, S=CNR, RN=CNR, CH_2=CO, RC≡N, RN=O, and RN=SO. Simple olefins or disubstituted acetylenes, e.g., PhC≡CMe (*99*), are unreactive, but powerfully electron-withdrawing groups encourage addition to occur, as with CH_2=CHCN or PhC≡CCl; the 1,2-dipole must evidently be susceptible to nucleophilic attack (*37*, *61*). However, 1,1-, 1,3-, and 1,4- insertions are also known.

$$\geqslant Sn-N\lessgtr + Y \longrightarrow \geqslant Sn-Y-N\lessgtr \qquad (68)$$

A. Reactions with Protic Species

The Sn—N bond in aminostannanes is readily cleaved by hydrogen chloride (*73*, *75*, *164*), and organotin cyanides (*105*, *106*) and azides (*104*, *106*) have conveniently been prepared, using hydrogen cyanide or hydrazoic acid. Acetic acid is also effective (*38*):

$$\geqslant Sn-N\lessdot + 2\,HX \longrightarrow \geqslant Sn-X + HN\lessdot \cdot HX \qquad (69)$$

Tin-nitrogen compounds are usually (but see Sec. II.D) easily hydrolyzed to the hydroxide [in case of $(Me_3SnOH)_2$] or oxide (*73, 75, 109, 165*):

$$\geqslant Sn-N\lessdot + H_2O \longrightarrow \rangle NH + \geqslant Sn-OH \longrightarrow (\geqslant Sn)_2O \qquad (70)$$

Compounds containing larger substituents, such as *n*-butyl or phenyl, on tin or nitrogen are somewhat more resistant to attack by moisture. In the series $(Me_3SiN(Me)\text{-})_nSnMe_{4-n}$, the water-sensitivity and steric hindrance are related (*145*).

Exposure of aminostannanes to the atmosphere, actually affords the carbonate (*75, 118, 164, 165*):

$$2 \geqslant SnN\lessdot + CO_2 + H_2O \longrightarrow (\geqslant Sn)_2CO_3 + 2 \rangle NH \qquad (71)$$

probably in a stepwise manner (*165*):

$$2\,R_3SnNR'_2 + H_2O \longrightarrow 2\,R'_2NH + (R_3Sn)_2O \xrightarrow{CO_2} (R_3Sn)_2CO_3 \qquad (72)$$

Aminostannanes react readily with alcohols, giving the corresponding alkoxides (*73, 75, 107*):

$$R_{4-n}Sn(NR'_2)_n + n\,R''OH \longrightarrow R_{4-n}Sn(OR'')_n + n\,R'_2NH \qquad (73)$$

This procedure was particularly convenient for the preparation of pure trimethylstannyl methoxide (*5*). Similarly, reactions with silanols provide routes to $\geqslant Sn-O-Si\leqslant$ compounds (*181*):

$$\geqslant Sn-N\lessdot + \rangle Si-OH \longrightarrow \geqslant Sn-O-Si\leqslant + \rangle NH \qquad (74)$$

Transamination reactions have already been discussed (Sec. II.B). In summary, displacements proceed from left to right in the series $-NH_2 > -NHR > -NR_2$ ($-NBu^n_2 > -NEt_2 > -NMe_2$), Eq. (22) (*74*). Similarly, reaction of Me_3SnNMe_2 with $Ph_2C{=}NH$ yielded Me_2NH and the stannylketimine $Me_3Sn-N{=}CPh_2$ (*43a*).

Examples of other Sn—N compounds which readily react with protic species such as HCl, H_2O, ROH, H_2S, and RNH_2 are *N*-stannyltriphenylphosphineimines (themselves obtained from $R'_3SnNR_2/HN{=}PPh_3$) (*106*), $R_3SnN{=}C{=}C(CN)CH_2R'$ (*166*), $R_3SnN(R')COOR''$ (*21*), $R_3SnN(R')COOSnR_3$ (*19, 22*), and *N*-stannyl-ureas and -biurets (*20, 24*). An explanation for the catalytic effect of organotins (such as R_3SnOR') on the conversion of isocyanates into urethanes by alcohols invokes tin carbamate alcoholysis as an integral step (*21*); a similar proposal has been made for the corresponding $R''N{=}C{=}NR''/R'''OH$ reaction (*25*) (for further details, see Chap. 4).

(Dimethylamino)trialkylstannanes react with phosphines, Eq. (75), and arsines, Eq. (76) (*52, 73, 75*); this is general, Eq. (77), for the other group V hydrides (M = P, As, Sb) (*64*):

$$R_3SnNMe_2 + Ph_2PH \longrightarrow R_3SnPPh_2 + Me_2NH \tag{75}$$

$$Me_3SnNMe_2 + Ph_2AsH \longrightarrow Me_3SnAsPh_2 + Me_2NH \tag{76}$$

$$3\,R_3SnNR'_2 + MH_3 \longrightarrow (R_3Sn)_3M + 3\,R'_2NH \tag{77}$$

Of interest are the displacement orders: (i) phosphine > amine, and (ii) arsine > amine. Evidently, tin(IV) is a class "*b*" acceptor (or soft acid), with preference for Sn—P and Sn—As rather than Sn—N bonds (*75*); this is confirmed by chemical observations on Sn—O and Sn—S compounds (*58*), and by preliminary thermochemical data (*11*). By contrast, Si appears to be a class "*a*" acceptor (or hard acid), as aminosilanes do not react with phosphines (*75*).

Upon addition of phenylacetylene to (dimethylamino)trimethylstannane at room temperature, dimethylamine was instantly evolved, and the residual tin compound was (phenylethynyl)trimethylstannane (*73, 75*). This reaction, Eq. (78), was found to be general. It has potential as a synthetic route to alkynylstannanes and, by using amido-derivatives of other electropositive elements, to other alkynylmetallanes. The advantages are the quantitative reactions, and the volatility of the only other product—Me_2NH (if R′ = Me). Other examples are in the reactions of Me_3SnNMe_2 with $HC{\equiv}C(CH_2)_2CMe_2$ and $HC{\equiv}CC(Me){=}CH_2$, respectively (*70a*):

$$R_{4-n}Sn(NR'_2)_n + n\,R''C{\equiv}CH \longrightarrow R_{4-n}Sn(C{\equiv}CR'')_n + n\,R'_2NH \tag{78}$$

When acetylene was used, the reaction proceeded directly to the second stage (*73, 75*):

$$2\,R_3SnNR'_2 + HC{\equiv}CH \longrightarrow R_3SnC{\equiv}CSnR_3 + 2\,R'_2NH \tag{79}$$

this was also the case with $HC{\equiv}C(CH_2)_3C{\equiv}CH$ (*70a*). Other unsaturated hydrocarbons, which react similarly are cyclopentadiene, Eq. (80), and indene, Eq. (81) (*73, 75*). In all these cases, the parent hydrocarbon is acidic and has a stable anion:

$R_3SnNR'_2$ + [cyclopentadiene] ⟶ R_3Sn–[cyclopentadienyl] + R'_2NH (80)

$R_3SnNR'_2$ + [indene] ⟶ R_3Sn–[indenyl] + R'_2NH (81)

A recent development has been the demonstration that some halogenated hydrocarbons react in a similar sense, Eq. (82), to afford R_3SnCCl_3 (*42, 52*), R_3SnCBr_3 (*54*), and $Me_3SnCCl{=}CCl_2$ (*41, 42*). Especially surprising is the ready formation in this way of $Me_3SnC_6F_5$ and $Et_3SnC_6F_5$ (*70a*) ,since C_6F_5H is certainly not an "acidic" hydrocarbon; C_6Cl_5H, however, proved unreactive. The C_6F_5H experiment suggests that kinetic acidity, (c.f. *174*), is an important criterion (hydrogen isotope exchange in C_6F_5H is facile in basic media). With $(CF_3)_2CFH$ and Me_3SnNMe_2 at 25°C, the Sn—C compound if formed was not isolated; Me_3SnF was obtaned (*41*):

$$R_3SnNR'_2 + HC\lessgtr \longrightarrow R_3SnC\lessgtr + R'_2NH \tag{82}$$

The diazomethane reaction, Eq. (83), has proved a little confusing. Initially, the volatile stannyldiazomethane was regarded as Me_3SnCHN_2 (*93*), formulated (on the basis of ^1H NMR) as a tautomer with the nitrile-imine form, $Me_3Sn\overset{+}{C}{=}N{-}\overset{-}{N}H$, predominating. More recent evidence [Eq. (63)] establishes $(Me_3Sn)_2CN_2$ (*95, 95b*):

$$Me_3SnNMe_2 + CH_2N_2 \xrightarrow{-Me_2NH} Me_3SnCHN_2 \xrightarrow{-Me_2NH} (Me_3Sn)_2CN_2 \tag{83}$$

Similarly, there have been obtained $(R_3Sn)_2CN_2$ (R = Et or *n*-Bu) (*93a*), $Me_3Sn(Me_3Si)CN_2$ (*93a, 95a*), and $R_3SnC(COOEt)N_2$ (*105a*) from R_3SnNMe_2 and CH_2N_2, Me_3SiCHN_2, or $CH(COOEt)N_2$. In the aminometallane/diazoalkane system reactivities increased in the sequences (i) Si $\ll$ Ge $<$ Pb and (ii) $MeCHN_2 < CH_2N_2 < CH(COOEt)N_2$ (*93a*). Reference (*95b*) is a review dealing with α-heterodiazoalkanes and the reactions of diazoalkanes with derivatives of metals and metalloids (*95b*).

Although (dimethylamino)trimethylstannane reacts with benzonitrile as a 1,2-dipolarophile, nitriles having an α-hydrogen atom appeared to behave as protic species (*75*):

$$R_3SnNR'_2 + R''R'''CHCN \longrightarrow R_3SnC(R'')(R''')\text{—}CN + R'_2NH \tag{84}$$

The amine was always isolated in high yield, but the residue invariably afforded a mixture of nitrile-containing products, suggesting thermal redistribution (*75*).

Nitromethane also afforded a high yield of amine, but again the tin products were not characterized (*75*); however, with Me_2CHNO_2, the β-nitrostannane was obtained (*104*):

$$R_3SnNR'_2 + Me_2CHNO_2 \longrightarrow R_3SnCMe_2NO_2 + R'_2NH \tag{85}$$

Similarly, (dialkylamino)trialkylstannanes reacted with acetone and isopropenyl acetate (*75*):

$$2\,R_3SnNR'_2 + 2\,Me_2CO \longrightarrow (R_3Sn)_2O + 2\,R'_2NH + (CH_3)_2C{=}CHCOCH_3 \quad (86)$$

$$2\,R_3SnNR'_2 + 2\,CH_2{=}\underset{\displaystyle Me}{\underset{|}{C}}OCOMe \longrightarrow$$

$$(R_3Sn)_2O + 2\,R'_2NCOCH_3 + (CH_3)_2C{=}CHCOCH_3 \quad (87)$$

It appears that both reactions proceed through the common intermediate $R_3SnCH_2COCH_3$, forming the organotin oxide and mesityl oxide by base catalysis.

Although decaborane is a protic acid, it forms a 1:1 complex with Me_3SnNMe_2, which appears to be $[Me_3Sn]^+\ [B_{10}H_{13}\cdot Me_2NH]^-$ (*39*).

In conclusion, the great facility of these Sn—N cleavage reactions suggests that the mechanism involves proton abstraction, rather than hydride ion or hydrogen atom loss, from HA. This is further evident from the non-reactivity of the Si—H bond, as demonstrated in the Me_3SnNMe_2/Et_3SiH (*75*), $Me_3SnNMe_2/NaBH_4$ or $LiAlH_4$ (*151*) systems (but see Sec.III.C).

B. Reactions with Metal Hydrides

The reaction of aminostannanes (*47*, *123–125*, *167*, *168*), and of hydrazino- (*128*) and amido- (*46*, *128*) stannanes, with organotin hydrides, Eq. (88), is another of great promise. It is probably the most important method for the synthesis of compounds containing Sn—Sn bonds (*45*, *46*, *167*), especially because of its unambiguous nature:

$$\geq Sn{-}N\langle \;+\; \geq Sn{-}H \longrightarrow \geq Sn{-}Sn\leq \;+\; \rangle NH \quad (88)$$

The reactivity of hydrides $R_3Sn{-}H$ increases in the order R = s-Bu < *i*-Bu < *n*-Bu < Ph (*125*), and increases in the series $R_3SnSnR_2H < R_3SnH < R_2SnH_2$ (*168*). Amide $R'_3SnNR''_2$ reactivity increases with increasing basicity of nitrogen, i.e., $R'_3SnNHR'' < R'_3SnNR''_2$ (*125*); $(Me_3Sn)_2NPr\text{-}i$ was also highly reactive. A polar mechanism, in the sense $\geq \overset{\delta-}{Sn}{-}\overset{\delta+}{H}$, is therefore indicated (*44*, *47*). In polar solvents, an S_E2 mechanism and in non-polar media, an S_F2 (four-center) mechanism was proposed (*47*); possible transition states are shown below.

$$\begin{array}{c} R'_3\overset{\delta+}{Sn}{-}\overset{\delta-}{N}\langle \\ \vdots \\ H^{\delta+} \\ | \\ R_3Sn^{\delta-} \end{array} \qquad \begin{array}{c} R'_3\overset{\delta+}{Sn}{-}\overset{\delta-}{N}\langle \\ \vdots \qquad \vdots \\ R_3\overset{\delta-}{Sn}{-}\overset{\delta+}{H} \end{array}$$

Tin hydrides are obviously very versatile, being also capable of participating in radical reactions (*167*); and in other polar ones, where the hydrogen is hydridic [as in Eq. (42)] (*103*).

Exchange reactions of the type shown in Eq. (89*a*) may sometimes be competing processes for the Sn—Sn forming reactions, Eq. (89*b*), but are not (*47*), (cf. *125*, *168*), important in the reaction sequence leading to Sn—Sn bond formation:

$$Et_3Sn{-}N\langle + R_3Sn{-}H \begin{cases} \xrightarrow{(a)} Et_3Sn{-}H + R_3Sn{-}N\langle \\ \xrightarrow{(b)} Et_3Sn{-}SnR_3 + HN\langle \end{cases} \quad (89)$$

For the synthesis of assemblies of more than two tin atoms, the use of dihydrides R_2SnH_2 rather than diamides $R'_2Sn(NR''_2)_2$ is preferred, as in Eqs. (90) and (91) (*168*). This is partly due to a question of relative reactivities, but also due to a competing exchange reaction in the $R_3SnH/R'_2Sn(NEt_2)_2$ system, leading to $R'_2Sn(H)NEt_2 + R_3SnNEt_2$:

$$2\,R'_3SnNEt_2 + R_2SnH_2 \longrightarrow R'_3Sn{-}\underset{\substack{|\\R}}{\overset{\substack{R\\|}}{Sn}}{-}SnR'_3 + 2\,Et_2NH \quad (90)$$

$$2\,R'_3SnNEt_2 + HSnR_2SnR_2H \longrightarrow R'_3Sn\left(\underset{\substack{|\\R}}{\overset{\substack{R\\|}}{Sn}}\right)_2SnR'_3 + 2\,Et_2NH \quad (91)$$

The tin amide ($R'_3SnNR''_2$)-metal hydride reaction has been extended to Ph_3GeH; R_3GeH and Ar_3SiH proved unreactive (*125*).

The metal amide/metal hydride reaction has been generalized, Eq. (92) (*31a*, *32*) (where L and L′ represent the sum of all ligands other than >N or H attached to the metal or metalloid M and M′); for tin compounds it is summarized by Eq. (65):

$$L'M'{-}N\langle + H{-}ML \longrightarrow L'M'{-}ML + HN\langle \quad (92)$$

It has been used for compounds containing Sn bonded to Ta (*76a*), Cr (*31a*), Mo (*30*, *31a*, *76a*), W (*31a*, *32*, *76a*), Mn (*76a*), Re (*76a*), Rh (*31a*, *32*), and Pt. The hydride need not necessarily be protic, but the reaction conditions are less severe if this is the case (*32*). The hydrides include π-$C_5H_5(CO)_3M{-}H$ (M = Cr, Mo, W), $(\pi\text{-}C_5H_5)_2(H_2)Ta{-}H$, $(\pi\text{-}C_5H_5)_2HM{-}H$ (M = Mo, W), and $(\pi\text{-}C_5H_5)_2M{-}H$ (M = Mn, Re). In the amide/hydride system, reactivities decreased in the series (i) Cr > Mo > W and (ii) SnN > GeN ~ SiN(*31a*). When the metal hydride also contains halogen, further elimination of tin halide is possible. Thus, $(F_3P)_4RhH$ initially affords $(F_3P)_4Rh{-}SnMe_3$ but subsequently Me_3SnF (*32*); while *trans*-$(Ph_3P)_2Pt(Cl)SnMe_3$, and Me_3SnCl with $[(Ph_3P)_2Pt]_n$, are formed successively from *trans*-$(Ph_3P)_2PtClH$ and Me_3SnNMe_2 (*33*). These reactions are therefore closely related to those described in Sec. III.E.

C. Metathetical Exchange Reations; Tin Amides as Aminating Reagents

The metathetical exchange reaction is represented by Eq. (66) and more specifically by Eq. (93). Its interest lies mainly as providing a convenient synthesis for a wide variety of amido derivatives >N—Q (*59*). In the sense of Eq. (93), (i) fluorine can be substituted (*59, 60*) in C_6F_6, C_6F_5Br (*38a*), C_6F_5CN (*38a*), $ClFC{=}CF_2$ (*59, 60*), $(CF_3)_2NH$, $BF_3 \cdot OEt_2$, PF_3, AsF_3, SbF_3, and TiF_4; (ii) chlorine in BCl_3, Me_3SiCl, Cl_2, and Ph_3OsCl_3 (*34*); (iii) bromine in $BrMn(CO)_5$ (*59, 60*); (iv) iodine in R′I (with $R_3SnN{=}PX_3$) (*106*); (v) hydrogen in $Et_3N \cdot BH_3$ (*59, 60*), B_2H_6 (*87*), $(Bu_2AlH)_2$ Eq. (100a), $Ph_2Si(Cl)H$ (*33*), and R_3SnH (*188*); methyl (*146*), ethyl (*59, 60*), butyl, and phenyl groups are displaced from $(MeLi)_n$ (*146*), $(Et_3Al)_2$ (*59, 60*), Bu_3B, and Ph_3B; (vi) methoxy, ethoxy, and phenoxy groups from $B(OMe)_3$, Ph_2BOMe, and most carboxylic esters, e.g., $PhOCOCH_3$ (*37, 59, 60*); and (vii) acetate can also be displaced (*21, 22, 38, 59, 60*) from vinyl acetate and from acetic anhydride (*59, 60*):

$$\geq Sn{-}N\lt \; + \; U{-}Q \longrightarrow \; \geq Sn{-}U \; + \; QN\lt \tag{93}$$

The reaction is further illustrated by:

$$Me_3SnNMe_2 + C_6F_5CN \xrightarrow{(99)} Me_3SnF + NC{-}C_6F_4{-}NMe_2 \tag{94}$$

(ring: 2,3,5,6-tetrafluoro, F F above and F F below)

$$Me_3SnNMe_2 + CF_2{=}CFCl \xrightarrow{(59,\,60)} Me_3SnF + CFNMe_2{=}CFCl \tag{95}$$

$$3\,Me_3SnNMe_2 + BF_3 \cdot OEt_2 \xrightarrow{(59,\,60)} 3\,Me_3SnF + B(NMe_2)_3 + OEt_2 \tag{96}$$

$$Me_3SnNMe_2 + Ph_2Si(Cl)H \xrightarrow{(33)} Me_3SnCl + Ph_2Si(NMe_2)H \tag{97}$$

$$2\,R_{4-n}Sn(NEt_2)_n + n\,B_2H_6 \xrightarrow{(87)} 2\,R_{4-n}SnH_n + 2n\,H_2BNEt_2 \tag{98}$$

$$(Me_3Sn)_3N + MeLi \xrightarrow{(146)} (Me_3Sn)_2NLi \text{ and } Me_3SnNLi_2 \tag{99}$$

$$2\,Me_3SnNMe_2 + (AlEt_3)_2 \xrightarrow{(59,\,60)} 2\,Me_3SnEt + (Et_2AlNMe_2)_2 \tag{100}$$

$$Me_nSn(NEt_2)_{4-n} + n\,Bu^n_2AlH \xrightarrow{(110)} Me_nSnH_{4-n} + n\,Bu^n_2AlNEt_2 \tag{100a}$$

$$3\ Me_3SnNMe_2 + 3\ B(OMe)_3 \xrightarrow{(59,\,60)} 3\ Me_3SnOMe + B(NMe_2)_3 + 2\ B(OMe)_3 \quad (101)$$

$$Me_3SnNMe_2 + PhOCOMe \xrightarrow{(37)} Me_3SnOPh + MeCONMe_2 \quad (102)$$

$$Me_3SnNMe_2 + CH_2{=}C(Me)CO_2Me \xrightarrow{(37)} Me_3SnOMe + CH_2{=}C(Me)CONMe_2 \quad (103)$$

$$Me_3SnNMe_2 + CH_2{=}CHOCOMe \xrightarrow{(37,\,38)} Me_3SnOCOCH_3 + Me_2NCH{=}CH_2 \quad (104)$$

Reactions (102), (103), and (104) have some mechanistic interest. With vinyl acetate, acyl-oxygen fission results; phenyl and alkyl acetates react by alkyl-oxygen fission. Methyl methacrylate gives, Eq. (103), MeO/NMe_2 exchange, while methyl acrylate undergoes insertion (*37*, *61*):

$$Me_3SnNMe_2 + CH_2{=}CHCO_2Me \longrightarrow Me_2NCH_2{-}CH(SnMe_3)CO_2Me \quad (105)$$

N-stannylketimines behave similarly, Eq. (106) (where UQ = $CH_2{=}CHCH_2Br$, $CH_3CH{=}CHCH_2Cl$, $PhCH_2Cl$, Br_2, and R″COCl) (*166*):

$$RCH_2C(CN){=}C{=}N{-}SnR'_3 + U{-}Q \longrightarrow R'_3Sn{-}U + RCH_2C(CN){=}C{=}N{-}Q \quad (106)$$

D. Reactions as Lewis Acids and Bases

In this section, we are concerned solely with those few reactions in which the integrity of the Sn—N bond is maintained; with the Sn—N compound behaving simply as a Lewis acid or base. Physical data relating to base strengths are discussed in Sec. IV.

In principle, an aminostannane has both acceptor (the tin atom) and donor (the nitrogen atom) sites, and is therefore capable of behaving as a Lewis acid or base. However, the number of such reactions that have been observed is quite small, probably because of the ready cleavage of the Sn—N bond.

N-Tributylstannylimidazole, which exists as a coordination polymer, forms monomeric complexes with basic ligands in which the tin is five-coordinate (*70*). The viscosity of the polymer solution drops sharply when a more powerful donor is added to the system, indicating depolymerization to be taking place:

$$\left[-N\langle C_3H_3\rangle N{-}Sn(Bu_3)- \right]_n + n\,B \longrightarrow n\ N\langle C_3H_3\rangle N{-}Sn(Bu_3){\cdot}B \quad (107)$$

Other Sn—N compounds with secondary donor sites may also exist with tin in a five-coordinate environment. A case in point is the nitramine, believed to have the chelate structure shown below, on the basis of electrical conductivity in nitrobenzene, molecular wight, and infrared and Mössbauer spectra (*188*):

$$Me_3\overset{-}{Sn}\langle O, N(Me) \rangle \overset{+}{N}O$$

The following complexes, in which a Sn—N compound behaves as a Lewis base, as illustrated by Eqs. (108)–(110), have been characterized: $(Me_2N)_4Sn \cdot VOCl_3$ (decomp. 45°C) (*98*), $[(Me_3Sn)_4N]^+Br^-$ (*1*), and *trans*-$(Bu_3P)PtCl_2(Me_3SnNMe_2)$ (*34*):

$$(Me_2N)_4Sn + VOCl_3 \longrightarrow (Me_2N)_4Sn \cdot VOCl_3 \quad (108)$$

$$(Me_3Sn)_3N + Me_3SnBr \longrightarrow [(Me_3Sn)_4N]^+Br^- \quad (109)$$

$$[(Bu_3P)PtCl(\mu\text{-}Cl)]_2 + 2\,Me_3SnNMe_2 \longrightarrow (Bu_3P)ClPtCl(NMe_2SnMe_3) \xrightarrow{\text{heat}} Me_3SnCl \quad (110)$$

E. Tin Amides as Dehydrohalogenating Reagents

It has been seen that tin amides react both with hydrogen (Secs. III. A and III. B) and halogen (Sec. III. C) compounds. It is therefore not surprising that they are among the most powerful dehydrohalogenating reagents, Eq. (67) (*33*), perhaps as a result of this synergic effect. This is further illustrated by reactions (111), (112), and (113). Neither of the iridium(I) products was isolated, but evidence for their existence was furnished by reaction products with other reagents [e.g., CO afforded *trans*-$(Ph_3P)_2Ir(CO)Cl$] (*33a*); in the absence of such added components, the ultimate iridium species appeared to be a complex mixture (*92a*). The efficacy of Me_3SnNMe_2 as a dehydrochlorinating reagent was attributed to a combination of factors which include (i) the weak and highly polar Sn—N bond, (ii) the high basicity of Me_3SnNMe_2, (iii) the high value for the heat of formation of crystalline $Me_3SnCl \cdot HNMe_2$, and (iv) for inorganic compounds, the intermediate formation of compounds having metal-tin bonds (see Sec. III. C for Pt—Sn and Rh—Sn). The reaction of Me_3SnNMe_2 and $(Ph_3P)_3IrHCl_2$ has been investigated under milder conditions (ambient temperature, 30 min) when $(Ph_3P)_2Ir(Cl_2)SnMe_3$ could be isolated (*92a*). This may represent the first stage in the reaction [see (iv)], although neither elimination of Me_3SnCl nor isolation of $(Ph_3P)_3IrCl$ has been established.

$$(Ph_3P)_3IrHCl_2 + Me_3SnNMe_2 \xrightarrow[\text{2h}]{\text{xylene (80–100°C)}} (Ph_3P)_3IrCl + Me_3SnCl \cdot HNMe_2 \quad (111)$$

$$(Ph_3P)_3IrH_2Cl + Me_3SnNMe_2 \xrightarrow[1\frac{1}{2}\text{ h}]{\text{xylene (80–100°C)}} (Ph_3P)_3IrH + Me_3SnCl \cdot HNMe_2 \quad (112)$$

$$MeCH_2CH_2CH_2Cl + Me_3SnNMe_2 \xrightarrow[4h]{40°C} \textit{cis-}(5.2\%) \text{ and } \textit{trans-}(88.5\%)\ MeCH{=}CHMe + (6.3\%)\ MeCH_2CH{=}CH_2 \quad (113)$$

It is interesting that in Eq. (113), there is a preponderance of the thermodynamically less stable olefins.

Other cases in which dehydrochlorination by Me_3SnNMe_2 probably occurred (in that $Me_3SnCl \cdot HNMe_2$ was formed) are with Cl_3CCHCl_2 (*42*) and $CH{\equiv}CCH_2Cl$ (*39*). Dehydrochlorination is obviously not invariably observed, e.g., see Eq. (97).

F. Reactions with 1,2-Dipoles, and Their Retrogressions

Those addition reactions which produce further compounds containing a tin-nitrogen bond (for example those with phenyl isocyanate) have been discussed already (Sec. II.C). Other examples of this general reaction which have been reported are with carbon dioxide and carbon disulfide (*58*, *72*):

$$R_3SnNR'_2 + CX_2 \longrightarrow R_3Sn{-}X{-}\underset{\underset{X}{\|}}{C}{-}NR'_2 \quad (114)$$

When $R = R' = Me$ and $X = O$, the product is actually a polymeric carbamate involving a 5-coordinate tin atom and planar trimethyltin group:

$$\cdots O\cdots \overset{\overset{Me}{|}}{\underset{Me\ Me}{Sn^{-}}}\cdots O\overset{\overset{NMe_2}{|}}{\overset{C}{\ \ \overset{+}{\cdots}\ \ }}O\cdots \overset{\overset{Me}{|}}{\underset{Me\ Me}{Sn^{-}}}\cdots O\overset{\overset{NMe_2}{|}}{\overset{C}{\ \ \overset{+}{\cdots}\ \ }}O\cdots$$

When $R = R' = Me$ and $X = S$, the product is predominantly monomeric but may still contain a 5-coordinate tin atom:

$$Me_3\overset{-}{Sn}\langle{}^{S}_{S}\rangle\overset{+}{C}{-}NMe_2$$

Other 1,2-dipolar insertion reactions which have similarly been demonstrated are as follows:

$$R_3SnNR'_2 + CH_2{=}C{=}O \xrightarrow{(58)} R_3SnCH_2CONR'_2 \quad (115)$$

$$R_3SnNR'_2 + CH_2{=}\underset{CH_2-C=O}{\overset{}{C-O}} \xrightarrow{(66)} R_3Sn{-}\langle{}^{O-CHMe}_{O=\overset{+}{C}(NR'_2)}\rangle{=}CH \quad (116)$$

$$R_3SnNR'_2 + (CF_3)_2CO \xrightarrow{(3)} R_3SnOC(CF_3)_2NR'_2 \quad (117)$$

$$R_3SnNR'_2 + SO_2 \xrightarrow{(58)} R_3SnOSONR'_2 \quad (118)$$

$$R_3SnNR'_2 + PhNO \xrightarrow{(61)} R_3SnONPhNR'_2 \quad (119)$$

$$R_3SnNR'_2 + R''NCS \xrightarrow{(58, 61, 72)} R_3SnSC(NR'')NR'_2 \quad (120)$$

$$R_3SnNR'_2 + PhC{\equiv}CCl \xrightarrow{(37, 61)} R_3SnCCl{=}CPhNR'_2 \quad (121)$$

$$R_3SnNR'_2 + EtOOCC{\equiv}CCOOEt \xrightarrow{(37, 61)} R_3Sn\underset{\displaystyle COOEt}{\underset{|}{C}}{=}C(COOEt)NR'_2 \quad (122)$$

$$R_3SnNR'_2 + CH_2{=}CHCN \xrightarrow{(37, 38a, 61)} R_3Sn\underset{\displaystyle CN}{\underset{|}{C}}H{-}CH_2NR'_2 \quad (123)$$

$$R_3SnNR'_2 + CH_2{=}C(Me)CN \xrightarrow{(37, 61)} R_3Sn\underset{\displaystyle CN}{\underset{|}{C}}(Me){-}CH_2NR'_2 \quad (124)$$

$$R_3SnNR'_2 + CH_2{=}C(Cl)CN \xrightarrow{(99)} R_3Sn\underset{\displaystyle CN}{\underset{|}{C}}(Cl){-}CH_2NR'_2 \quad (125)$$

$$R_3SnNR'_2 + CH_2{=}CHCHO \xrightarrow{(37,61)} R_3Sn\left(\underset{\displaystyle CHO}{\underset{|}{C}}H{-}CH_2\right)_n NR'_2 \quad (126)$$

$$R_3SnNR'_2 + CH_2{=}CHCO_2Me \xrightarrow{(37, 61)} R_3Sn\underset{\displaystyle CO_2Me}{\underset{|}{C}}H{-}CH_2NR'_2 \quad (127)$$

$$R_3SnNR'_2 + MeCH{=}CHCHO \xrightarrow{(37, 61)} R_3Sn\underset{\displaystyle CHO}{\underset{|}{C}}H{-}CH(Me)NR'_2 \quad (128)$$

$$R_3SnNR'_2 + PhCH{=}CHCHO \xrightarrow{(37, 61)} R_3Sn\underset{\displaystyle CHO}{\underset{|}{C}}H{-}CH(Ph)NR'_2 \quad (129)$$

The insertion of isothiocyanates, Eq. (120), was originally (*58*, *72*) thought to proceed by addition across the $-N{=}C\langle$ rather than the C=S bond (*25*, *61*).

The contrasting behavior of ketene, Eq. (115), and diketene, Eq. (116), is noteworthy.

The possibility of insertion of more than 1 mole of the addendum was noted, Eq. (126), with $CH_2{=}CHCHO$ (*37*, *61*); at low temperature, only 1 mole is introduced.

The reactions of some distannyl- and tristannyl-amines with dipoles containing a C=S group have been studied (*67, 68, 182*). The insertion products were not isolated, as shown in Eq. (130) (*49, 182*) and Eq. (131) (*49, 67, 68*):

$$(Me_3Sn)_3N + CS_2 \longrightarrow Me_3SnNCS + (Me_3Sn)_2S \tag{130}$$

$$(Me_3Sn)_2NMe + {>}C{=}S \longrightarrow (Me_3Sn)_2S + {>}C{=}NMe \tag{131}$$

The first example of a 1,1-insertion of a stable compound into a non-transition metal M—X bond was observed in Sn—N chemistry, Eq. (132) (*61*). Reaction (50) represents a related 1,1-insertion of a nitrene (*156*):

$$Me_3SnNMe_2 + CN{-}C_6H_4{-}Me \longrightarrow (Me_3Sn)(Me_2N)C{=}N{-}C_6H_4{-}Me \tag{132}$$

Insertion in a formal 1,3-manner has been observed with epoxides, Eq. (133) (*183*); however, a trace of $LiNEt_2$ was required to catalyze the process; and a chain process, with $RCH(OLi){-}CH_2NEt_2$ as a propagating species, may be involved. Similarly, ethylene and propene sulfides afforded $Me_3SnSCH_2CH_2NMe_2$ and $Et_3SnSCH(Me)CH_2NMe_2$, respectively (*38a*); a catalyst was not required. Reaction (83) may also involve initial 1,3-insertion of $\overset{-}{C}H_2{-}N{=}\overset{+}{N}$ (*93, 93a*):

$$Bu_3SnNEt_2 + \underset{\diagdown O \diagup}{RCH{-}CH_2} \xrightarrow{LiNEt_2} Bu_3SnOCH(R)CH_2NEt_2 \tag{133}$$

The cleavage of γ-butyrolactone, Eq. (135), is an example of a 1,4-insertion reaction (*69*). It is interesting that while the Sn—N compound induces acyl-oxygen cleavage of the lactone, analogous Si—N or Ge—N compounds produce alkyl-oxygen cleavage:

$$Me_3MNMe_2 + H_2C\underset{O}{\overset{CH_2}{\diamond}}C{=}O \begin{cases} \xrightarrow{M = Sn} Me_3SnOCH_2CH_2CONEt_2 \\ \xrightarrow[M = Si \text{ or } Ge]{} Me_3MOCOCH_2CH_2NEt_2 \end{cases} \tag{134}$$

Related insertion reactions are also exhibited by some Sn—O (see Chap. 4), Sn—H (see Chap. 1), and many other compounds (*96*).

Many of these insertion reactions, which are often dipolar in character and may involve cyclic transition states,

$$\begin{matrix} R_3Sn-NR'_2 \\ \curvearrowleft\curvearrowright \\ A{=}B \end{matrix}$$

are probably thermodynamically controlled processes (*58*, *72*). Their retrogressions have been observed in a number of cases; for example during attempted distillation of the adducts $Ph_3SnOPh \cdot PhNCO$ (*21*) and $Bu_3SnOMe \cdot ArN{=}C{=}NAr$ (*25*).

IV. Physical Properties

The physical properties which are associated with synthetic chemistry (mp, bp, d_4^{20}, n_D^{20}) have been recorded for a large number of tin-nitrogen derivatives (see Appendices). The results of systematic studies of other properties (ir, Raman, nmr, and Mössbauer spectra, dipole moments basicity, thermochemistry, ionization potentials, and viscosity) are also increasing, although there are, as yet, no data on bond lengths or angles or electronic spectra; and there is little information on reaction mechanisms.

Infrared spectroscopy has been useful in following the rate of reactions, particularly those involving $\geq Sn-H$ (*128*), and in elucidating the structures of the products of addition reactions (*58*, *61*, *131*). The Me_3Sn group in tris-(trimethylstannyl)amine, and in other aliphatic aminotrimethylstannanes, is not planar, as indicated by the presence of both the symmetric and antisymmetric SnC_3 stretching modes (*52*, *165*), whereas in the imidazole derivatives, the presence of only one SnC_3 stretch suggests a planar configuration (*78*, *109*). The identification of both ν_{asym} and ν_{sym} (SnC_3) in $R_3SnN(NO_2)R'$ strengthened the proposal of a five-coordinate monomeric structure for the nitraminostannane (*188*). Infrared spectral data have also been used to discuss the question of basicity (see below).

The question of assignment of SnN stretching frequencies is still not settled. For $(R_3Sn)_3N$, ν_{asym} (Sn_3N) was initially assigned to the 712–728 cm^{-1} region, with the symmetrical vibration falling below 400 cm^{-1} (*165*), while the band at 510 cm^{-1} in $Me_3SnNPhCONMe_2$ was tentatively proposed as ν(SnN) (*58*). In $Me_3SnNHPh$, the use of ^{14}N and ^{15}N isotopomers has led to the suggestion that $\nu(Sn{-}^{14}N)$ lies at 843 ± 1 cm^{-1} and $\nu(Sn{-}^{15}N)$ at 835 cm ± 1 cm^{-1}; the bands at 535 and 508 cm^{-1} were assigned as ν_{asym} and ν_{sym} (SnC_3), respectively (*134*). Analysis of the infrared and Raman spectra of $(Me_3Sn)_3N$ showed that the skeletal geometry could not be D_{3h}, but was pyramidal around nitrogen (*64*); moreover, the strongly Raman polarized band at 514 cm^{-1} was confidently attributed to ν_{sym} (Sn_3N), while the band at 672 cm^{-1} was assigned as ν_{asym} (Sn_3N). The weak Raman line at 538 cm^{-1} in

$Sn(NMe_2)_4$ was identified as ν_{asym} (SnN_4) (*64*); this has been confirmed (536 cm^{-1}) and the strongly-polarized Raman line at 516 cm^{-1} was attributed to the symmetrical counterpart (*28*). Other assignments of tin-nitrogen stretching frequencies have been to bands at 596 and 592 cm^{-1} in $Ph_3SnN(Ph)P(NPh)Ph_2$ and $[Ph_3SnN(Ph)]_2P(NPh)Ph$, respectively (*156*); $\sim$880cm^{-1} for ν_{asym} (SnN_2) in $R_2Sn(NR'_2)_2$ (*190*); and 479, 480, and 472 cm^{-1} in $Me_3C(Me_3Si)NSnMe_3$, $[Me_3C(Me_3Si)]_2N$—, and $Me_3C(Me_3Si)NSnMe_2Cl$, respectively (*158*).

Compounds on which proton nmr data have been reported (*83, 86, 100, 107, 108, 136–139, 145, 147–150, 152, 154, 157, 158*) or Mössbauer spectra (*51, 63, 188*) are indicated in the Appendices. In all compounds so far examined, the spectra are consistent with the predicted structures.

Some attention has been devoted to spin-spin coupling [$J(^{119}Sn—H)$ and $J(^{117}Sn—H)$] in Sn—N—C—H systems (*86, 107, 108, 137, 145*). In certain cases [R = Me, Ph and $0 \leqslant n \leqslant 3$ in $R_nSn(NEt)_{4-n}$] such coupling was not detected (*86, 107*), presumably because of fast exchange; spectra taken at $-50°C$ failed to reveal satellites (*108*). However, in the ^1H nmr spectrum of Et_3SnNMe_2 the ^{117}Sn and ^{119}Sn satellites, observable at moderate temperatures, disappeared at 150°C (*137*); while Me_3SnNMe_2 showed no satellites even at $-68°C$. Evidently, exchange is favored by either a decrease in the number (n) or size of the groups R attached to Sn, and by a decrease in the size of the groups R′ on N. In summary, for exchange rates: (i) $n\text{-}Bu_3SnNMe_2 < Et_3SnNMe_2 < Me_3SnNMe_2$, and (ii) $n\text{-}Bu_3SnNMe_2 < n\text{-}Bu_2Sn(NMe_2)_2 < n\text{-}BuSn(NMe_2)_3$. By examining mixtures of $R_nSn(NR'R'')_{4-n}/R'R''NH$, exchange and transamination have been examined (see Sec. II.B) (*137*). $J(^{15}N—H)$ in each of ($Me_3M—^{15}NPh$) (M = C, Si, or Sn) and

$$\begin{array}{c} Me_3M—^{15}NPh \\ | \\ H \end{array}$$

$[Ph^{15}NH_3]^+$ was roughly similar, suggesting that M⤺N̈ π-bonding was not likely to be important (*138*). More qualitative examination of ^1H nmr spectra in $Me_3MN(GeMe_3)Me$ (M = Si, Ge, Sn, or Pb) led to the suggestion that both π-bonding and the electronegativity of M decreased in the series Si > Ge > Sn > Pb (*157*); while π-bonding to Sn in $Me_3SnN<$ was said to decrease from $(CH_2)_4N > (CH_2)_2N > Me_2N$, with the converse situation valid for Si analogs (*83*). For a series of pentafluorophenyl compounds of the type C_6F_5—X (F F / F⟨benzene ring⟩—X / F F), a plot of $J_{2,4}$ (the $o\text{-}^{19}F/p\text{-}^{19}F$ coupling constant) against δ_4 (the $p\text{-}^{19}F$ chemical shift, relative to $CFCl_3$) was essentially linear (*65*); the position on the line for X = $NHSnMe_3$ is at high δ_4 and low $J_{2,4}$, suggesting that π-donation from N to the aromatic ring is considerable, and is greater than for related Si or B compounds [this is consistent both with (i) Sn amides being stronger bases than Si or B analogs, and (ii) M—N π-bonding being less important for the Sn than the other two compounds].

The relative basicities of stannyl-, germyl-, and silylamines of the type Me_3MNMe_2 and $(Me_3M)_3N$ (M = Sn, Ge, or Si) have been examined with $CHCl_3$ or $CDCl_3$ as reference acids (*1*). From measurements of the difference $\Delta\nu$(C—D), as between $CDCl_3$ and $CDCl_3$ with base it was concluded that base strengths decreased from Sn > Ge > Si, with Me_3SnNMe_2 comparable in basicity with *tert*-BuNHEt; it was therefore concluded that it is unlikely that there is significant Sn—$\ddot{N}$ π-bonding (*1*). The problem of basicity in the same systems was referred to in terms of the five parameters: the "interaction" parameters $\Delta\nu$(C—D), $\Delta J(^{13}C—H)$ in $^{13}CHCl_3$, and $\Delta\delta_\infty$ (the $CHCl_3$ ^1H nmr chemical shift difference at infinite dilution); and the "intrinsic" parameters $J(^{13}C—H)$ in —M—N—^{13}C—H and ν(CH) of $\rangle NCH_3$ (*136*); it was concluded that the various methods gave mutually inconsistent results. Electron impact measurements of first ionization potentials (clearly an "intrinsic" parameter) for the series Me_3MNMe_2, show decreasing values (eV in parentheses) for the series M = Si (7.75 ± 0.10) > Ge (7.47 ± 0.08) > Sn (7.11 ± 0.02), consistent with the inverse order of nitrogen lone-pair availability (*94*).

Thermochemical data are available for the series $(Me_3Sn)_nNMe_{3-n}$ (n = 1, 2, or 3) (*11*). Heats of hydrolysis, $-\Delta H_{\text{hydrolysis}}$, in 1 *M* HCl at 25°C are 38.2, 45.4, and 70.2 kcal/mole, for Me_3SnNMe_2 (liq.), $(Me_3Sn)_2NMe$ (liq.), and $(Me_3Sn)_3N$ (cryst.), respectively. Calculating from the measured $-\Delta H_{\text{hydrolysis}}$ of Me_3SnCl (cryst.), ΔH_f° is 55 kcal/mole. Using these values, ΔH_f° for Me_3SnNMe_2 (liq.), $(Me_3Sn)_2NMe$ (liq.), and $(Me_3Sn)_3N$ (cryst.) are −55.6, −24.3, and −21.6 kcal/mole, respectively. Taking reasonable estimates for the heats of vaporization or sublimation of the SnN compounds, and assuming the thermochemical contribution of the Me_3Sn and Me groups to be constant, $\bar{E}$(Sn—N) is 43.5, 48.7, and 40 kcal/mole, respectively. It is too early to say whether the differences in $\bar{E}$(Sn—N) between the three compounds are significant in terms of bonding (after all, the geometry and hence hybridization around N is unknown, but is unlikely to be the same, for the three compounds), but clearly the Sn—N bond is relatively weak. For comparison E(Si—N) and E(Ge—N) in Me_3MNMe_2 (M = Si or Ge) are 79.9 (*10*) and 64.3 (*11*) kcal/mole, respectively.

Dipole moments of a series of (diethylamino) stannanes have been recorded for benzene solutions, and the partial moment of Sn—NEt_2 is 1.7 D (*104*).

The molecular refraction of the Sn—N bond in (tributylstannyl)succinimide was calculated to be 4.02. Using this value for calculations on organotin pseudohalides, the iso structures gave the best agreement between calculated and observed molar refractivity (*48*). That the compounds $R'_2CHC(CN){=}C{=}NSnR_3$ are polymeric, with five-coordinate Sn, was deduced from their vicosity (*127*).

V. Tin Pseudohalides with Sn—N Bonds

A. Azides

Trimethyl-, tri-*n*-butyl-, and triphenyl-tin azides were obtained from the corresponding tin chlorides and sodium azide (*106, 109*); such reactions are frequently accomplished in aqueous solution (*169*), so the reactions are really between hydrazoic acid and the tin oxide or hydroxide (*177*). Reaction of organotin amides with hydrazoic acid affords a route to corresponding azides (*106*).

The tin azides R_3SnN_3 are polymeric in the solid state with tin in a pyramidal five-coordinate environment and azide bridging groups, on the basis of infrared (*109, 176, 178*) and Mössbauer (*63*) spectral data; the hindered $(PhCMe_2CH_2)_3SnN_3$, however, is monomeric (*141*).

The compounds have high thermal stability; pyrolysis of Ph_3SnN_3 gave Ph_4Sn (*140*). They react with phosphines, either to form 1:1 adducts (*177*) or stable phosphinimines (*106, 153, 177, 186*). Me_3SnN_3 with R_3P (R = Me, Et, or Ph) gives Me_4Sn and $Me_2Sn(N_3){-}N{=}PR_3$ (*153*), while $Ph_3Sn{-}N_3$ afforded $Ph_3Sn{-}N{=}PR_3$ (R = *n*-Bu, *n*-Oct, or Ph) (*101*). Trimethyltin azide behaved as a pseudohalide in its reactions with the metal hydrides π-$C_5H_5(CO)_3M{-}H$ [M = Mo (*95a*), W (*92b*)] and $Fe_2(CO)_9$ to give metal-tin bonded products. Reaction with acetylenedicarboxylic acid and ester proceeded according to Eq. (135) (*109*) and Eq. (61) (*112*). Lewis acid properties of Ph_3SnN_3 were demonstrated by the isolation of BBr_3 and $SnCl_4$ 1 : 1 adducts (*179*):

$$2\ n\text{-}Bu_3SnN_3 + HOOCC{\equiv}CCOOH \longrightarrow n\text{-}Bu_3SnOOCC{\equiv}CCOOSnBu_3 + 2\ HN_3 \qquad (135)$$

B. Isocyanides

There is no clear evidence that the tin isocyanides $>Sn{-}\overset{+}{N}{\equiv}\overset{-}{C}$ exist; it is more likely that compounds so described are in fact the cyanides $>Sn{-}C{\equiv}N$ (*97, 180*). The distinction is perhaps somewhat artificial, since the trimethylstannyl compound $(Me_3SnCN)_n$ has been shown by single crystal x-ray diffraction to be built of linear polymeric chains having Me_3Sn groups linked by CN bridges (*150a*).

C. Isocyanates

These are compounds of the general formula $\geq Sn-N=C=O$ (*97*). The compounds are generally available from tin halides (or, less generally, R_3SnH) and a metal (e.g., Ag, Na, or K) salt of cyanic acid (*171, 176*).

The reaction of urea with R_3SnOH or $(R_3Sn)_2O$ (R = Et or *i*-Bu) in the molten state (130–140°C) furnished R_3SnNCO in high yield (*172*); with R_3SnCl, the results were poorer, and $(n\text{-}Bu_2SnO)_n$ afforded $[n\text{-}Bu_2SnOSn(NCO)_2]_n$. The distannoxanes and isocyanic acid at 70–80°C gave R_3SnNCO (R = Et or Bu) (*173*). Interaction of EtO_2CNH_2 and R_3SnOMe gave R_3SnNCO and MeOH (*40*).

Infrared spectral data on Me_3SnNCO suggest a polymeric structure for the solids; ν_{asym} (NCO) shifts to higher frequency upon dissolution (*176*). The 1 : 1 adduct of $(Bu_3Sn)_2O$ with CCl_3CN (*25, 55*), upon being heated, yielded some Bu_3SnNCO.

Dimethyltin diisocyanate (*142*) was readily hydrolyzed, Eq. (136). The diisocyanates appear to be particularly sensitive to moisture; hydrolysis products of the type $n\text{-}Bu_8Sn_4(NCO)_4O_2$, $n\text{-}Bu_8Sn_4(NCO)_2(OH)_2O_2$, and $Ph_4Sn_2(NCO)_2(OH)_2$ have been isolated (*116*):

$$Me_2Sn(NCO)_2 + 3\,H_2O \longrightarrow \frac{1}{n}(Me_2SnO)_n + CO_2 + 2\,NH_3 \qquad (136)$$

D. Isothiocyanates

These are compounds of the general formula $\geq Sn-N=C=S$ (*97*). The interaction of organotin halides and metal thiocyanates has provided synthetic routes, under the conditions of Eqs. (137)–(139) (*48*). For the reactants of Eq. (138), liquid sulfur dioxide or acetonitrile have been proposed as more suitable solvents (*62*); these media dissolve appreciable amounts of ammonium or potassium thiocyanates, while the chlorides are only sparingly soluble. Similarly $(n\text{-}Bu_3Sn)_2O$ and KCNS furnished $n\text{-}Bu_3SnNCS$ in 60% yield, after $\frac{1}{2}$ h at 80°C in 2*M* HCl (*48*), and the interaction of organotin oxides, hydroxides, or alkoxides with ammonium thiocyanate, Eq. (140), is a convenient synthetic route (*132*). Triaryltin derivatives, from the tin iodide and AgNCS (*170, 171*), were obtained analogously:

$$Et_3SnI + AgNCS \xrightarrow{C_6H_6} Et_3SnNCS + AgI \qquad (137)$$

$$Me_2SnCl_2 + 2\,NaCNS \xrightarrow{Me_2CO} Me_2Sn(NCS)_2 + 2\,NaCl \qquad (138)$$

$$n\text{-}Bu_3SnCl + KCNS\ (\text{or } NH_4CNS) \xrightarrow{ROH;\ reflux} n\text{-}Bu_3SnNCS + KCl \qquad (139)$$

$$\geq Sn-OH + NH_4SCN \longrightarrow \geq SnNCS + NH_3 + H_2O \qquad (140)$$

The metathetical exchange reaction, Eq. (141), has been employed (*7*). The position of NCS (as well as some other pseudohalides) in a "conversion" series was placed as follows: $Et_3SnSMe = (Et_3Sn)_2S \rightarrow Et_3SnI \rightarrow Et_3SnBr \rightarrow Et_3SnCN \rightarrow Et_3SnNCS \rightarrow Et_3SnCl \rightarrow (Et_3Sn)_2O \rightarrow Et_3SnNCO \rightarrow Et_3SnOCOMe \rightarrow Et_3SnF$ (*8*); thus, using silver salts, any compound may be converted to one on its right in the series. Thiocyanic acid, Eq. (142) (*8*), sulfur, Eq. (143) (*48*, *160*), and thiourea, Eq. (144) (*48*) have also been employed as starting materials. Cleavage of certain tin amides with compounds containing $\rangle C{=}S$ gave isothiocyanates (see Sec. III.E) (*49*, *67*, *68*, *182*):

$$2\ Me_3SiNCS + (Et_3Sn)_2O \longrightarrow 2\ Et_3SnNCS + (Me_3Si)_2O \qquad (141)$$

$$(Et_3Sn)_2O + 2\ HNCS \longrightarrow 2\ Et_3SnNCS + H_2O \qquad (142)$$

$$R_3SnCN + S \longrightarrow R_3SnNCS \qquad (143)$$

$$n\text{-}Bu_3SnCl + (NH_2)_2CS \longrightarrow n\text{-}Bu_3SnNCS\ (16\%\text{ in 2 h at }170°C) \qquad (144)$$

n-$Bu_2Sn(NCS)_2$ formed a 1:1 complex with 2,2′-dipyridyl (*4*), and similar complexes of the Me, Et, and n-Pr homologs are known (*184*, *185*). The anions $[MeSn(NCS)_5]^{2-}$, $[Me_2Sn(NCS)_4]^{2-}$, and $[Me_3Sn(NCS)_3]^{2-}$ are known (*35*).

E. Miscellaneous Pseudohalides

Triphenyltin isoselenocyanate was prepared by reaction (145) (*9*) or from Ph_3SnCl and KSeCN. It is stable in air and soluble in water and alcohol:

$$Ph_6Sn_2 + Se(SeCN)_2 \longrightarrow 2\ Ph_3SnNCSe + Se \qquad (145)$$

Triphenyltin hyponitrite (*15*) has been prepared from the silver salt Eq. (146), and dicyanamide and dicyanoketenimine derivatives have also been prepared (*82*) by conventional methods Eqs. (147), (148):

$$2\ Ph_3SnI + Ag_2N_2O_2 \longrightarrow (Ph_3Sn)_2N_2O_2 + 2\ AgI \qquad (146)$$

$$R_2SnI_2 + 2\ AgN(CN)_2 \longrightarrow R_2Sn[N(CN)_2]_2 + 2\ AgI \qquad (147)$$

$$Ph_2SnO + 2\ HN{=}C{=}(CN)_2 \longrightarrow Ph_2Sn[N{=}C{=}C(CN)_2]_2 + H_2O \qquad (148)$$

VI. Addendum

Since this review was submitted, further papers have appeared on tin-nitrogen chemistry, and the bibliography is now brought up to June 1970; contributions of the authors and their collaborators for this period are found in the main text. Elementoorganic amines and imines, including those of Sn, have been reviewed (*219*).

Aminostannanes have been prepared by transmetallation reactions (*195*, *211*) and during the ammonolysis of organostannanes (*222*), organodistannanes (*221*), and organotin halides (*204*), and some physical data have been recorded (*194*, *209*, *212*, *213*). The synthesis and structure of *N*-trialkylstannyl keteneimines have been detailed (*215*, *225*). The preparations of a tin(II)-nitrogen compound (*202*), tin homologs of cyclotrisiloxadiazanes (*229*), and *N*-trialkylstannyl sulfonamides (*196*, *197*), ureas (*192*), and percarbamates $Ph_3SnN(R)COOOBu^t$ (*191*) have been reported. Organotin-nitrogen compounds are involved in the reactions of organotin arsines with phenyl azide (*226*) and in various insertion reactions (*191*, *192*, *196*, *201*, *206–208*, *217*, *218*, *220*). Further examples have been published of reactions of styannylamines with phenylalkynes (*210*), chloramine and *N,N*-dimethylchloramine (*203*), acetone (*218*), aliphatic and cycloaliphatic ketones (*217*), benzoyl isocyanate (*208*), benzoyl heterocumulenes (*207*), alkylating agents (*209*), and bis(trimethylsilyl)mercury (*214*). Reactions of stannyliminophosphoranes toward alkyl metals (*210*) have been investigated.

The structural parameters of $Sn(NMe_2)_4$ have been determined by electron diffraction (*228*). The tetrahedral molecule has the bond lengths Sn—N, 2.045; C—N, 1.45; and C—H, 1.10 Å; while the bond angles are CNC 119 $\pm$ 3 and SnNC 117 $\pm$ 1.5° when NSnN 109.5 and HCN 109.5° are assumed. The crystal and molecular structures of the carbon disulfide addition product with Me_3SnNMe_2, dimethyldithiocarbamatotrimethylstannane, which occurs in two different modifications, have been determined (*223*, *224*).

With regard to organotin pseudohalides, tetramethylammonium hexa-azidostannate has been prepared (*216*); the reactions of triphenyltin azide with tin hydrides (*227*), and the formation of alkyltin isothiocyanate complexes with ligands containing oxygen and nitrogen donor atoms (*205*), have been reported. The crystal structures have been determined for trimethyltin isothiocyanate (*200*), dimethyltin diisothiocyanate (*193*, *198*), and $(Me_3Sn)_2N_2C$ (*199*).

VII. Appendices

Appendices 1–16 provide a summary of reported data on individual compounds containing a Sn—N bond. The numbers listed under the headings "preparations" and "reactions" refer to the equations in Secs. I–V. Similarly, the letters under the heading "physical properties" are interpreted as follows: a = infrared spectral data; b = nuclear magnetic resonance data; c = mass spectroscopic data; d = dipole moment measurements; e = ultraviolet spectral data; f = Mössbauer spectral data; and g = base-strength measurements.

APPENDIX 1

Sn—N Compounds Derived from Simple Amines

	$\rangle$Sn—N$\langle$	Method of prep.	bp (deg/mm) (mp)	d_4^{20}	n_D^{20}	Reported reactions	Physical properties	Refs.
1	Me_3SnNMe_2	3	126/760	1.2173	1.4572	32		(*74*)
		3	126/760			29, 32, 33, 114		(*72*)
		3	128/760				b	(*108*, *187*)
						117		(*3*)
						65, 67, 92		(*32*)
						67, 92, 97, 111–113		(*31*, *31a*, *76a*)
						102–104, 120, 121, 123, 126–129		(*37*)
						104		(*38*)
						82		(*41*, *42*)
						22, 25		(*50*)
						28		(*51*)
						77		(*52*)
						33–35, 114, 115, 118, 120		(*58*)
						18, 66, 95, 96, 100, 101		(*59*)
						95, 96, 100, 101		(*60*)
						119, 121, 123, 124, 126–129, 132		(*61*)
						77		(*64*)
						116		(*66*)
						34, 75–81, 120		(*73*)
						69–71, 73, 75–81, 84–86		(*75*)

(*continued*)

APPENDIX 1 (*continued*)

Sn—N Compounds Derived from Simple Amines

	$\geq Sn-N\leq$	Method of prep.	bp (deg/mm) (mp)	d_4^{20}	n_D^{20}	Reported reactions	Physical properties	Refs.
1	Me_3SnMe_2 (continued)					27		(*83*)
						98		(*87*, *110*)
						83		(*93*, *93a*)
						79, 94, 125		(*38a*, *70a*, *146*)
							b	(*137*)
						133		(*93a*)
2	$Me_3SnNHEt$	16	153/760		1.4689			(*2*)
3	$Me_3SnN<(CH_2)_2$ (aziridine ring: CH_2—CH_2)	27	53–55/16					(*83*)
4	$Me_3SnNHPh$	6, 28	77/0.05	1.4255	1.5721			(*72*, *74*)
		6					a	(*134*)
							b	(*139*)
5	$Me_3SnNHC_6F_5$						b	(*65*)
6	$Me_3SnNHC_6H_4Cl$-*p*	6						(*51*)
7	$Me_3SnNHC_6H_4Me$-*p*	6	106/0.5	1.3536	1.5622			(*74*)
8	$Me_3SnNMePh$	3	83/0.1	1.3645	1.5757			(*72*, *74*)
9	Me_3SnNEt_2	3	162/760		1.4651			(*2*)
		3, 22	43/8		1.4618			(*74*)
		3	140/720				b	(*86*)
		3	156–162/760			70		(*165*)
		3	36/6				b	(*189*)
		3					g	(*1*)

		3				73		(*5*)
						77		(*52*)
						64		(*54*)
						36		(*58*, *73*)
						34		(*69*)
		3				87		(*75*)
		3				69	d	(*104*, *105*)
		3				100*a*		(*110*)
		3				88		(*123*)
		3				65, 88		(*125*, *168*)
10	Me_3SnNPr_2	3	63/8	1.1539	1.4645			(*72*, *74*)
11	$Me_3SnNBu^n_2$	3, 22	74/2.5	1.1068	1.4559			(*72*, *74*)
12	Me_3SnNPh_2	3	108/0.1	1.3176	1.6096			(*72*, *74*)
		3					f	(*51*)
13	$Me_3SnN(C_6H_{11})_2$	3	96/0.2	1.1972	1.5055			(*74*)
14	Et_3SnNMe_2	3	76/9		1.4783			(*72*, *74*)
		3				71, 73, 75, 78, 87		(*75*)
		3					b	(*137*)
15	Et_3SnNEt_2	3	72/2	1.1692	1.4724			(*74*)
		3	40/0.1			73	b	(*107*)
		15	114–117/23			69, 72	a	(*164*)
		3	225/720					(*187*)
		22				88		(*47*)
		3				64		(*54*)
		3				84		(*75*)
		3				69	c	(*104*, *105*)
		3				88		(*124*, *167*)
		3				88		(*125*, *168*)
16	$Et_3SnNHPh$	6	100/0.2					(*74*)
17	Et_3SnNPh_2	22				88		(*47*)

(*continued*)

APPENDIX 1 (*continued*)

Sn—N Compounds Derived from Simple Amines

	$\geq Sn-N \leq$	Method of prep.	bp (deg/mm) (mp)	d_4^{20}	n_D^{20}	Reported reactions	Physical properties	Refs.
18	$Et_3SnN(Ph)CH_2Ph$	45	149–151/0.002		1.5907			(*120*)
19	$Et_3SnN(C_6H_4Me\text{-}p)CH_2Ph$	45	142/0.001		1.5843	71		(*118*, *122*)
								(*120*)
20	$Pr^n_3SnNEt_2$	15	118–120/13			69, 72	a	(*164*)
21	$Bu^n_3SnNHBu^t$	28	124/1	1.0577	1.4773			(*74*)
22	$Bu^n_3SnNMe_2$	3	86/0.1		1.4737			(*72*, *74*)
		3				73, 79, 80, 84, 86		(*75*)
		3					b	(*137*)
23	$Bu^n_3SnNEt_2$	3	95/0.1			73	b	(*107*)
		14	115–120/0.1					(*151*)
		15	124–134/8			69, 72	a	(*164*)
		3	95–100/0.16			133		(*183*)
		3				64		(*54*)
		3				69	c	(*104*, *105*)
		3				98		(*110*)
		3				88		(*125*)
		3				88		(*167*, *168*)
24	$Bu^n_3SnN(Ph)CH_2Ph$	45	160/0.002					(*120*)
25	$Bu^i_3SnNEt_2$	3	75–77/0.2	1.08		88		(*125*, *168*)
								(*167*)
26	Ph_3SnNMe_2	3	166/0.1					(*72*, *74*)
		3	(62)			70, 78, 80, 81		(*75*)
27	Ph_3SnNEt_2	3	165–170/0.1				bc	(*86*)
		3	(40)			69		(*104*, *105*)

		3				88		(*168*)
28	(cyclo-C_6H_{11})$_3SnNEt_2$	3	158–161/0.001			88		(*125*)
			(79–80)					(*168*)
29	$Me_2Sn(NH_2)Na$	54					a	(*80*)
30	$Me_2Sn(NMe_2)_2$	3	138/760	1.1482	1.4463	31		(*72, 74*)
		3	45/0.1				b	(*108*)
		22					b	(*137*)
31	$Me_2Sn(NMeEt)_2$	3					b	(*137*)
32	$Me_2Sn(NEt_2)_2$	3	78/4					(*74*)
		3	65/0.1				b	(*86*)
		3				69	c	(*104, 105*)
		3				98, 100*a*		(*110*)
		3					b	(*136*)
		3				22	b	(*137*)
		3				23, 24		(*147*)
		3				88		(*167, 168*)
		3					a	(*190*)
33	$Me_2Sn(NPr^i_2)_2$	3	66/0.05	1.1060	1.4685			(*72, 74*)
34	Me, N, Me_2Sn, N, Me [ring structure]	23	88–90/18				b	(*147*)
			(1–3)					
35	$Et_2Sn(NEt_2)_2$	3	77/0.1			73	b	(*107*)
			108/1.0					
		3	58–60/0.2			88		(*168*)
		3				69		(*104, 105*)
		3				88		(*167*)
		3						(*124, 187*)

(*continued*)

APPENDIX 1 (*continued*)

Sn—N Compounds Derived from Simple Amines

	$>Sn-N<$	Method of prep.	bp (deg/mm) (mp)	d_4^{20}	n_D^{20}	Reported reactions	Physical properties	Refs.
36	$Bu^n_2Sn(NMe_2)_2$	3	72/0.05	1.1247	1.4747			(*72*, *74*)
		3				73		(*75*)
37	$Bu^n_2Sn(NEt_2)_2$	3	98/0.1			78	b	(*107*)
		14	95–105/0.1					(*151*)
		3				69	c	(*104*, *105*)
						88		(*167*)
38	$Bu^i_2Sn(NEt_2)_2$	3	83–86/0.001			88		(*168*)
		3				88		(*167*)
39	$Ph_2Sn(NMe_2)_2$	3	128/0.2					(*72*, 74)
40	$Ph_2Sn(NEt_2)_2$	3	145–150/0.1				b	(*86*)
		3				69	c	(*104*, *105*)
		3				98		(*110*)
41	$MeSn(NMe_2)_3$	3	50/0.1			26	b	(*108*)
		3	45–50/0.1					(*87*)
		3						(*187*)
42	$MeSn(NEt_2)_3$	3	92/0.1				b	(*86*)
		3				100*a*		(*87*)
		3					c	(*104*)
43	$EtSn(NEt_2)_3$	3	76/0.1			73	b	(*107*)
		3					c	(*104*)
44	$Bu^nSn(NMe_2)_3$	3	67/0.1					(*72*, *74*)
45	$Bu^nSn(NEt_2)_3$	3	96/0.1			73	b	(*107*)
		3	115/1.0					(*151*)
		3					c	(*104*)

46	$PhSn(NMe_2)_3$	3	80/0.1					*(74)*
		3						*(187)*
47	$PhSn(NEt_2)_3$	3	130/0.1				b	*(86)*
		3					c	*(104)*
48	$Sn(NMe_2)_4$	3	51/0.15	1.687	1.4774			*(72, 74)*
		3				108		*(98)*
		3					f	*(51)*
		3				68		*(49)*
		3					a	*(28, 64)*
49	$Sn(NEt_2)_4$	3	90/0.1			74		*(181)*
		3	90/0.05	1.1042	1.4800			*(74)*
		3	116/1.0				b	*(86)*
		3	116/0.1					*(187)*
		3				98		*(110)*
		3					f	*(51)*

APPENDIX 2

Sn—N Compounds Derived from Cyclic Amines

	$\gt Sn—N$ (ring)	Method of prep.	bp (deg/mm) (mp)	d_4^{20}	n_D^{20}	Reported reactions	Physical properties	Refs.
50	Me_3SnN (pyrrole ring)	14	101.5–102/17		1.5302		b	(*83*, *111*)
51	Me_3SnN (imidazole ring)	14 15	(234–236) (235–238)			70	a	(*78*, *109*) (*111*)
52	Me_3SnN (triazole ring)	14 14, 52	(277–278) (277–278d)			70	a	(*78*, *109*) (*111*)
53	Me_3SnN (piperidine ring)	3, 22	48/1					(*74*)
54	Me_3SnN (benzimidazole ring)	14	(sub > 200)				a	(*78*, *109*) (*111*)
55	Me_3SnN (benzotriazole ring)	14	(221.5–223d)				a	(*78*, *109*) (*111*)

56	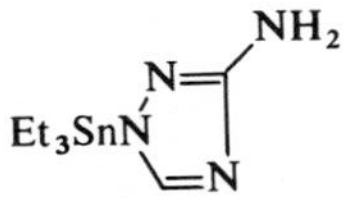 Et_3SnN (3-amino-1,2,4-triazol-1-yl)	14	(240d)		(*111*)
57	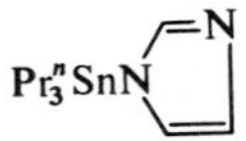 Pr^n_3SnN (imidazol-1-yl)	14 52	(152–154) (149–150)		(*78*, *109*) (*111*)
58	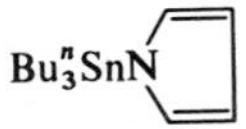 Bu^n_3SnN (pyrrol-1-yl)	14	139–141/0.62		(*111*)
59	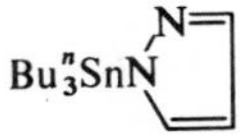 Bu^n_3SnN (pyrazol-1-yl)	52	98–102/0.04	1.4995	(*111*)
60	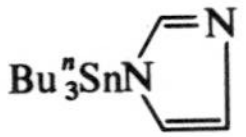 Bu^n_3SnN (imidazol-1-yl)	52 52	(65–67) (64–64.5)		(*78*, *109*) (*111*)
61	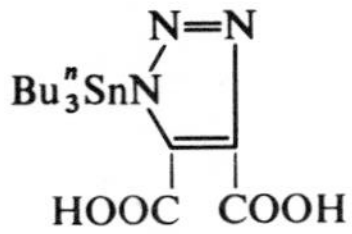 Bu^n_3SnN (4,5-bis(COOH)-1,2,3-triazol-1-yl; HOOC, COOH)	61	(64.5–67)		(*112*)
62	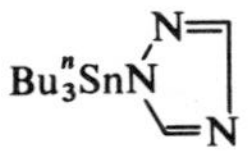 Bu^n_3SnN (1,2,4-triazol-1-yl)	52	(66–71)		(*78*, *109*) (*111*)
63	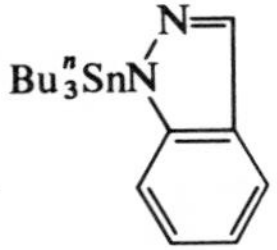 Bu^n_3SnN (indazol-1-yl)	14	124–128/0.002	1.5376	(*111*)

(*continued*)

APPENDIX 2 (*continued*)

Sn—N Compounds Derived from Cyclic Amines

	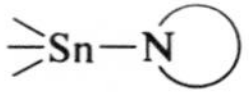 $\geq$Sn—N (ring)	Method of prep.	bp (deg/mm) (mp)	d_4^{20}	n_D^{20}	Reported reactions	Physical properties	Refs.
64	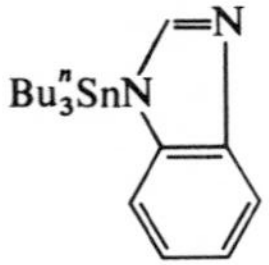 Bu^n_3SnN (benzimidazole)	52	(137.5–139)					(*78*, *109*) (*111*)
65	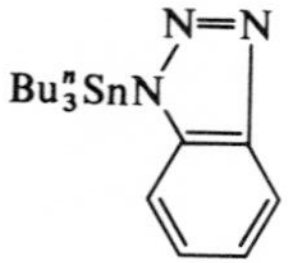 Bu^n_3SnN (benzotriazole)	52	(78–88)					(*78*, *109*) (*111*)
66	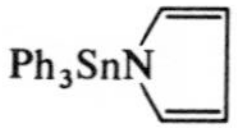 Ph_3SnN (pyrrole)	14 15	(205–206) (203.2–204)					(*111*) (*111*)
67	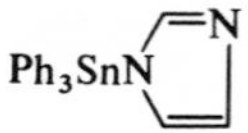 Ph_3SnN (imidazole)	52 52	(304–305.5) (310–311)					(*109*) (*111*)
68	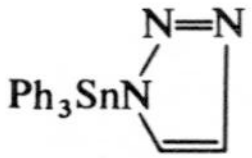 Ph_3SnN (1,2,3-triazole)	14	(311–313d)					(*111*)
69	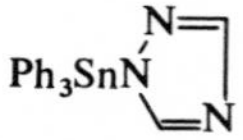 Ph_3SnN (1,2,4-triazole)	52	(294–295.5)					(*109*, *111*)

70	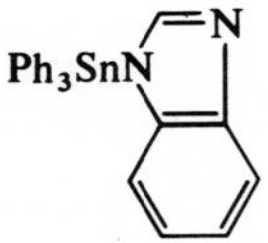 Ph_3SnN (benzimidazol-1-yl)	52	(298d)		(*109*, *111*)
71	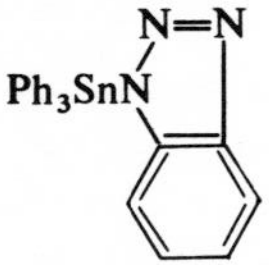 Ph_3SnN (benzotriazol-1-yl)	52	(270.5–272)		(*109*, *111*)
72	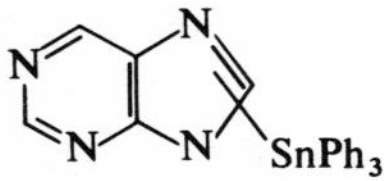 (purine, N–$SnPh_3$)	52	(304–307d)	a, e	(*92*)
73	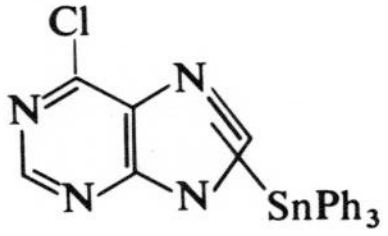 (6-Cl-purine, N–$SnPh_3$)	52	(101)	a, e	(*92*)
74	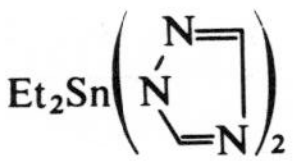 $Et_2Sn(C_2H_2N_3)_2$ (1,2,4-triazolyl)	14			(*111*)
75	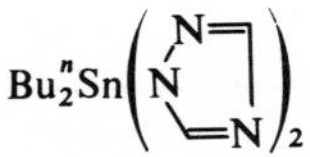 $Bu^n_2Sn(C_2H_2N_3)_2$ (1,2,4-triazolyl)	14	(240d)		(*111*)

APPENDIX 3

Sn—N Compounds Derived from Hydrazine

	$\rangle$Sn—N(—)—N$\langle$	Method of prep.	bp (deg/mm) (mp)	d_4^{20}	n_D^{20}	Reported reactions	Physical properties	Refs.
76	$(Me_3Sn)_2NNMe_2$	22	60/0.3				a, b	(*50*)
77	$Me_3Sn(Me)N—N(Me)SnMe_3$	22	52/0.3				a, b	(*50*)
78	$Me_3Sn(Ph)N—N(Ph)SnMe_3$	22	(196)				a, b	(*50*)
79	$(Me_3Sn)_2N—N(Ph)SnMe_3$	25	120/0.5				a, b	(*50*)
80	$(Me_3Sn)_2NN(SnMe_3)_2$	22					a, b	(*50*)
81	$Et_3SnNPhNHPh$	43				70, 88		(*128*)
		43				88		(*47*)
82	$Ph_3SnNPhNHPh$	43	(130)			88		(*128*)
83	$Et_3SnN(COOEt)—N(COOEt)SnEt_3$	44	156/0.1		1.4998	70, 88		(*128*)
84	$Et_3Sn(SO_2Ph)NNHPh$	43	(116)					(*121*)
85	$Bu^n_3SnN(COOEt)—N(COOEt)SnBu^n_3$	52	125–135/0.07					(*55*)

APPENDIX 4

Sn—N Compounds Containing More Than One Tin Atom per Molecule

	$\gt Sn—N(—)—Sn\lt$	Method of prep.	bp (deg/mm) (mp)	d_4^{20}	n_D^{20}	Reported reactions	Physical properties	Refs.
86	$(Me_3Sn)_2NMe$	7, 29	64/3	1.4794	1.4901	68		*(72, 74)*
		29				131		*(49)*
		29					f	*(51)*
						131		*(67, 68)*
87	$(Me_3Sn)_2NEt$	29	93/15	1.4805	1.4968			*(72, 74)*
88	$(Me_3Sn)_2NPr^i$	29	95–97/11			65		*(125)*
89	$(Me_3Sn)_2NPh$	51	99–100/1				a	*(156)*
90	$(Me_3Sn)_2NLi$	99						*(146)*
91	$(Me_3Sn)_3N$	32	70/0.2	1.5084	1.5331			*(72, 74)*
		9	84/0.4				b	*(100)*
		10	133–134/20 (22–24)			70	a	*(165)*
		9	130/14 (26–28)			130		*(182, 49)*
							g	*(1)*
		32					f	*(51)*
		32					a	*(64)*
		53						*(27)*
92	$(Et_3Sn)_3N$	10	86–88/9			71[g]	a	*(165)*
93	$(Pr^n_3Sn)_3N$	10	122–123/10			71	a	*(165)*
94	$(Me_3Sn)_4N^+Br^-$	109					g	*(116)*

(continued)

APPENDIX 4 (*continued*)

Sn—N Compounds Containing More Than One Tin Atom per Molecule

	$>Sn-N(-)-Sn<$	Method of prep.	bp (deg/mm) (mp)	d_4^{20}	n_D^{20}	Reported reactions	Physical properties	Refs.
95	cyclo-$[Me_2Sn-N(Me)]_3$ (ring: Me_2Sn, N–Me, $SnMe_2$, N–Me, $SnMe_2$, N–Me)	31	114/0.2					(*74*)
96	cyclo-$[Me_2Sn-N(Et)]_3$ (ring: Me_2Sn, N–Et, $SnMe_2$, N–Et, $SnMe_2$, N–Et)	31	104/0.05					(*72, 74*)

APPENDIX 5

Sn—N Compounds Containing Other Metals

No.	$>Sn-N(-)-M$	Method of prep.	bp (deg/mm) (mp)	d_4^{20}	n_D^{20}	Reported reactions	Physical properties	Refs.
97	$(Me_3Sn)(Me_3Si)NMe$	3	79–81/30				b	(*150*)
		3	59–61/11				b	(*145*)
98	$(Me_3Sn)(Me_3Ge)NMe$	3	28/2					(*154*)
		3					b	(*157*)
99	$(Me_3Sn)(Me_3Si)NCMe_3$	3	50/0.25					(*158*)
100	$(Me_2SnCl)(Me_3Si)NCMe_3$	3	74/0.1					(*158*)
101	$Me_3SnN(SiMe_3)_2$	25	58/0.1				b	(*108, 152*)
		10	58–59/1			99		(*146*)
		13	(20–22)			70	a, b	(*148, 149*)

(*continued*)

APPENDIX 5 (*continued*)

Sn—N Compounds Containing Other Metals

	$>Sn-N(-)-M$	Method of prep.	bp (deg/mm) (mp)	d_4^{20}	n_D^{20}	Reported reactions	Physical properties	Refs.
102	$(Me_3Sn)(Me_3Ge)NSiEt_3$	3	138–141/10					(*144*)
103	$(Me_3Sn)(Me_3Ge)N-SiMe_2-N(SiMe_3)Me$	3	84–86/0.2					(*143*)
104	$Bu^n_3SnN(SiMe_3)_2$	13	140–145/1				a	(*148*)
105	$Me_2Sn(NMe_2)N(SiMe_3)_2$	26	58/0.1				b	(*108*)
106	$Me_2Sn(N(Me)SiMe_3)_2$	3	61–63/0.5				b	(*145*, *150*)
107	$Me_2Sn(N(SiMe_3)(CMe_3))_2$	3	133/0.15					(*158*)
108	$Me_2Sn(N(SiEt_3))_2GeMe_2$ (cyclic)	20	124–128/0.2					(*144*)

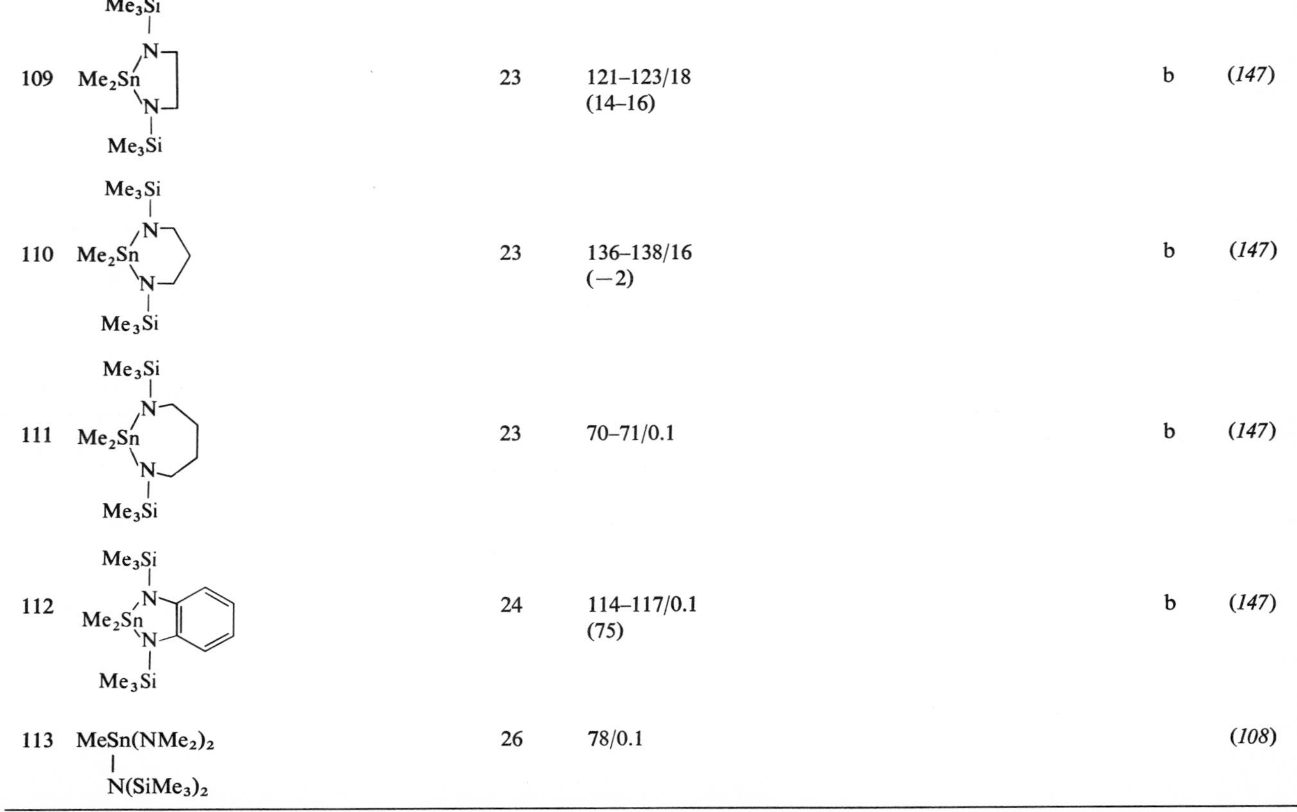

109		23	121–123/18 (14–16)	b	(*147*)
110		23	136–138/16 (−2)	b	(*147*)
111		23	70–71/0.1	b	(*147*)
112		24	114–117/0.1 (75)	b	(*147*)
113	$MeSn(NMe_2)_2$ with $N(SiMe_3)_2$	26	78/0.1		(*108*)

(*continued*)

APPENDIX 5 (*continued*)

Sn—N Compounds Containing Other Metals

	>Sn—N(—)—M	Method of prep.	bp (deg/mm) (mp)	d_4^{20}	n_D^{20}	Reported reactions	Physical properties	Refs.
114	$MeSn(N(Me)SiMe_3)_3$	3	83–85/0.5 (22–23)				b	(*145*)
115	$Sn(N(Me)SiMe_3)_4$	3	129–132/0.5 (90–92)				b	(*145*)

APPENDIX 6

Sn—N Compounds Derived from Formamide

	$\geq$Sn—N(—)—C(=O)—H	Method of prep.	bp (deg/mm) (mp)	d_4^{20}	n_D^{20}	Reported reactions	Physical properties	Refs.
116	$Me_3SnNPhCHO$	42						(*46*)
117	Et_3SnNBu^nCHO	22				88		(*47*)
118	Et_3SnNBu^tCHO	22				88		(*47*)
119	$Et_3SnN(CHO)C_6H_4Cl$-*p*	42	120–122/0.1 (77–79)				a	(*131*)
						88		(*47*)
120	$Et_3SnN(CHO)C_6H_4NO_2$-*p*	42	(116–119)				a	(*131*)
						88		(*47*)
121	$Et_3SnNPhCHO$	42	171–172/12 (50–53)					(*130*)
		52	97/0.1 (49.5–52.5)					(*55*)
		42	110/0.3 (56–68)					(*129*)
		42	171–172/13 (50–53)				a	(*131*)
						88		(*47, 128*)
122	$Et_3SnN(C_6H_{13})CHO$	52	105–109/0.3		1.4918			(*129*)
		42	105–106/0.1 (28)		1.4910		a	(*130, 131*)
						88		(*47*)

(*continued*)

APPENDIX 6 (*continued*)

Sn—N COMPOUNDS DERIVED FROM FORMAMIDE

	>Sn—N(—)—C(=O)—H	Method of prep.	bp (deg/mm) (mp)	d_4^{20}	n_D^{20}	Reported reactions	Physical properties	Refs.
123	$Et_3SnNCHO$ – $Et_3SnNCHO$ (N–N linked)	42	(191–195)				a	(*131*)
124	$Bu^n_3SnNMeCHO$	52	130–140/0.4					(*55*)
			(37–37.5)					
125	$Bu^n_3SnNPhCHO$	42	170/0.1				a	(*16*, *131*)
			(64)					
			(64–67)					(*55*)
126	$Bu^n_3SnN(C_6H_{13})CHO$	52	167–170/4					(*55*)
			(35–37)					
127	$Et_2Sn(NPhCHO)_2$	42					a	(*131*)

APPENDIX 7

Sn—N Compounds Derived from Carbamic Acid

	$\rangle Sn-N(-)-C(=O)-O-$	Method of prep.	bp (deg/mm) (mp)	d_4^{20}	n_D^{20}	Reported reactions	Physical properties	Refs.
128	Et_3SnNBu^nCOOEt	39	57.5/0.01					(*16*, *21*)
		39						(*17*)
129	$Et_3SnNPhCOOMe$	52	105–108/0.4		1.5388			(*129*)
130	$Et_3SnNPhCOOBu^i$	39	(45–47)			70		(*119*)
131	$Et_3SnNPhCOOSnEt_3$	39	(58–60)				a	(*19*, *22*)
132	$Et_3SnN(C_6H_{13})COOMe$	52	92–93/0.2					(*129*)
133	$Et_3SnN(1\text{-}C_{10}H_7)COOEt$	39	104–108/0.05					(*16*, *21*)
134	$Pr_3^nSnNPhCOOSnPr_3^n$	39						(*17*)
135	$Bu_3^nSnNMeCOOMe$	39	101.5/0.05					(*21*)
136	$Bu_3^nSnNMeCOOSnBu_3^n$	39					a	(*20*, *22*)
137	$Bu_3^nSnNEtCOOMe$	39	89–90/0.05					(*21*)
		52	90–95/0.2					(*55*)
138	$Bu_3^nSnNEtCOOSnBu_3^n$	39	oil d.					(*19*)
		39				39		(*20*)
						66, 73		(*22*)
139	$Bu_3^nSnNPhCOOMe$	39	99–100/0.1			40		(*18*)
		39				64, 66		(*21*)
		39				41		(*23*)
		52	118–120/0.1					(*55*)
140	$Bu_3^nSnNPhCOOPh$	39	solid d.					(*21*)
141	$Bu_3^nSnNPhCOOSnBu_3^n$	39	150/0.02			59, 66, 73	a	(*19*, *25*)
		39				41		(*22*, *23*)

(*continued*)

APPENDIX 7 (*continued*)

Sn—N Compounds Derived from Carbamic Acid

	$>Sn-N(-)-C(=O)-O-$	Method of prep.	bp (deg/mm) (mp)	d_4^{20}	n_D^{20}	Reported reactions	Physical properties	Refs.
142	$Bu^n_3SnN(1\text{-}C_{10}H_7)COOMe$	39	120/0.01			69, 70, 72		(*16*, *21*)
143	$Bu^n_3SnN(1\text{-}C_{10}H_7)COOSnBu^n_3$	39	oil d.				a	(*19*, *22*)
144	$Et_2Sn(Cl)NPhCOOMe$	58						(*53*)
145	$Bu^n_2Sn(Cl)NEtCOOMe$	58						(*53*)
146	$Bu^n_2Sn(Br)NEtCOOMe$	58						(*53*)
147	$Bu^n_2Sn(I)NEtCOOMe$	58						(*53*)
148	$Bu^n_2Sn(OAc)NEtCOOMe$	58						(*53*)
149	$Bu^n_2Sn(OCOC_{11}H_{23})NEtCOOMe$	58						(*53*)
150	$Bu^n_2Sn(NCS)NEtCOOMe$	58						(*53*)
151	$Bu^n_2Sn(Cl)NPhCOOMe$	58						(*53*)
152	$Bu^n_2Sn(NPhCOOMe)_2$	39	154/0.2					(*53*)
153	$Bu^n_2Sn(NEtCOOMe)_2$	39	d.					(*53*)

APPENDIX 8

Sn—N Compounds Derived from Urea

	$>Sn-N(-)-C(=O)-N<$	Method of prep.	bp (deg/mm) (mp)	d_4^{20}	n_D^{20}	Reported reactions	Physical properties	Refs.
154	$Me_3SnNPhCONMe_2$	33	103/0.5				a	(*58, 72*)
155	$Bu^n_3SnNMeCONPhSnBu^n_3$	40	149–150/0.05				b	(*24*)
156	$Bu^n_3SnNEtCONPhSnBu^n_3$	40	167/0.05				b	(*24*)
157	$(Bu^n_3SnNPr^i)_2CO$	39	150/0.05					(*25*)
158	$(Bu^n_3SnNPh)_2CO$	40, 41	130/0.05					(*19*)
		59						
		39	200/0.1					(*22*)
		52	130/0.05					(*55*)
159	$(Bu^n_3SnNC_6H_{11})_2CO$	39	155/0.1					(*25*)
160	$[Bu^n_3SnN(1\text{-}C_{10}H_7)]_2CO$	40	>220/0.05					(*19*)
		41						
		59						(*22*)
		39						(*25*)
		52						(*55*)
161	$(Bu^n_3SnNPhCO)_2NEt$	40	oil			70, 73	a	(*20, 24*)
162	$[Bu^n_3SnN(1\text{-}C_{10}H_7)CO]_2NEt$	40	oil			70	a	(*20, 24*)
163	$Bu^n_3SnNPhCONMeCON(1\text{-}C_{10}H_7)SnBu^n_3$	40	oil			40, 70		(*20, 24*)
164	$Bu^n_3SnNPhCONEtCON(1\text{-}C_{10}H_7)SnBu^n_3$	40	oil			40, 70		(*20, 24*)
165	Cyclic ring: Bu^n_3Sn-N, $N-SnBu^n_3$, $N-SnBu^n_3$ linked by $C=O$ groups (three $C=O$, three N; Bu^n_3Sn on each N)	52	300/0.04					(*55*)

APPENDIX 9

Sn—N Compounds Derived from Simple Amides

	$>Sn-N(-)-C(=O)-C<$	Method of prep.	bp (deg/mm) (mp)	d_4^{20}	n_D^{20}	Reported reactions	Physical properties	Refs.
166	$Et_3SnN(CO)_2C_6H_4$ (phthalimide)	52	(71–73)					(*77*, *100*)
167	$Et_3SnN(CO)(SO_2)C_6H_4$ (saccharin)	52	(113.5–114)					(*100*)
168	$Et_3SnNPhCOMe$	41	111–112/0.15 (46–48)					(*129*)
169	$Bu^n_3SnN(CO)_2(CH_2)_2$ (succinimide)	52	144/0.2 (21)	1.23	1.5086			(*48*)
170	$Bu^n_3SnNPhCOMe$	52	104–106/0.03					(*55*)
171	$Bu^n_3SnNHCOMe$	52	142–144/0.1					(*55*)

APPENDIX 10

Sn—N Compounds Containing Sulfur

	$\gt Sn—N\cdots S$	Method of prep.	bp (deg/mm) (mp)	d_4^{20}	n_D^{20}	Reported reactions	Physical properties	Refs.
172	$Me_3SnNPhCSNMe_2$	120					a	*(58, 73)*
173	$Me_3SnNPhSONMe_2$	35					a	*(58)*
174	$Et_3SnNHSO_2Me$	14	(38)					*(77, 110)*
175	$Et_3SnNHSO_2C_6H_4Me$-*p*	52	(69.5–71)					*(77, 110)*
176	$Et_3SnNPhCSH$	42	115/0.2		1.5910		a	*(130, 131)*
177	$Bu^n_3SnNEtSO_2Ph$	14						*(115)*
178	$Bu^n_3SnNPhC(=S)Me$	41	68/0.02					*(16, 17)*
179	$Bu^n_2Sn(NHSO_2Ph)_2$	14	(135–137)					*(114, 115)*
180	$Bu^n_2Sn(NEtSO_2Ph)_2$	14						*(115)*
181	$Bu^n_2Sn(NBu^nSO_2Ph)_2$	14						*(114, 115)*
182	$Bu^n_2Sn(NHSO_2C_6H_{11})_2$	14						*(114, 115)*
183	$Bu^n_2Sn(NHSO_2C_6H_4Me$-*p*$)_2$	14						*(114, 115)*

APPENDIX 11

Sn—N Compounds Containing Phosphorus

	>Sn—N---P	Method of prep.	bp (deg/mm) (mp)	d_4^{20}	n_D^{20}	Reported reactions	Physical properties	Refs.
184	$Et_3SnNMeP(=O)(OMe)_2$	62					b	(*106*)
185	$Et_3SnN{=}P(NMe_2)_3$	62	81–84/0.1			69	b	(*106*)
186	$Et_3SnN{=}PBu^n_3$	22, 62	101–105/0.1			69		(*106*)
187	$Et_3SnN{=}PPh_3$	62	(49–52) 180–182/0.1			69		(*106*)
188	$Bu^n_3SnN{=}P(NMe_2)_3$	62	138–140/0.1			69		(*106*)
189	$Bu^n_3SnN{=}PBu^n_3$	62	162–165/0.1			69		(*106*)
190	$Ph_3SnNPhCOPPh_2$	38						(*155*)
191	$Ph_3SnNPhCSPPh_2$	38						(*155*)
192	$Ph_3SnNPh(Ph_2){=}NPh$	49	(170d)				a	(*156*)
193	$(Ph_3SnNPh)_2PPh{=}NPh$	50	(160)				a	(*156*)

APPENDIX 12

ORGANOTIN NITRAMINES

	$>Sn-N(-)-NO_2$	Method of prep.	bp (deg/mm) (mp)	d_4^{20}	n_D^{20}	Reported reactions	Physical properties	Refs.
194	$Me_3SnN(NO_2)Me$	19	(155–156)				a	(*188*)
195	$Et_3SnN(NO_2)Me$	19, 57	(109.5–110.5)				a, f	(*188*)
196	$Et_3SnN(NO_2)Et$	19	(108–109.5)				a	(*188*)
197	$Et_3SnN(NO_2)Pr^i$	19, 57	(144–144.5)				a	(*188*)
198	$Et_3SnN(NO_2)Bu^t$	19	(179–180.5)				a	(*188*)
199	$Et_3SnN(NO_2)Ph$	19	(138–139)				a	(*188*)
200	$Pr^i_3SnN(NO_2)Me$	52	(69–72)				a	(*188*)
201	$Pr^i_3SnN(NO_2)Bu^t$	52	(77–81)				a	(*188*)
202	$Ph_3SnN(NO_2)Me$	19	(142–144)				a	(*188*)
203	$Ph_3SnN(NO_2)Ph$	19	(138–139d)				a	(*188*)

APPENDIX 13

ORGANOTIN AZIDES

	$\geq Sn-N=N=N$	Method of prep.	bp (deg/mm) (mp)	d_4^{20}	n_D^{20}	Reported reactions	Physical properties	Refs.
204	Me_3SnN_3	64	(115–117)					*(92b, 95a, 106)*
			(119.5–121.5)					*(109)*
			(121–122)				a	*(176)*
		14	(120–121)				a, e	*(178)*
						62		*(153)*
							f	*(63)*
205	$Me_3SnN_3 \cdot NEt_3$		(125–128)					*(176)*
206	$Me_3SnN_3 \cdot NC_5H_5$		(136–137)					*(176)*
207	Et_3SnN_3	64	63–65/1			62		*(106)*
			(37–39)					
208	$Bu^n_3SnN_3$	64	126/1			62		*(106)*
		52	118–120/0.18		1.5745		a	*(109)*
		52				135		*(112)*
209	$(PhCMe_2CH_2)_3SnN_3$	14	(96–96.5)				f	*(141)*
210	Ph_3SnN_3	64	(118–120)					*(106)*
			(115–116)				a	*(109)*
		14	(111.5–112.5)					*(140)*
		14	(115)					*(177)*
						62		*(101)*
							f	*(63)*

211	$Ph_3SnN_3 \cdot SnCl_4$				(*179*)
212	$(o\text{-}MeC_6H_4)_3SnN_3$	14	(81)		(*169*)
213	$(p\text{-}MeC_6H_4)_3SnN_3$	14	(115)		(*169*)
214	$(PhCH_2)_3SnN_3$	14	(119)		(*169*)
215	$Me_2Sn(N_3)N{=}PMe_3$	62	(238–242)	a	(*153*)
216	$Me_2Sn(N_3)N{=}PEt_3$	62	(218–224)	a, b	(*153*)
217	$Me_2Sn(N_3)N{=}PPh_3$	62	(225–230)	a, b	(*153*)
218	$Me_2Sn(N_3)_2$	64	(151–153d)		(*106*)
219	$Et_2Sn(N_3)_2$	64	(134–137d)		(*106*)
220	$Bu^n_2Sn(N_3)_2$	64	(63–65) 170–180/0.1		(*106*)

APPENDIX 14

ORGANOTIN ISOCYANATES

	$>Sn{-}N{=}C{=}O$	Method of prep.	bp (deg/mm) (mp)	d_4^{20}	n_D^{20}	Reported reactions	Physical properties	Refs.
221	Me_3SnNCO	14	(105–107)				a	*(176)*
								(49)
222	Et_3SnNCO	14	239					*(8)*
			(48)					
			(51–53)					*(40)*
		52	120–121/11				a	*(172)*
223	Pr^n_3SnNCO							*(40)*
224	Bu^n_3SnNCO	14	112–115/0.5					*(116)*
			144–147/1.3		1.490		a	*(172)*
								(55)
225	$(Bu^n_3SnNCO)_3$	52			1.510 (22°C)			*(173)*
226	Bu^i_3SnNCO	52	103/0.3		1.489 (21°C)			*(172)*
227	Ph_3SnNCO	14	(100)					*(171)*
		14	(98–99)					*(172)*
228	$(Ph_3SnNCO)_3$	52	(>360)					*(173)*
229	$Me_2Sn(NCO)_2$	14				136		*(142)*
230	$Bu^n_2Sn(NCO)_2$	14	(48–51)					*(116)*
231	$(Bu^n_2SnNCO)_2O$		(195–215)					*(172)*
232	$(Ph_2SnNCO)_2O$		(158–162)					*(116)*
233	$Ph_2Sn[OSn(OH)Ph_2]NCO$		(300–301)					*(116)*
234	$[Ph_2Sn(NCO)_2]_2 \cdot bipy$		(204–206d)					*(116)*

APPENDIX 15

Organotin Isothiocyanates

	$\geq Sn-N=C=S$	Method of prep.	bp (deg/mm) (mp)	d_4^{20}	n_D^{20}	Reported reactions	Physical properties	Refs.
235	Me_3SnNCS	140	(105–108)					*(132)*
		143	(108.5)					*(160)*
		14	(107–110)				a	*(176)*
			(108)				a	*(182)*
								(49)
236	$Me_3SnNCS \cdot PhNH_2$		(56–57.5)					*(182)*
237	$Me_3SnNCS \cdot NC_5H_5$		(112–114)					*(132)*
238	Et_3SnNCS	141	130/1					*(7)*
		142	282d/760		1.5825			*(8)*
			(33)					
239	Pr^n_3SnNCS	140	126–128/0.2					*(132)*
240	Bu^n_3SnNCS	137	160–162/0.8	1.2350	1.543			*(48)*
		140	150–152/0.3					*(132)*
			155/0.4					
241	$Bu^n_2Sn(NEtCOOMe)NCS$							*(53)*
242	$(C_6H_{11})_3SnNCS$	140	(123)					*(132)*
243	Ph_3SnNCS	140	(171–172)					*(132)*
		14	(172–173)					*(171)*
			(168.5–169.5)				a	*(182)*
244	$(o\text{-}MeC_6H_4)_3SnNCS$	14	(120)					*(170)*
245	$(p\text{-}MeC_6H_4)_3SnNCS$	14	(128)					*(170)*
246	$Me_2Sn(NCS)_2$	14	(198.6–199.4)					*(162)*
			(194–196)					*(185)*
			(201–203d)				a	*(182)*

(continued)

APPENDIX 15 (*continued*)

Organotin Isothiocyanates

	$\geq Sn-N=C=S$	Method of prep.	bp (deg/mm) (mp)	d_4^{20}	n_D^{20}	Reported reactions	Physical properties	Refs.
247	$Me_2Sn(NCS)_2 \cdot bipy$		(219–220.5)					(*185*)
248	$Me_2Sn(OC_9H_6N)NCS$	14	(123–124)					(*184*)
249	$(CH_2{=}CH)_2Sn(NCS)_2$		(163.5–165)					(*163*)
250	$Et_2Sn(NCS)_2$	14	(188.5–190)					(*29, 185*)
251	$Et_2Sn(NCS)_2 \cdot bipy$		(220–222)					(*185*)
252	$(Et_2SnNCS)_2O$		(178–179)					(*185*)
253	$Et_2Sn[OSn(OH)Et_2]NCS$	14	(170–176d)					(*185*)
254	$Pr^n_2Sn(NCS)_2$	14	(135–136)					(*185*)
255	$Pr^n_2Sn(NCS)_2 \cdot bipy$		(158–159)					(*185*)
256	$(Pr^n_2SnNCS)_2O$	14	(108)					(*185*)
257	$Pr^n_2Sn[OSn(OH)Pr^n_2]NCS$	14	(162–167d)					(*185*)
258	$Pr^n_2Sn(OC_9H_6N)NCS$		(144)					(*184*)
259	$Bu^n_2Sn(NCS)_2$	14	(144–145)					(*62*)
		140	(143–145)					(*132*)
								(*162*)
		14	(142–142.5)					(*185*)
260	$Bu^n_2Sn(NCS)_2 \cdot bipy$		(152–153)					(*132*)
			(152.5–153)					(*4*)
			(150–150.5)					(*185*)
261	$(Bu^n_2SnNCS)_2O$	14	(83.5–84.5)					(*185*)
262	$Bu^n_2Sn[OSn(OH)Bu^n]_2NCS$	14	(123–124)					(*185*)
263	$Ph_2Sn(NCS)_2$		(186–187)				a	(*182*)
264	$[Et_4N]_2[MeSn(NCS)_5]$		(120–121d)					(*35*)
265	$[Et_4N]_2[Me_2Sn(NCS)_4]$		(113–115)					(*35*)
266	$[Et_4N][Me_3Sn(NCS)_2]$		(163–169)					(*35*)

APPENDIX 16

MISCELLANEOUS Sn—N COMPOUNDS

	$\geqslant$Sn—N$\lt$	Method of prep.	bp (deg/mm) (mp)	d_4^{20}	n_D^{20}	Reported reactions	Physical properties	Refs.
267	$Me_3SnNCSe$							*(180)*
268	$Me_3SnN{=}CPhNMe_2$	34	79/0.2				a	*(58, 73)*
269	$Me_3SnN{=}CFe(CO)_4$	60	(sub. 110–115/0.2)					*(161)*
270	$Me_3SnN(C_6H_4Me\text{-}p)C{=}NC_6H_4Me\text{-}p$ with NEt_2 on C	36	168/0.1				a	*(58, 73)*
271	$(Me_3Sn)_2N_2C$	147						*(199)*
272	*trans*-$(Bu^n_3P)PtCl_2(Me_3SnNMe_2)$	110						*(34)*
273	$Et_3SnN{=}C{=}CMe_2$	2						*(126)*
274	$Et_3SnN{=}C{=}C(CN)CH_2R$	47						*(127, 166)*
275	$Et_3SnN(C_6H_{11})CH{=}NC_6H_{11}$	46	126/0.0001		1.5233			*(120)*
276	$Bu^n_3SnN{=}C{=}C(CN)CH_2R$	47						*(127, 166)*
277	$Bu^n_3SnN{=}C(CCl_3)OMe$	41	80d/0.05					*(17, 25)*
278	$Bu^n_3SnN{=}C(CCl_3)OSnBu^n_3$	41	115d/0.05					*(25)*
279	$Bu^n_3SnN(1\text{-}C_{10}H_7)C{=}N(1\text{-}C_{10}H_7)$ with OMe on C	41	140/0.02					*(17, 25)*
280	$Bu^n_3SnN(C_6H_4Me\text{-}p)C{=}NC_6H_4Me\text{-}p$	41						*(25)*
281	$Ph_3SnNCSe$	145						*(9)*
282	$(Ph_3Sn)_2N_2O_2$	146					a	*(15)*
283	$Me_2Sn[N(CN)_2]_2$	147						*(82)*
284	$Ph_2Sn[N(CN)_2]_2$	147						*(82)*

(continued)

APPENDIX 16 (*continued*)

MISCELLANEOUS Sn—N COMPOUNDS

	$\gtrless Sn-N \lessgtr$	Method of prep.	bp (deg/mm) (mp)	d_4^{20}	n_D^{20}	Reported reactions	Physical properties	Refs.
285	$Ph_2Sn[N=C=C(CN)_2]_2$	148						*(82)*
286	$(Me_2N)_4Sn \cdot VOCl_3$	108						*(98)*
287	$(Me_3Sn)_2CN_2 \cdot$ *p*-$MeC_6H_4N=C=NC_6H_4Me$-*p*	63	(132)					*(95)*
288	$(Me_3Sn)_2CN_2 \cdot PhNCS$	63	(124–125)					*(95)*
289	$(Me_3Sn)_2CN_2 \cdot CS_2$	63	(96)					*(95)*
290	$(Me_3Sn)_2CN_2 \cdot MeOOCC{\equiv}CCOOMe$	63	(120–125d)					*(95)*
291	$(Me_3Sn)_2CN_2 \cdot CH_2=CHCN$	63	(120–125d)					*(95)*
292	$(Me_3Sn)_2CN_2 \cdot CH_2=CMeCN$	63	(60d)					*(95)*
293	$Me_3SnN=CPh_2$	3	121/0.15				e	*(35a)*
294	$Et_3SnN=CPh_2$	22	~150/0.002				a, b	*(43a)*
295	$Ph_3SnN=CPh_2$	3	(78–88d)				e	*(35a)*

REFERENCES

1. E. W. Abel, D. A. Armitage, and D. B. Brady, *Trans. Faraday Soc.*, **62**, 3459 (1966).
2. E. W. Abel, D. B. Brady, and B. R. Lerwill, *Chem. Ind., London*, 1333 (1962).
3. E. W. Abel and J. P. Crow, *J. Chem. Soc. A*, 1361 (1968).
4. D. L. Alleston and A. G. Davies, *J. Chem. Soc.*, 2050 (1962).
5. E. Amberger, M-R. Kula, and J. Lorberth, *Angew. Chem. Int. Ed.*, **3**, 138 (1964).
6. H. H. Anderson, *J. Am. Chem. Soc.*, *74*, 1421 (1952).
7. H. H. Anderson, *J. Org. Chem.*, **19**, 1766 (1954).
8. H. H. Anderson and J. A. Vasta, *J. Org. Chem.*, **19**, 1300 (1954).
9. E. E. Aynsley, N. N. Greenwood, G. Hunter, and M. J. Sprague, *J. Chem. Soc. A*, 1344 (1966).
10. J. C. Baldwin, M. F. Lappert, J. B. Pedley, and J. A. Treverton, *J. Chem. Soc. A*, 1980 (1967).
11. J. C. Baldwin, M. F. Lappert, J. B. Pedley, J. S. Poland, and J. A. Treverton, unpublished work, 1967.
12. E. Bannister and G. W. A. Fowles, *J. Chem. Soc.*, 751 (1958).
13. E. Bannister and G. W. A. Fowles, *J. Chem. Soc.*, 4374 (1958).
14. G. Baum, W. L. Lehn, and C. Tamborski, *J. Org. Chem. 29*, 1264 (1964).
15. W. Beck, H. Engelmann, and H. S. Smedal, *Z. Anorg. Allgem. Chem.*, **357**, 134 (1968).
16. A. J. Bloodworth and A. G. Davies, *Proc. Chem. Soc.*, 264 (1963).
17. A. J. Bloodworth and A. G. Davies, *Proc. Chem. Soc.*, 315 (1963).
18. A. J. Bloodworth and A. G. Davies, *Chem. Commun.*, p. 24 (1965).
19. A. J. Bloodworth and A. G. Davies, *Chem. Ind., London*, 900 (1965).
20. A. J. Bloodworth and A. G. Davies, *Chem. Ind., London*, 1868 (1965).
21. A. J. Bloodworth and A. G. Davies, *J. Chem. Soc.*, 5238 (1965).
22. A. J. Bloodworth and A. G. Davies, *J. Chem. Soc.*, 6245 (1965).
23. A. J. Bloodworth and A. G. Davies, *J. Chem. Soc.*, 6858 (1965).
24. A. J. Bloodworth and A. G. Davies, *J. Chem. Soc., C*, 299 (1966).
25. A. J. Bloodworth, A. G. Davies, and S. C. Vasishtha, *J. Chem. Soc. C*, 1309 (1967).
26. A. B. Bruker, L. D. Balashova, and L. Z. Soborovskii, *Dokl. Akad. Nauk SSSR*, **135**, 843 (1960); through *CA*, **55**, 13301*a* (1961).
27. R. H. Bullard and W. R. Robinson, *J. Am. Chem. Soc.*, **49**, 1368 (1927).
28. H. Bürger and W. Sawodny, *Inorg. Nucl. Chem. Letters*, **2**, 209 (1966).
29. A. Cahours, *Ann. Chem.*, **122**, 48 (1862).
30. D. J. Cardin, S. A. Keppie, B. M. Kingston, and M. F. Lappert, *Chem. Commun.*, 1035 (1967).
31. D. J. Cardin, S. A. Keppie, and M. F. Lappert, *Inorg. Nucl. Chem. Letters*, **4**, 365 (1968).

31a. D. J. Cardin, S. A. Keppie, and M. F. Lappert, *J. Chem. Soc. A*, 2594 (1970).

32. D. J. Cardin and M. F. Lappert, *Chem. Commun.*, 506 (1966).
33. D. J. Cardin and M. F. Lappert, *Chem. Commun.*, 1034 (1967).
34. D. J. Cardin and M. F. Lappert, unpublished work, 1967.
35. A. Cassol, R. Portanova, and R. Barbieri, *J. Inorg. Nucl. Chem.*, **27**, 2275 (1965).

35a. L. H. Chan and E. G. Rochow, *J. Organometal Chem.*, **9**, 231 (1967).

36. G. Chandra, T. A. George, and M. F. Lappert, *Angew. Chem. Int. Ed.*, **5**, 514 (1966).
37. G. Chandra, T. A. George, and M. F. Lappert, *Chem. Commun.*, 116 (1967).
38. G. Chandra, T. A. George, and M. F. Lappert, *J. Chem. Soc. C*, 2565 (1969).

38a. G. Chandra, A. D. Jenkins, M. F. Lappert, and R. C. Srivastava, *J. Chem. Soc. A*, 2550 (1970).
39. G. Chandra and M. F. Lappert, *J. Chem. Soc. A*, 1940 (1968).
40. V. A. Chauzov, O. V. Litvinova, and Yu I. Baukov, *Zhur. Obshchei. Khim.*, **36**, 952 (1966); through *CA*, **65**, 8955 (1966).
41. T. Chivers and B. David, *J. Organometal. Chem.*, **10**, P35 (1967).
42. T. Chivers and B. David, *J. Organometal. Chem.*, **13**, 177 (1968).
43. G. E. Coates and K. Wade, *Organometallic Compounds*, 3rd ed., Vol. I, Methuen, London, 1967.
43a. M. R. Collier and M. F. Lappert, unpublished work, 1969.
44. H. M. J. C. Creemers and J. G. Noltes, *Rec. trav. chim.*, **84**, 590 (1965).
45. H. M. J. C. Creemers and J. G. Noltes, *Rec. trav. chim.*, **84**, 382 (1965).
46. H. M. J. C. Creemers, J. G. Noltes, and G. J. M. van der Kerk, *Rec. trav. chim.*, **83**, 1284 (1964).
47. H. M. J. C. Creemers, F. Verbeek, and J. G. Noltes, *J. Organometal. Chem.*, **8**, 469 (1967).
48. R. A. Cummins and P. Dunn, *Australian J. Chem.*, **17**, 411 (1964).
49. R. F. Dalton and K. Jones, *J. Chem. Soc. A*, 590 (1970).
50. R. F. Dalton and K. Jones, unpublished work, 1967.
51. R. F. Dalton and K. Jones, *Inorg. Nucl. Chem. Letters*, **5**, 785 (1969).
52. A. G. Davies, *Ann. N.Y. Acad. Sci.*, [2], **26**, 923 (1964).
53. A. G. Davies and P. G. Harrison, *J. Chem. Soc. C*, 1313 (1967).
54. A. G. Davies and T. N. Mitchell, *J. Organometal. Chem.*, **6**, 568 (1966).
55. A. G. Davies, T. N. Mitchell, and W. R. Symes, *J. Chem. Soc. C*, 1311 (1966).
56. E. A. V. Ebsworth, J. R. Hall, M. J. Mackillop, D. C. McKean, N. Sheppard, and L. A. Woodward, *Spectrochim. Acta* **13**, 202 (1958).
57. M. V. George, P. B. Talukdar, and H. Gilman, *J. Organometal. Chem.*, **5**, 397 (1966).
58. T. A. George, K. Jones, and M. F. Lappert, *J. Chem. Soc.*, 2157 (1965).
59. T. A. George and M. F. Lappert, *Chem. Commun.*, 463 (1966).
60. T. A. George and M. F. Lappert, *J. Chem. Soc. A*, 992 (1969).
61. T. A. George and M. F. Lappert, *J. Organometal. Chem.*, **14**, 327 (1968).
62. B. S. Green, D. B. Sowerby, and K. J. Wihksne, *Chem. Ind. London*, 1306 (1960).
63. R. H. Herber, H. Stöckler, and W. T. Reichle, *J. Chem. Phys.*, **42**, 2447 (1965).
64. R. E. Hester and K. Jones, *Chem. Commun.*, 317 (1966).
65. M. G. Hogben, A. J. Oliver, and W. A. G. Graham, *Chem. Commun.*, 1183 (1967).
66. J. R. Horder and M. F. Lappert, *Chem. Commun.*, 485 (1967); *J. Chem., Soc. A*, 173 (1969).
67. K. Itoh, Y. Fukumoto, and Y. Ishii, *Tetrahedron Letters*, 3199 (1968).
68. K. Itoh, I. K. Lee, I. Matsuda, S. Sakai, and Y. Ishii, *Tetrahedron Letters*, 2667 (1967).
69. K. Itoh, S. Sakai, and Y. Ishii, *Tetrahedron Letters*, 4941 (1966).
70. M. J. Janssen, J. G. A. Luijten, and G. J. M. van der Kerk, *J. Organometal. Chem.*, **1**, 286 (1964).
70a. A. D. Jenkins, M. F. Lappert, and R. C. Srivastava, *J. Organometal. Chem.*, **23**, 165 (1970).
71. K. Jones, unpublished work, 1968.
72. K. Jones and M. F. Lappert, *Proc. Chem. Soc.*, 358 (1962).
73. K. Jones and M. F. Lappert, *Proc. Chem. Soc.*, 22 (1964).
74. K. Jones and M. F. Lappert, *J. Chem. Soc.*, 1944 (1965).
75. K. Jones and M. F. Lappert, *J. Organometal. Chem.*, **3**, 295 (1965).
76. K. Jones and M. F. Lappert, *Organometal. Chem. Rev.*, **1**, 67 (1966).
76a. S. A. Keppie and M. F. Lappert, unpublished work, 1970.

77. G. J. M. van der Kerk and J. G. A. Luitjten, *J. Appl. Chem.*, **6**, 49 (1956).
78. G. J. M. van der Kerk, J. G. A. Luijten, and M. J. Janssen, *Chimia*, **16**, 10 (1962).
79. G. J. M. van der Kerk, J. G. Noltes, and J. G. A. Luijten, *Rec. trav. chim.*, **81**, 853 (1962).
80. S. F. A. Kettle, *J. Chem. Soc.*, 2936 (1959).
81. P. E. Koenig, J. M. Morris, E. J. Blanchard, and P. S. Mason, *J. Org. Chem.*, **26**, 4777 (1961).
82. H. Köhler and B. Seifert, *J. Organometal. Chem.*, **12**, 253 (1968).
83. R. G. Kostyanovskii and A. K. Prokofiev, *Izvest. Akad. Nauk SSSR Ser. Khim.*, **2**, 473 (1967); through *CA*, **67**, 21982 (1967).
84. C. A. Kraus and A. M. Neal, *J. Am. Chem. Soc.*, **52**, 695 (1930).
85. H. G. Kuivila, A. K. Sawyer, and A. G. Armour, *J. Org. Chem.*, **26**, 1426 (1961).
86. M-R. Kula, C. Kreiter, and J. Lorberth, *Chem. Ber.*, **97**, 1294 (1964).
87. M-R. Kula, J. Lorberth, and E. Amberger, *Chem. Ber.*, **97**, 2087 (1964).
88. E. J. Kupchik and P. J. Calabretta, *Inorg. Chem.*, **3**, 905 (1964).
89. E. J. Kupchik and P. J. Calabretta, *Inorg. Chem.*, **4**, 973 (1965).
90. E. J. Kupchik and R. E. Connolly, *J. Org. Chem.*, **26**, 4747 (1961).
91. E. J. Kupchik and T. Lanigan, *J. Org. Chem.*, **27**, 3661 (1962).
92. E. J. Kupchik and E. F. McInerney, *J. Organometal. Chem.*, **11**, 291 (1968).
92a. M. F. Lappert and T. E. Levitt, unpublished work, 1970.
92b. M. F. Lappert, T. Nile, and K. H. Pannell, unpublished work, 1970.
93. M. F. Lappert and J. Lorberth, *Chem. Commun.*, 836 (1967).
93a. M. F. Lappert, J. Lorberth, and J. S. Poland, *J. Chem. Soc. A*, 2954 (1970).
94. M. F. Lappert, J. B. Pedley, and P. N. K. Riley, unpublished work, 1967.
95. M. F. Lappert and J. S. Poland, *Chem. Commun.*, 156 (1969).
95a. M. F. Lappert and J. S. Poland, *Chem. Commun.*, 1061 (1969).
95b. M. F. Lappert and J. S. Poland, in *Advances in Organometallic Chemistry* (F. G. A. Stone and R. West, eds.), Vol. 9, Academic, New York, 1970, in press.
96. M. F. Lappert and B. Prokai, in *Advances in Organometallic Chemistry* (F. G. A. Stone and R. West, eds.), Vol. 5, Academic, New York, 1967, p. 225.
97. M. F. Lappert and H. Pyszora, in *Advances in Inorganic Chemistry and Radiochemistry* (H. J. Emeleus and A. G. Sharpe, eds.), Vol. 9, Academic, New York, 1966, p. 133.
98. M. F. Lappert and G. Srivastava, *Inorg. Nucl. Chem. Letters*, **1**, 53 (1965).
99. M. F. Lappert and R. C. Srivastava, unpublished work, 1969.
100. W. L. Lehn, *J. Am. Chem. Soc.*, **86**, 305 (1964).
101. W. L. Lehn, *Inorg. Chem.*, **6**, 1061 (1967).
102. A. J. Leusink, H. A. Budding, and J. G. Noltes, *Rec. trav. chim.*, **85**, 151 (1966).
103. A. J. Leusink and J. G. Noltes, *Rec. trav. chim.*, **84**, 585 (1965).
104. J. Lorberth, Dissertation, München, 1965.
105. J. Lorberth, *Chem. Ber.*, **98**, 1201 (1965).
105a. J. Lorberth, *J. Organometal. Chem.*, **15**, 251 (1968).
106. J. Lorberth, H. Krapf, and H. Nöth, *Chem. Ber.*, **100**, 3511 (1967).
107. J. Lorberth and M-R. Kula, *Chem. Ber.*, **97**, 3444 (1964).
108. J. Lorberth and M-R. Kula, *Chem. Ber.*, **98**, 520 (1965).
109. J. G. A. Luijten, M. J. Janssen, and G. J. M. van der Kerk, *Rec. trav. chim.*, **81**, 202 (1962).
110. J. G. A. Luijten and G. J. M. van der Kerk, *Investigations in the Field of Organotin Chemistry*, Tin Research Institute, Greenford, 1955.
111. J. G. A. Luijten and G. J. M. van der Kerk, *Rec. trav. chim.*, **82**, 1181 (1963).
112. J. G. A. Luijten and G. J. M. van der Kerk, *Rec. trav. chim.*, **83**, 295 (1964).

113. J. G. A. Luijten, F. Rijkens, and G. J. M. van der Kerk, in *Advances in Organometallic Chemistry* (F. G. A. Stone and R. West, eds.), Vol. 3, Academic, New York, 1965, p. 397.
114. G. P. Mack and E. Parker, U.S. Pat. 2,618,625 (1952); through *CA*, **47**, 1977 (1953).
115. G. P. Mack and E. Parker, U.S. Pat. 2,634,281 (1953); through *CA*, **48**, 1420 (1954).
116. A. S. Mufti and R. C. Poller, *J. Chem. Soc.*, 5055 (1965).
117. W. P. Neumann, *Die Organische Chemie des Zinns*, Ferdinand Enke, Stuttgart, 1967.
118. W. P. Neumann and E. Heymann, *Angew. Chem. Int. Ed.*, **2**, 100 (1963).
119. W. P. Neumann and E. Heymann, *Ann. Chem.*, **683**, 11 (1965).
120. W. P. Neumann and E. Heymann, *Ann. Chem.*, **683**, 24 (1965).
121. W. P. Neumann and H. Lind, *Angew. Chem. Int. Ed.*, **6**, 76 (1967).
122. W. P. Neumann, H. Niermann, and R. Sommer, *Angew. Chem.*, **73**, 768 (1961).
123. W. P. Neumann, E. Petersen, and R. Sommer, *Angew. Chem. Int. Ed.*, **4**, 599 (1965).
124. W. P. Neumann and B. Schneider, *Angew. Chem. Int. Ed.*, **3**, 751 (1964).
125. W. P. Neumann, B. Schneider, and R. Sommer, *Ann. Chem.*, **692**, 1 (1966).
126. W. P. Neumann, R. Sommer, and H. Lind, *Ann. Chem.*, **688**, 14 (1965).
127. W. P. Neumann, R. Sommer, and E. Müller, *Angew. Chem. Int. Ed.*, **5**, 545 (1966).
128. J. G. Noltes, *Rec. trav. chim.*, **83**, 515 (1964).
129. J. G. Noltes, *Rec. trav. chim.*, **84**, 799 (1965).
130. J. G. Noltes and M. J. Janssen, *Rec. trav. chim.*, **82**, 1055 (1963).
131. J. G. Noltes and M. J. Janssen, *J. Organometal. Chem.*, **1**, 346 (1964).
132. K. C. Pande, *J. Organometal. Chem.*, **13**, 187 (1968).
133. R. C. Poller, *J. Organometal. Chem.*, **3**, 321 (1965).
134. E. W. Randall, J. J. Ellner, and J. J. Zuckerman, *Inorg. Nucl. Chem. Letters*, **1**, 109 (1966).
135. E. W. Randall, C. H. Yoder, and J. J. Zuckerman, *Inorg. Nucl. Chem. Letters*, **1**, 105 (1966).
136. E. W. Randall, C. H. Yoder, and J. J. Zuckerman, *Inorg. Chem.*, **6**, 744 (1967).
137. E. W. Randall, C. H. Yoder, and J. J. Zuckerman, *J. Am. Chem. Soc.*, **89**, 3438 (1967).
138. E. W. Randall and J. J. Zuckerman, *Chem. Commun.*, 732 (1966).
139. E. W. Randall and J. J. Zuckerman, *J. Am. Chem., Soc.*, **90**, 3167 (1968).
140. W. T. Reichle, *Inorg. Chem.*, **3**, 402 (1964).
141. W. T. Reichle, *Inorg. Chem.*, **5**, 87 (1966).
142. E. G. Rochow, D. Seyferth, and A. C. Smith, *J. Am. Chem. Soc.*, **75**, 3099 (1953).
143. O. J. Scherer and D. Biller, *Angew. Chem. Int. Ed.*, **6**, 446 (1967).
144. O. J. Scherer and D. Biller, *Z. Naturforsch.*, **22b**, 1079 (1967).
145. O. J. Scherer and P. Hornig, *J. Organometal. Chem.*, **8**, 465 (1967).
146. O. J. Scherer, J. F. Schmidt, and M. Schmidt, *Z. Naturforsch.*, **19b**, 447 (1964).
147. O. J. Scherer, J. Schmidt, J. Wokulat, and M. Schmidt, *Z. Naturforsch.*, **20b**, 183 (1965).
148. O. J. Scherer and M. Schmidt, *Angew. Chem. Int. Ed.*, **2**, 478 (1963).
149. O. J. Scherer and M. Schmidt, *J. Organometal. Chem.*, **1**, 490 (1964).
150. O. J. Scherer and M. Schmidt, *J. Organometal. Chem.*, **3**, 156 (1965).
150a. E. O. Schlemper and D. Britton, *Inorg. Chem.*, **5**, 507 (1966).
151. D. Schmid, Dissertation, München, 1963.
152. H. Schmidbaur, *J. Am. Chem. Soc.*, **85**, 2336 (1963).
153. H. Schmidbaur and W. Wolfsberger, *Chem. Ber.*, **101**, 1664 (1968).
154. M. Schmidt and I. Ruidisch, *Angew. Chem. Int. Ed.*, **3**, 637 (1964).
155. H. Schumann and P. Jutzi, *Chem. Ber.*, **101**, 24 (1968).
156. H. Schumann and A. Roth, *J. Organometal. Chem.*, **11**, 125 (1968).
157. I. Schumann-Ruidisch and B. Jutzi-Mebert, *J. Organometal. Chem.*, **11**, 77 (1968).

158. I. Schumann-Ruidisch, W. Kalk, and R. Brüning, *Z. Naturforsch.*, **23b**, 307 (1968).
159. R. Schwarz and A. Jeanmaire, *Ber.*, **64**, 1442 (1932).
160. D. Seyferth and N. Kahlen, *J. Org. Chem.*, **25**, 809 (1960).
161. D. Seyferth and N. Kahlen, *J. Am. Chem. Soc.*, **82**, 1080 (1960).
162. D. Seyferth and E. G. Rochow, *J. Am. Chem. Soc.*, **77**, 1302 (1955).
163. D. Seyferth and F. G. A. Stone, *J. Am. Chem. Soc.*, **79**, 515 (1957).
164. K. Sisido and S. Kozima, *J. Org. Chem.*, **27**, 4051 (1962).
165. K. Sisido and S. Kozima, *J. Org. Chem.*, **29**, 907 (1964).
166. R. Sommer and W. P. Neumann, *Angew. Chem. Int. Ed.*, **5**, 515 (1966).
167. R. Sommer, W. P. Neumann, and B. Schneider, *Tetrahedron Letters*, 3875 (1964).
168. R. Sommer, B. Schneider, and W. P. Neumann, *Ann. Chem.*, **692**, 12 (1966).
169. T. N. Srivastava and S. N. Bhattacharya, *J. Inorg. Nucl. Chem.*, **28**, 1480 (1966).
170. T. N. Srivastava and S. N. Bhattacharya, *J. Inorg. Nucl. Chem.*, **28**, 2445 (1966).
171. T. N. Srivastava and S. K. Tandon, *Indian J. Appl. Chem.*, **26**, 171 (1963).
172. W. Stamm, *J. Org. Chem.*, **30**, 693 (1965).
173. Stauffer Chem. Co., Neth. Applic., 6,411,318; through *CA*, **63**, 13316 (1965).
174. A. Streitwieser, J. A. Hudson, and F. Mares, *J. Am. Chem. Soc.*, **90**, 648 (1968).
175. J. S. Thayer, *Organometal. Chem. Rev.*, **1**, 157 (1966).
176. J. S. Thayer and D. P. Strommen, *J. Organometal. Chem.*, **5**, 383 (1966).
177. J. S. Thayer and R. West, *Inorg. Chem.*, **3**, 406 (1964).
178. J. S. Thayer and R. West, *Inorg. Chem.*, **3**, 889 (1964).
179. J. S. Thayer and R. West, *Inorg. Chem.*, **4**, 114 (1965).
180. J. S. Thayer and R. West, in *Advances in Organometallic Chemistry* (F. G. A. Stone and R. West, eds.), Vol. 5, Academic Press, New York, 1967, p. 169.
181. I. M. Thomas, *Can. J. Chem.*, **39**, 1386 (1961).
182. T. T. Tsai, A. J. Sicree, and W. L. Lehn, *Tech. Rept. AFML-TR-66-108*, Air Force Materials Laboratory, Wright-Patterson AFB, Ohio, 1966.
183. A. Tzchach and E. Reiss, *J. Organometal. Chem.*, **8**, 255 (1967).
184. M. Wada, K. Kawakami, and R. Okawara, *J. Organometal. Chem.*, **4**, 159 (1965).
185. M. Wada, M. Nishino, and R. Okawara, *J. Organometal. Chem.*, **3**, 70 (1965).
186. R. M. Washburn and R. A. Baldwin, U.S. Pat. 3,112,311 (1963).
187. E. Wiberg and R. Rieger, German Pat. 1,121,050 (1960); through *CA*, **56**, 14328*b* (1962).
188. L. J. Winters and D. T Hill, *Inorg. Chem.*, **4**, 1433 (1965).
189. C. M. Wright and E. L. Muetterties, in *Inorganic Syntheses* (E. L. Muetterties, ed.), Vol. 10, McGraw-Hill, New York, 1967, p. 137.
190. C. H. Yoder and J. J. Zuckerman, *Inorg. Chem.*, **5**, 2055 (1966).
191. A. J. Bloodworth, *J. Chem. Soc. C*, 2380 (1968).
192. A. J. Bloodworth, A. G. Davies, and S. C. Vasishta, *J. Chem. Soc. C*, 2640 (1968).
193. Y. M. Chow, *Inorg. Chem.*, **9**, 794 (1970).
194. T. Cuvigny and H. Normant, *Compt. Rend. C*, **268**, 834 (1969).
195. T. Cuvigny and H. Normant, *Compt. Rend. C*, **269**, 1398 (1969).
196. A. G. Davies and J. D. Kennedy, *J. Chem. Soc. C*, 2630 (1968).
197. A. G. Davies and T. N. Mitchell, *J. Organometal. Chem.*, **17**, 158 (1969).
198. R. A. Forder and G. M. Sheldrick, *J. Organometal. Chem.*, **21**, 115 (1970); 22, 611 (1970).
199. R. A. Forder and G. M. Sheldrick, *Chem. Commun.*, 1023 (1970).
200. R. A. Forder and G. M. Sheldrick, *Chem. Commun.*, 1125 (1969).
201. M. Fukui, Y. Ishii, and K. Itoh, *Tetrahedron Letters*, 3867 (1968).
202. P. G. Harrison and J. J. Zuckerman, *Inorg. Nucl. Chem. Letters*, **5**, 545 (1969).
203. R. E. Highsmith and H. H. Sisler, *Inorg. Chem.*, **8**, 1029 (1969).

204. R. E. Highsmith and H. H. Sisler, *Inorg. Chem.*, **8**, 996 (1969).
205. J. H. Holloway, G. P. McQuillan, and D. S. Ross, *J. Chem. Soc. A*, 2505 (1969).
206. K. Itoh, Y. Ishii, and Y. Fukumoto, *Tetrahedron Letters*, 3199 (1968).
207. K. Itoh, I. Matsuda, and Y. Ishii, *Tetrahedron Letters*, 2675 (1969).
208. K. Itoh, I. Matsuda, T. Katsuura, and Y. Ishii, *J. Organometal. Chem.*, **19**, 347 (1969).
209. J. Lorberth, *J. Organometal. Chem.*, **16**, 235 (1969).
210. J. Lorberth, *J. Organometal. Chem.*, **16**, 327 (1969).
211. J. Lorberth, *J. Organometal. Chem.*, **19**, 435 (1969).
212. J. Lorberth and H. Nöth, *J. Organometal. Chem.*, **19**, 203 (1969).
213. J. Mack and C. H. Yoder, *Inorg. Chem.*, **8**, 278 (1969).
214. T. N. Mitchell and W. P. Neumann, *J. Organometal. Chem.*, **22**, C25 (1970).
215. E. Müller, R. Sommer, and W. P. Neumann, *Ann. Chem.*, **718**, 1 (1968).
216. F. Petillon, M. T. Youinou, and J. E. Guerchais, *Bull. Soc. Chim. France*, 4293 (1969).
217. J. C. Pommier and A. Roubineau, *J. Organometal. Chem.*, **16**, 23P (1969).
218. J. C. Pommier and A. Roubineau, *J. Organometal. Chem.*, **17**, 25P (1969).
219. O. J. Scherer, *Angew. Chem. Int. Ed.*, **8**, 861 (1969).
220. O. J. Scherer and P. Hornig, *Chem. Ber.*, **101**, 2533 (1968).
221. O. Schmitz-Dumont and H. J. Götze, *Z. Anorg. Allgem. Chem.*, **371**, 38 (1969).
222. O. Schmitz-Dumont, H. J. Götze, and H. Götze, *Z. Anorg. Allgem. Chem.*, **366**, 180 (1969).
223. G. M. Sheldrick and W. S. Sheldrick, *J. Chem. Soc. A*, 490 (1970).
224. G. M. Sheldrick, W. S. Sheldrick, R. F. Dalton, and K. Jones, *J. Chem. Soc. A*, 493 (1970).
225. R. Sommer, E. Müller, and W. P. Neumann, *Ann. Chem.*, **718**, 11 (1968).
226. H. Schumann and A. Roth, *Chem. Ber.*, **102**, 3731 (1969).
227. T. T. Tsai, W. L. Lehn, and C. J. Marshall, *J. Organometal. Chem.*, **22**, 387, (1970).
228. L. V. Vilkov, N. A. Tarasenko, and A. K. Prokof'ev, *Zhur. Strukt. Khim.*, **11**, 129 (1970); through *CA*, **72**, 126014*n* (1970).
229. U. Wannagat and F. Rabet, *Inorg. Nucl. Chem. Letters*, **6**, 155 (1970).
230. W. Wolfsberger and H. Schmidbaur, *J. Organometal. Chem.*, **17**, 41 (1969).

8. ORGANOTIN COMPOUNDS WITH Sn–P, Sn–As, Sn–Sb, AND Sn–Bi BONDS

HERBERT SCHUMANN, INGEBORG SCHUMANN-RUIDISCH

Institute for Inorganic and Analytical Chemistry of the Technical University of Berlin
Berlin, Germany

AND MAX SCHMIDT

Institute for Inorganic Chemistry of the University of Würzburg
Würzburg, Germany

I. Introduction

Contrary to the organotin halides and organotin chalcogens, which had been known for many years, until very recently no one had worked on the chemistry of organotin compounds in which a heavy element of the fifth main group of the periodic table is bonded directly to tin. A 1936 patent (*P1*) as well as several papers by Arbuzov et al. (*4*, *5*, *7*) and Malatesta et al. (*33*) from 1947 to 1950 did indeed describe the synthesis of organotin phosphines, arsines, stibines, and bismuthines. But later examinations (*6*) showed that these compounds did not contain direct bonds between tin and the group V element. While, thereafter, the first real organotin phosphine was prepared in 1959 by Kuchen and Buchwald (*30*), it was not until 1964 that organotin arsines (*26*), stibines (*63*), and bismuthines (*63*) also became known.

These are the reasons why even in the excellent summary by Ingham et al. (*23*), one cannot find any information about the compounds discussed here. The syntheses and properties as well as the chemical reactions of covalent tin group V compounds are thus dealt with only in some newer monographs and summaries (*1f, 15, 16, 21b, c, 25, 28, 28a, 31, 31a, 35, 37b, 38–41, 45a, 64, 70a–d*). The present article takes into consideration all the existing literature about this branch of organotin chemistry. The findings, which in a way are still very incomplete, show that the investigation of this class of compounds has just begun. Hopefully, the interest in these compounds will increase considerably during the coming years.

II. Organotin-Phosphorus Compounds

A. Preparation

As expected, organotin phosphines are very similar to their homologous nitrogen compounds. Accordingly, there are certain parallels in the methods of preparation for these two classes of compounds, both of which were discovered at about the same time. Organotin halides are also the most important starting materials for the synthesis of organotin-phosphorus compounds.

Organotin halides react smoothly with alkali organophosphides with the formation of alkali halides and organotin phosphines. Using this method, namely by reacting triethyltin bromide with sodium diphenylphosphide in ether, Kuchen and Buchwald (*30*) in 1959 obtained and isolated the first organotin phosphine:

$$(C_2H_5)_3SnBr + (C_6H_5)_2PNa \longrightarrow (C_2H_5)_3Sn{-}P(C_6H_5)_2 + NaBr$$

Shortly afterwards, in a similar manner, Bruker et al. (*9, P4*), prepared the first organotin phosphines in which two or even three triorganotin groups were bonded to phosphorus, namely by reacting triethyltin chloride or trimethyltin bromide with sodium methylphosphide or monosodium phosphide, respectively:

$$2\,(C_2H_5)_3SnCl + 2\,CH_3P\begin{smallmatrix}\diagup H\\ \diagdown Na\end{smallmatrix}$$

$$\big\downarrow$$

$$(C_2H_5)_3Sn{-}\underset{\displaystyle CH_3}{\underset{|}{P}}{-}Sn(C_2H_5)_3 + CH_3PH_2 + 2\,NaCl$$

$$3\,(CH_3)_3SnBr + 3\,NaPH_2 \longrightarrow P[Sn(CH_3)_3]_3 + 3\,NaBr + 2\,PH_3$$

They were unable to obtain any organotin phosphines of the type $R_3Sn{-}P(H){-}CH_3$ or R_3SnPH_2. This fact, unfortunately, was always confirmed in many later tests in which the synthesis of such compounds was attempted. Lithium diphenylphosphide also reacts in tetrahydrofuran with triphenyltin chloride to form triphenylstannyldiphenylphosphine (*11*, *44*, *45*). The lithium diphenylphosphide used for this synthesis was impure due to its formation from triphenylphosphine and lithium because phenyllithium is formed simultaneously. Therefore, difficulties arose in separating the organotin phosphine and tetraphenyltin, because both were formed at the same time:

$$(C_6H_5)_3P + 2\,Li \longrightarrow (C_6H_5)_2PLi + C_6H_5Li$$

$$(C_6H_5)_2PLi + C_6H_5Li + 2\,(C_6H_5)_3SnCl \longrightarrow (C_6H_5)_3Sn{-}P(C_6H_5)_2 + (C_6H_5)_4Sn + 2\,LiCl$$

The reaction of diphenylphosphinomagnesium bromide with triphenyltin chloride in ether–benzene (*11*) or the reaction of sodium diorganophosphide with organotin halides in liquid ammonia (*11*) turned out to be a better method. The latter reaction makes possible not only the synthesis of trialkylstannyldiorganophosphines, but also the isolation of diphenylstannylbis(diphenylphosphine). Besides that, this reaction proved itself mainly because it was possible to remove organotin halides that had not reacted from the reaction mixture, as their solid ammonia complexes:

$$(C_6H_5)_3SnCl + (C_6H_5)_2PMgBr \longrightarrow (C_6H_5)_3Sn{-}P(C_6H_5)_2 + MgClBr$$

$$(C_2H_5)_3SnBr + (C_6H_5)(C_2H_5)PNa \longrightarrow (C_2H_5)_3Sn{-}P(C_6H_5)(C_2H_5) + NaBr$$

$$(C_6H_5)_2SnCl_2 + 2\,(C_6H_5)_2PNa \longrightarrow (C_6H_5)_2Sn[{-}P(C_6H_5)_2]_2 + 2\,NaCl$$

It also is possible to obtain organotin phosphines by the removal of alkali halides from alkali organotin compounds and organochlorophosphines. If, for instance, triphenyltin lithium, dissolved in tetrahydrofuran, is added dropwise to a benzene solution of diphenylchlorophosphine, phenyldichlorophosphine, or phosphorus trichloride, it reacts with the formation of triphenylstannyldiphenylphosphine, bis(triphenylstannyl)phenylphosphine, or tris(triphenylstannyl)phosphine, respectively. However the yield is low (*45*, *53*, *57*):

$$n\,(C_6H_5)_3SnLi + (C_6H_5)_{3-n}PCl_n \longrightarrow [(C_6H_5)_3Sn]_nP(C_6H_5)_{3-n} + n\,LiCl$$

In the same manner tributylstannyldiphenylphosphine (*44*), bis(trimethylstannyl)methylphosphine (*8*, *P5*) and bis(triethylstannyl)methylphosphine (*8*, *P5*) were successfully synthesized. But if these methods of synthesis are reversed, i.e., by dropping the organochlorophosphine or phosphorus trichloride into a solution of triphenyltin lithium in tetrahydrofuran, it then is possible to isolate the trimeric, cyclic compounds diphenylstannylphenylphosphine or diphenylstannyltriphenylstannylphosphine besides hexaphenylditin, triphenylphosphine, and lithium chloride (*53*, *57*). The reaction mechanism so far is unknown:

$$9\,R_3SnLi + 9\,R_2PCl \longrightarrow \text{cyclo-}[R_2Sn{-}PR]_3 + 3\,R_3Sn{-}SnR_3 + 6\,R_3P + 9\,LiCl$$

$$12\,R_3SnLi + 6\,RPCl_2 \longrightarrow \text{cyclo-}[R_2Sn{-}P(SnR_3)]_3 + 3\,R_3Sn{-}SnR_3 + 3\,R_3P + 12\,LiCl$$

$$(R = C_6H_5)$$

Organotin halides in benzene solution do not react with di- or monoorganophosphines even when refluxed. Only in the presence of an equivalent amount of a base, e.g., triethylamine, is it possible under a nitrogen atmosphere to split out hydrogen chloride:

$$R_{4-n}SnCl_n + n\,R_2PH + n\,(C_2H_5)_3N \longrightarrow R_{4-n}Sn[PR_2]_n + n\,(C_2H_5)_3N\cdot HCl$$

$$n\,R_3SnCl + R_{3-n}PH_n + n\,(C_2H_5)_3N \longrightarrow (R_3Sn)_nPR_{3-n} + n\,(C_2H_5)_3N\cdot HCl$$

This way turned out to be the most convenient one for the synthesis of organotin phosphines. In this way a large number of mono-, bis-, and tris(organotin)phosphines were successfully prepared, with alkyl and/or aryl groups bonded to tin and phosphorus. The yield was always very high (*45*, *46*, *46a*, *53*, *56*, *60*, *66*, *66a*).

In this connection, special emphasis has to be given to some oligomeric organotin phosphines which were prepared in yields between 5 and 80%, besides polymeric products, by simultaneous dropwise addition of organotin halides and phenylphosphine (or, instead of phenylphosphine, simultaneous passing in of phosphine) to a benzene solution of triethylamine (*47, 48, 48a–c*):

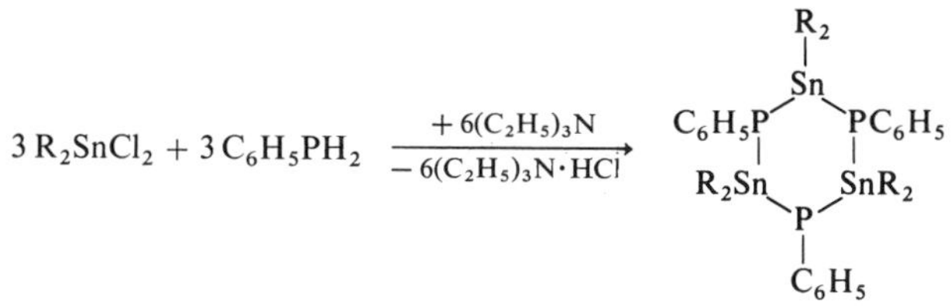

$$3\,R_2SnCl_2 + 3\,C_6H_5PH_2 \xrightarrow[-\,6(C_2H_5)_3N\cdot HCl]{+\,6(C_2H_5)_3N} [R_2SnPC_6H_5]_3 \text{ (six-membered ring: } R_2Sn, C_6H_5P, R_2Sn, C_6H_5P, SnR_2, PC_6H_5)$$

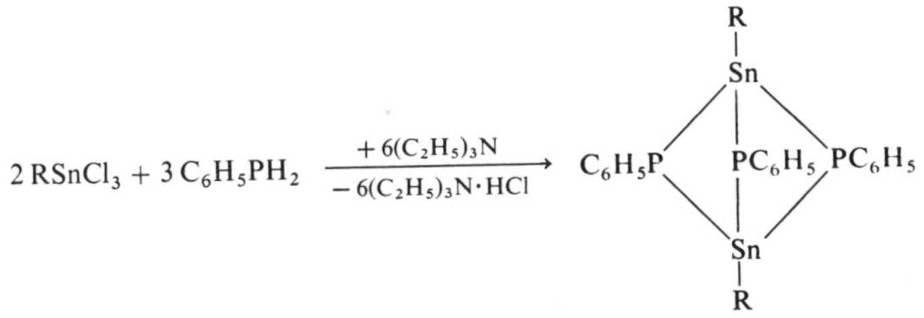

$(R = CH_3, C_4H_9, C_6H_5)$

$$2\,RSnCl_3 + 3\,C_6H_5PH_2 \xrightarrow[-\,6(C_2H_5)_3N\cdot HCl]{+\,6(C_2H_5)_3N} (RSn)_2(PC_6H_5)_3$$

$(R = CH_3, C_4H_9, C_6H_5)$

$$4\,C_6H_5SnCl_3 + 4\,PH_3 \xrightarrow[-\,12(C_2H_5)_3N\cdot HCl]{+\,12(C_2H_5)_3N} (C_6H_5Sn)_4P_4 \text{ (cube)}$$

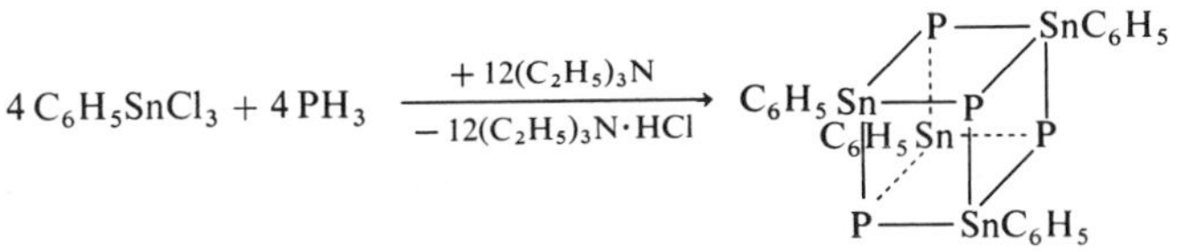

Dibutylphosphine and diphenylphosphine, in reactions with hexabutyldistannoxane, liberate water and yield the corresponding organotin phosphines nearly quantitatively and in a high degree of purity (*24, 36*):

$$(C_4H_9)_3Sn{-}O{-}Sn(C_4H_9)_3 + 2\,R_2PH \longrightarrow 2\,(C_4H_9)_3Sn{-}PR_2 + H_2O$$

The organotin amines are equally suited as starting materials for the synthesis of organotin phosphines. Trimethylstannyldimethylamine, e.g., reacts not only with many other proton-active substances, but also with diphenylphosphine or di(*tert*-butyl)phosphine with the liberation of dimethylamine (*26, 27, 59a*):

$$(CH_3)_3Sn{-}N(CH_3)_2 + (C_6H_5)_2PH \longrightarrow (CH_3)_3Sn{-}P(C_6H_5)_2 + (CH_3)_2NH$$

Other distinct methods that allow the synthesis of simple organotin phosphines turned out to be the reaction of triorganotin halides with diethylphosphonate and sodium in ethanol (*P3*) as well as the reaction of triethyltin iodide with ethyl diethylphosphinite. This last reaction yields, as the authors believe, via a diethylethoxytriethylstannylphosphonium iodide, a "stable" organotin-substituted phosphine oxide (*37*):

$$(C_4H_9)_3SnCl + (C_2H_5O)_2P(=O)H + Na \longrightarrow (C_4H_9)_3Sn-P(=O)(OC_2H_5)_2 + NaCl + \tfrac{1}{2}H_2$$

$$(C_2H_5)_3SnI + (C_2H_5)_2POC_2H_5 \longrightarrow \left[(C_2H_5)_2\overset{\oplus}{P}(-OC_2H_5)-Sn(C_2H_5)_3\right]I^{\ominus}$$

$$\downarrow$$

$$(C_2H_5)_3Sn-P(=O)(C_2H_5)_2 + C_2H_5I$$

Organotin phosphines (*65, 66a*) in which, besides the organotin groups, organogermanium or organolead moieties are attached to the phosphorus atom are also easy to prepare. For instance, butyllithium in benzene solution cleaves bis(triphenylstannyl)phenylphosphine to form triphenylbutyltin and lithium triphenylstannylphenylphosphide. This last compound, which is not isolable, reacts just like lithium bis(triphenylstannyl)phosphide, which is prepared in the same way, with triphenylgermanium chloride or triphenyllead chloride with the formation of organotin-organogermanium phosphines or organotin-organolead phosphines, respectively:

$$[(C_6H_5)_3Sn]_2PC_6H_5 + LiC_4H_9 \longrightarrow (C_6H_5)_3SnC_4H_9 + (C_6H_5)_3Sn-P(Li)C_6H_5$$

$$(C_6H_5)_3Sn-P(Li)C_6H_5 + (C_6H_5)_3MCl \longrightarrow (C_6H_5)_3Sn-P(M(C_6H_5)_3)C_6H_5 + LiCl$$

$$(M = Ge, Pb)$$

$$[(C_6H_5)_3Sn]_3P + C_4H_9Li \longrightarrow [(C_6H_5)_3Sn]_2PLi + (C_6H_5)_3SnC_4H_9$$

$$[(C_6H_5)_3Sn]_2PLi + (C_6H_5)_3MCl \longrightarrow [(C_6H_5)_3Sn]_2PM(C_6H_5)_3 + LiCl$$

$$(M = Ge, Pb)$$

Unfortunately, until now it was not possible to synthesize a triphenylgermanyl-triphenylstannyl-triphenylplumbyl-phosphine in this manner (*45*).

A somewhat unusual way to synthesize organotin phosphines was found in the reaction of tetraphenyltin with elementary phosphorus at higher temperatures in a sealed tube (*64*). In this way, phosphorus-phosphorus bonds can be split by nucleophilic attack similar to sulfur-sulfur bonds because in the different phosphorus modifications the lone pair of electrons of each phosphorus atom is not localized as such, but is more or less included into the bonds by $(p\text{-}d)\pi$ interactions. Therefore, the final product of the degradation of phosphorus by tetraphenyltin must be triphenylphosphine, as well as alloy-like tin phosphides.

When a mixture of tetraphenyltin and white phosphorus is heated to 220–230°C in a sealed tube, the primary products are diphenyltin and triphenylphosphine (*54*). The "carbenoid" diphenyltin adds to the lone pair of electrons of the phosphorus in triphenylphosphine with the formation of the addition complex, diphenylstannyltriphenylphosphine:

$$3\,(C_6H_5)_4Sn + \tfrac{1}{2}\,P_4 \longrightarrow 3\,(C_6H_5)_2Sn + 2\,(C_6H_5)_3P$$
$$(C_6H_5)_2Sn + P(C_6H_5)_3 \longrightarrow (C_6H_5)_2Sn \leftarrow |P(C_6H_5)_3$$
$$(C_6H_5)_2Sn \leftarrow |P(C_6H_5)_3 \longrightarrow (C_6H_5)_3Sn{-}P(C_6H_5)_2$$

This compound, which cannot be isolated, but whose existence could be proved by the formation of its stable oxidation products, diphenyltin oxide and triphenylphosphine oxide, rearranges at higher reaction temperatures to the stable isomer triphenylstannyldiphenylphosphine. By an excess of phosphorus this rearrangement product becomes further dephenylized (*45*, *53*, *55*). Between 230 and 250°C, e.g., cyclic trimeric diphenylstannylphenylphosphine is formed, whereas between 250 and 280°C, highly polymeric organotin phosphines are obtained (*45*, *55*). These highly polymeric organotin phosphines can be identified from their oxidation products (Table 1).

In the whole temperature range up to 280°C small amounts of alloy-like tin phosphides, which are rich in phosphorus, are also formed. Above 280°C, all tin-carbon bonds are split and the final product of the reaction of phosphorus with tetraphenyltin is, besides triphenylphosphine, nearly pure metallic tin:

$$3\,(C_6H_5)_4Sn + P_4 \longrightarrow 3\,Sn + 4\,(C_6H_5)_3P$$

The unusual ways of synthesis for organotin phosphines described now lead to complicated molecules of this class of compounds.

From bis(triethylstannyl)methylphosphine and diethyldisulfide, e.g., one obtains triethylstannyl(ethylsulfido)methylphosphine (*10*):

$$[(C_2H_5)_3Sn]_2PCH_3 + (C_2H_5)_2S_2 \longrightarrow (C_2H_5)_3Sn{-}P\begin{matrix} \diagup CH_3 \\ \diagdown S{-}C_2H_5 \end{matrix} + (C_2H_5)_3Sn{-}SC_2H_5$$

TABLE 1

THE REACTION PRODUCTS ISOLATED FROM $(C_6H_5)_4Sn$ AND P_x BETWEEN 260 AND 280°C ($R = C_6H_5$)

Basic structure of the			
Organotin phosphine	Oxidation products	Oxidation products	Cleavage products
RSnP—	O RSnOPO— O	$(C_6H_5O_4PSn)_n$	RSn(O)OH H_3PO_4
R_2PSnP—	O O O $R_2POSnOPO$—	$(C_{12}H_{10}O_6P_2Sn)_n$	SnO_2 H_3PO_4 $R_2P(O)OH$
R R_2PSnPR	O R O $R_2POSnOPR$ O	$(C_{24}H_{20}O_5P_2Sn)_n$	RSn(O)OH $R_2P(O)OH$ $RP(O)(OH)_2$
R $(RPSn—)_2PR$	O R O $(RPOSnO)_2PR$ O	$(C_{30}H_{25}O_9P_3Sn_2)_n$	$RSn(O)OH_2$ RP(O)(OH)
R $(RPSn—)_3P$	O R O $(RPOSnO)_3P$ O	$(C_{36}H_{30}O_{13}P_4Sn_3)_n$	RSn(O)OH $RP(O)(OH)_2$ H_3PO_4

A phosphine-imine derivative that is organotin-substituted at the phosphorus atom apparently does not form by Arbuzov rearrangement of a *N*-organotin-substituted aminophosphine, but via a *P*-organotin substituted phosphonium salt intermediate (*41b*, *42*, *43*):

$$[(CH_3)_3C]_2P{-}N{<}^{H}_{Li} + (CH_3)_3SiCl \longrightarrow [(CH_3)_3C]_2P{-}N{<}^{H}_{Si(CH_3)_3} + LiCl$$

$$[(CH_3)_3C]_2P{-}N{<}^{H}_{Si(CH_3)_3} + C_4H_9Li \longrightarrow [(CH_3)_3C]_2P{-}N{<}^{Li}_{Si(CH_3)_3} + C_4H_{10}$$

$$[(CH_3)_3C]_2P{-}N{<}^{Li}_{Si(CH_3)_3} + (CH_3)_3SnCl \longrightarrow \left[[(CH_3)_3C]_2\overset{\oplus}{P}(Sn(CH_3)_3){-}N(Li){-}Si(CH_3)_3 \right] Cl^{\ominus}$$

$$\xrightarrow{-LiCl} [(CH_3)_3C]_2P(Sn(CH_3)_3){=}N{-}Si(CH_3)_3$$

Finally, very stable organotin phosphines with a quarternary phosphorus atom are formed in the reaction of some organotin phosphines with nickel tetracarbonyl (*1b*, *46a*, *59a*, *67*, *69a*) iron pentacarbonyl (*46a*, *59a*, *69*, *69c*) chromium hexacarbonyl (*46a*, *68*), or nitrosylcobalt tricarbonyl (*68*, *69b*) with the liberation of carbon monoxide.

$$[(CH_3)_3Sn]_3P + M(CO)_n \longrightarrow [(CH_3)_3Sn]_3P{\to}M(CO)_{n-1} + CO$$

$$M = Ni(n=4),\quad Fe(n=5),\quad Cr(n=6);$$

$$[(CH_3)_3Sn]_3P + CoNO(CO)_3 \longrightarrow [(CH_3)_3Sn]_3P{\to}CoNO(CO)_2 + CO$$

In the same way one can obtain spiroheterocyclic compounds with tin as the central atom (*18*):

$$Sn[P(C_6H_5)_2]_4 + 2\,M(CO)_6 \longrightarrow (CO)_4M{<}^{P(C_6H_5)_2}_{P(C_6H_5)_2}{>}Sn{<}^{P(C_6H_5)_2}_{P(C_6H_5)_2}{>}M(CO)_4 + 4\,CO$$

(M = Cr, Mo, W)

When using dinitrosyliron dicarbonyl or nitrosylcobalt tricarbonyl it is possible to isolate the intermediate tetrakis(diphenylphosphinodinitrosylcarbonyliron)tin or tetrakis(diphenylphosphinonitrosyldicarbonylcobalt) tin, respectively:

$$Sn[P(C_6H_5)_2]_4 + 4\,Fe(NO)_2(CO)_2 \longrightarrow Sn\left[\begin{matrix} C_6H_5 \\ | \\ -P\longrightarrow Fe(NO)_2CO \\ | \\ C_6H_5 \end{matrix}\right]_4 + 4\,CO$$

$$Sn[P(C_6H_5)_2]_4 + 4\,Co(NO)(CO)_3 \longrightarrow Sn\left[\begin{matrix} C_6H_5 \\ | \\ -P\longrightarrow Co(NO)(CO)_2 \\ | \\ C_6H_5 \end{matrix}\right]_4 + 4\,CO$$

B. Chemical Properties and Reactions

Under normal conditions, organotin phosphines are liquid or crystalline substances and, with the exception of the phenylstannyl compounds, they usually can be distilled without decomposing. They dissolve well without decomposition in aromatic hydrocarbons and tetrahydrofuran, and some compounds also dissolve in diethyl ether. The crystalline compounds can be easily recrystallized from aliphatic hydrocarbons like pentane or methylcyclohexane. The analytical determination of tin in the presence of phosphorus is especially difficult. Conventional methods do not give good results, but it can be accomplished very well by the method of x-ray fluorescence (*32*). Organotin phosphines, like all covalent group IV–V compounds, are sensitive to oxygen. The degree of that sensitivity is essentially dependent on the shielding by the organic groups linked to tin and phosphorus. In general, alkyl compounds are less stable than aryl compounds. Tin tetrakis(diphenylphosphine) and tris(triphenylstannyl)phosphine, e.g., react only slowly with oxygen of the air, while nearly all of the other organotin phosphines known so far can be handled only in an oxygen-free atmosphere. In principle, this oxidation occurs according to the following scheme either when exposed to air or, always quantitatively, with hydrogen peroxide in ethanol forming the corresponding organotin phosphinates (*11, 45, 50, 55–57, 65, 66, 66a*):

$$\geqslant Sn-\bar{P}\lessdot + O_2 \longrightarrow \;\geqslant Sn-O-\overset{O}{\overset{\|}{P}}\lessdot$$

Up to now it has been impossible to fix with certainty that stage of the reaction in which only one oxygen atom per tin-phosphorus bond has been added. Thus, tri-*n*-butylstannyldiphenylphosphine oxide (*44*) described in the literature certainly is tri-*n*-butylstannyldiphenylphosphinate. And the reaction of triphenylstannyldiphenylphosphine with equimolar amounts of *C*-phenyl-*N*-phenyl-nitrone, 1,2-dihydroisoquinoline-*N*-oxide, or benzonitril oxide leads only to a 1:1 mixture of triphenylstannyldiphenylphosphinate and unreacted triphenylstannyldiphenylphosphine (*50*):

$$(C_6H_5)_3Sn{-}P(C_6H_5)_2 + 2\,C_6H_5CH{=}\overset{\overset{O}{\uparrow}}{N}{-}C_6H_5 \longrightarrow 2\,C_6H_5CH{=}N{-}C_6H_5 + (C_6H_5)_3Sn{-}O{-}\overset{\overset{O}{\|}}{P}(C_6H_5)_2$$

Of the two possible intermediates of this oxidation reaction, only triphenylstannyldiphenylphosphenate was isolated as a substance which is extremely sensitive toward oxygen and which becomes oxidized immediately yielding triphenylstannyldiphenylphosphinate, while triphenylstannyldiphenylphosphine oxide rearranges immediately to the isomeric phosphenate (*50*):

$$(C_6H_5)_3Sn{-}O{-}Li + (C_6H_5)_2PCl \longrightarrow (C_6H_5)_3Sn{-}O{-}\bar{P}(C_6H_5)_2 + LiCl$$

$$(C_6H_5)_3SnLi + (C_6H_5)_2P{\overset{\displaystyle /\!\!/ O}{\underset{\displaystyle \backslash Cl}{}}} \longrightarrow (C_6H_5)_3Sn{-}\overset{\overset{O}{\|}}{P}(C_6H_5)_2 + LiCl$$

$$\downarrow$$

$$(C_6H_5)_3Sn{-}O{-}P(C_6H_5)_2$$

These facts, as well as the phenomenon that triorganotin phosphenates remove oxygen from triorganostannoles, thus forming triorganotin diorganophosphinates and triorganotin hydrides (*24*):

$$R_2P{-}O{-}SnR'_3 + R'_3Sn{-}OH \longrightarrow R'_3Sn{-}O{-}\overset{\overset{O}{\|}}{P}R_2 + R'_3SnH$$

suggest the following course of reaction for the oxidation of organotin phosphines. In the first step an organotin phosphine is converted to an organotin phosphine oxide by formal addition of one oxygen atom to the free electron-pair of the phosphorus (intermediates are probably undetectable peroxo-compounds). Now, the formerly free electron-pair of the phosphorus atom can no longer strengthen the bond between tin and phosphorus and therefore, the tin-phosphorus bond breaks and, by a reversed Arbuzov rearrangement, a phosphenate is formed from the phosphine oxide. In the second step, the electron-pair of the phosphorus, which is now available again adds another oxygen atom with formation of the corresponding phosphinate:

$$\gt Sn-\bar{P}\lt \xrightarrow{+\frac{1}{2}O_2} \gt Sn-\overset{O}{\overset{\|}{P}}\lt \longrightarrow \gt Sn-O-\bar{P}\lt \xrightarrow{+\frac{1}{2}O_2} \gt Sn-O-\overset{O}{\overset{\|}{P}}\lt$$

In this reaction, the second addition of oxygen is favored over the first one, which means that in the oxidation of organotin phosphines one always gets only the corresponding phosphinates.

Triphenylstannyldiphenylphosphine reacts with sulfur (S_8) in the same manner (*50*) giving triphenylstannyldiphenyldithiophosphinate:

$$4\,(C_6H_5)_3Sn-P(C_6H_5)_2 + S_8 \longrightarrow 4\,(C_6H_5)_3Sn-S-\overset{S}{\overset{\|}{P}}(C_6H_5)_2$$

Using bis(triphenylstannyl)phenylphosphine or tris(triphenylstannyl)phosphine in this reaction, only the cleavage products of organotinthiophosphinates are formed, namely bis(triphenylstannyl)sulfide and polymeric phenylphosphorus sulfides or P_2S_5, respectively (*50*).

In an inert atmosphere, perphenylized organotin phosphines are not hydrolyzed by water, probably because they are too hydrophobic (*45, 56, 57*). In alkyl-substituted compounds, the tin-phosphorus bond is split, and due to its polarization stannoxanes and phosphines are formed (*11*). An alcoholic solution of sodium hydroxide splits all known organotin phosphines. For complicated compounds it is easy to derive the structure of the original molecule (*45, 55–57*) from the kind of cleavage products formed:

$$\gt Sn-\bar{P}\lt \xrightarrow[OH^-]{+H_2O} \gt Sn-OH + H\bar{P}\lt$$

Organotin phosphines should react with methyl iodide, forming phosphonium salts, but it turns out that these phosphonium salts are unstable and decompose immediately, forming organotin iodides and phosphonium iodides (*11, 45, 56, 59a, 66a*):

$$R_3Sn-P(C_6H_5)_2 + 2\,CH_3I \longrightarrow R_3SnI + [(CH_3)_2\overset{\oplus}{P}(C_6H_5)_2]\overset{\ominus}{I}$$

When cleaving trialkylstannyldiphenylphosphine with sodium in liquid ammonia, one gets sodium diphenylphosphide and trialkyltin sodium (*11*). Butyllithium cleaves the tin-phosphorus bond of bis(triphenylstannyl)-phenylphosphine or tris(triphenylstannyl) phosphine with the formation of triphenylbutyltin and lithium triphenylstannylphenylphosphide or lithium bis(triphenylstannyl)phosphide (*65, 66a*). The two compounds mentioned last are suitable as starting materials for the synthesis of organometallic phosphines with mixed substituents:

$$[(C_6H_5)_3Sn]_2PC_6H_5 + LiC_4H_9 \longrightarrow \begin{matrix}(C_6H_5)_3Sn \\ C_6H_5\end{matrix}\!\!>\!P{-}Li + (C_6H_5)_3SnC_4H_9$$

$$[(C_6H_5)_3Sn]_3P + LiC_4H_9 \longrightarrow [(C_6H_5)_3Sn]_2P{-}Li + (C_6H_5)_3SnC_4H_9$$

Another cleavage reaction of organotin phosphines is one with organotin hydrides. Besides an exchange of the organotin groups between the organotin phosphine and the organotin hydride (*13, 14, 36*) one finds a side reaction that leads in only small yields to the formation of hexaorganoditin and phosphine (*13, 14, 36*):

$$(C_6H_5)_3SnH + (C_4H_9)_3Sn{-}P(C_4H_9)_2 \longrightarrow (C_6H_5)_3Sn{-}P(C_4H_9)_2 + (C_4H_9)_3SnH$$

$$(C_2H_5)_3Sn{-}P(C_6H_5)_2 + (CH_3)_3SnH \longrightarrow (C_2H_5)_3Sn{-}Sn(CH_3)_3 + (C_6H_5)_2PH$$

Trimethylstannyldiphenylphosphine reacts with butylthiodiphenylborane under fission of the Sn—P bond (*21a*):

$$(CH_3)_3Sn{-}P(C_6H_5)_2 + (C_6H_5)_2B{-}SC_4H_9 \longrightarrow (CH_3)_3Sn{-}SC_4H_9 + (C_6H_5)_2B{-}P(C_6H_5)_2$$

The very reactive tin-phosphorus bond of organotin phosphines can also be split by 1,2-dipolar agents (*49, 51*). For instance, carbon disulfide inserts between tin and phosphorus in triphenylstannyldiphenylphosphine with the formation of triphenylstannyldiphenylphosphinodithioformate:

$$(C_6H_5)_3Sn{-}P(C_6H_5)_2 + CS_2 \longrightarrow (C_6H_5)_3Sn{-}S{-}C\!\begin{matrix}{/\!\!/}S \\ \diagdown P(C_6H_5)_2\end{matrix}$$

As 1,2-dipoles, carbon oxisulfide, thiophosgene, thiourea, phenyl isocyanate, and phenylisothiocyanate also react with triphenylstannyldiphenylphosphine in the same manner, forming *O*-triphenylstannyldiphenylphosphinothioformate, dichloro(triphenylstannylmercapto) (diphenylphosphino)methane, diamino(triphenylstannylmercapto) (diphenylphosphino)methane, *N*-phenyl-*N*-triphenylstannylcarbamoyldiphenylphosphine, and *N*-phenyl-*N*-triphenylstannylthiocarbamoyldiphenylphosphine, respectively. But no cleavage was observed with carbon dioxide.

These reactions, which are analogous to the "insertion-reactions" at the Si—N and Sn—N bond, probably proceed by a polar four-centered mechanism:

$$\begin{matrix}(C_6H_5)_3Sn & - & \underline{P}(C_6H_5)_2 \\ \uparrow & & \downarrow \\ |\overline{S}^{\delta-} & - & C^{\delta+} = \overline{S}\end{matrix} \longrightarrow \begin{matrix}(C_6H_5)_3Sn & \overline{P}(C_6H_5)_2 \\ | & | \\ |\underline{S} & - C = \overline{S}\end{matrix}$$

Tetrakis(diphenylphosphino)tin reacts with carbon disulfide or phenyl isocyanate with splitting of all four tin-phosphorus bonds in the molecule. With tris(triphenylstannyl)phosphine these cleavage reactions do not take place (*49, 51*).

This reaction principle also does not work in the case of 1,3-dipolar reagents; nitrones and nitriloxides react only by oxidation (*50*), whereas nitrilimines and diazomethane do not react at all with organotin phosphines. Organic azides like phenyl azide split the tin-phosphorus bond with the formation of a new kind of phosphinimine (*60*). For instance, in the reaction with triphenylstannyldiphenylphosphine, in the first stage triphenylstannyl-diphenylphosphine-phenylimine is formed with elimination of nitrogen. This unstable compound rearranges as in a reversed Arbusov rearrangement, regenerating the free electron-pair and forming *N*-triphenylstannyl-*N*-phenyl-diphenylphosphino-amine. Then a second molecule of phenylazide is added immediately and, with repeated elimination of nitrogen, *N*-triphenylstannyl-*N*-phenylamino-diphenylphosphinophenylimine is formed:

$$(C_6H_5)_3Sn{-}P(C_6H_5)_2 \xrightarrow[-N_2]{+C_6H_5N_3} (C_6H_5)_3Sn{-}\overset{\overset{\displaystyle NC_6H_5}{\|}}{P}(C_6H_5)_2 \longrightarrow$$

$$(C_6H_5)_3Sn{-}\overset{\overset{\displaystyle C_6H_5}{|}}{N}{-}P(C_6H_5)_2 \xrightarrow[-N_2]{+C_6H_5N_3} (C_6H_5)_3Sn{-}\underset{\underset{\displaystyle C_6H_5}{|}}{N}{-}\overset{\overset{\displaystyle NC_6H_5}{\|}}{P}(C_6H_5)_2$$

In the same manner, bis(triphenylstannyl)phenylphosphine and bis(trimethylstannyl)phenylphosphine react with phenylazide, but in the last case the phosphinimine formed is unstable and decomposes in the following way:

$$[(CH_3)_3Sn]_2PC_6H_5 + 3\ C_6H_5N_3 \longrightarrow [(CH_3)_3Sn]_2NC_6H_5 + C_6H_5P(NC_6H_5)_2 + 3\ N_2$$

On the other hand, tris(triphenylstannyl)phosphine does not react with phenyl azide even in boiling benzene.

Triphenylstannyldiphenylphosphine reacts in boiling benzene with allyl chloride or styrene with the formation of 1-triphenylstannyl-2-diphenyl-phosphino-3-chloropropane and 1-triphenylstannyl-2-phenyl-2-diphenyl-phosphinoethane, respectively (*52*):

$$(C_6H_5)_3Sn{-}P(C_6H_5)_2 + ClCH_2{-}CH{=}CH_2 \longrightarrow (C_6H_5)_3Sn{-}CH_2{-}\overset{\overset{\displaystyle CH_2Cl}{|}}{C}H{-}P(C_6H_5)_2$$

$$(C_6H_5)_3Sn{-}P(C_6H_5)_2 + C_6H_5CH{=}CH_2 \longrightarrow (C_6H_5)_3Sn{-}CH_2{-}\underset{\underset{\displaystyle C_6H_5}{|}}{C}H{-}P(C_6H_5)_2$$

With phenylacetylene the reaction stops at the stage of 1-triphenylstannyl-2-phenyl-2-diphenylphospinoethylene (*52*). Catalytic amounts of azobisisobutyronitrile increase the rate of the reaction, which points to a radical mechanism. Contrary to that, vinylic carbonyl compounds like mesityl oxide, acrolein, or cinnamic aldehyde add trimethylstannyldiphenylphosphine only at the carbonyl function (*70*):

$$(CH_3)_3Sn{-}P(C_6H_5)_2 + CH_2{=}CH{-}C(=O)H \longrightarrow CH_2{=}CH{-}CH(O{-}Sn(CH_3)_3)(P(C_6H_5)_2)$$

A simple method for the synthesis of oligomeric phenylphosphines and phenylphosphinoarsines was found in the reaction of organotinphosphines with phenylphosphorus or phenyl arsenic halides (*21*, *62*, *62a*). Thus tris(trimethylstannyl)phosphine reacts in ether solution and at room temperature with diphenylphosphorus chloride or diphenylarsenic chloride with the elimination of trimethyltin chloride and the formation of tris(diphenylphosphino)-phosphine and tris(diphenylarsino)phosphine, respectively (*62*, *62a*):

$$[(CH_3)_3Sn]_3P + 3\,(C_6H_5)_2XCl \longrightarrow 3\,(CH_3)_3SnCl + [(C_6H_5)_2X]_3P$$
$$(X = P, As)$$

In the same manner, by reacting trimethylstannyldiphenylphosphine with arsenic trifluoride it was possible to synthesize tris(diphenylphosphino) arsine (*21*, *21a*) and by reactions of bis(trimethylstannyl)phenylphosphine with diphenylphosphorus chloride or diphenylarsenic chloride it was possible for the first time to synthesize bis(diphenylphosphino)phenylphosphine and bis(diphenylarsino)phenylphosphine, respectively (*62*, *62a*):

$$3\,(CH_3)_3Sn{-}P(C_6H_5)_2 + AsF_3 \longrightarrow [(C_6H_5)_2P]_3As + 3\,(CH_3)_3SnF$$
$$[(CH_3)_3Sn]_2PC_6H_5 + 2\,(C_6H_5)_2XCl \longrightarrow [(C_6H_5)_2X]_2PC_6H_5 + 2\,(CH_3)_3SnCl$$
$$(X = P, As)$$

Because of the described chemical properties and reactions of the organotin phosphines it was assumed that the "free" electron-pair at the phosphorus atom participates by way of $(p\text{-}d)\pi$ interaction between phosphorus and tin (*45*). Therefore, tris(triorganotin)phosphines should have no or only little tendency to act as σ donors in transition metal complexes. But as has already been mentioned on p. 589, dimethylstannyl bis(diphenylphosphine) (*1a*, *69a*), tris(trimethylstannyl)phosphine (*67–69*, *69a–c*) bis(trimethylstannyl)phenylphosphine (*69*, *69a*), bis(trimethylstannyl)methylphosphine (*46*, *46a*), trimethylstannyldiphenylphosphine (*1a*, *69*, *69a*), trimethylstannyldi(*tert*-butyl)phosphine (*59a*), and tetrakis(diphenylphosphino)tin (*18*) react with some transition metal carbonyls in tetrahydrofuran at room temperature with formation of remarkably stable transition metal complexes in which there are stable organotin-substituted phosphines with four bonds at the phosphorus atom. The stability of these compounds towards oxygen confirms the hypothesis that, in the oxidation of organotin phosphines, the oxygen attacks phosphorus at its "free" electron-pair. In the complexes, because of the coordination with the transition metal, this electron-pair is no longer available for electrophilic attack. The iron complexes are especially well suited as catalysts for different organic reactions (*69*).

C. Physical Properties

All organotin phosphines known so far are listed in Table 2. As far as their boiling points, melting points, refractive indices or densities are given in the literature, these values are included in the table. More detailed information about specific physical properties of organotin phosphines are, so far, found only scarcely in the literature. No X-ray structure analyses have yet been done and therefore no bond distances or bond angles have been determined.

The infrared, and in some cases, the Raman spectra of a number of organotin phosphines have been measured by Hester and Jones (*22*), Reich (*19*), Campbell et al. (*11*), Ellermann (*18*) and Schumann et al. (*19, 45, 46a, 47, 48, 48a, b, 50, 59a, 65, 66 66a, 67, 68, 69*). Table 3 shows the frequencies of the tin-phosphorus vibration bands found for these compounds. The vibrational spectra of the tris(triorganotin)phosphines show that, because of the agreement of the infrared and Raman spectra with each other and because of the measured depolarization ratios, the compounds have C_{3v} symmetry (*19, 22*). Since the tin-phosphorus valence vibrations of the other organotin phosphines are in the same frequency area, one certainly can expect similar bonding conditions in these compounds. The observed slight shift of the tin-phosphorus valence vibrations in the organotin-phosphine-substituted transition metal complexes compared to the valence vibrations in the free organotin phosphines shows that the tin-phosphorus bond is weakened only slightly by the complex formation (*18, 67, 68, 69a,b*).

Besides, one has to remember that, between the tin-phosphorus valence vibrations and the transition metal-phosphorus valence vibrations of the same symmetry, coupling is expected, mass-wise as well as bonding-wise. The νSn—P and the νM—P therefore do not represent pure valence vibrations.

In the uv-spectrum of triethylstannyldiethylphosphine a very intense band is found at 214 mμ, which is attributed to the tin-phosphorus bond (*11*). From the intensity of the band one can conclude that it is due to a $\pi \rightarrow \pi^*$-transition.

TABLE 2

ORGANOTIN PHOSPHINES

Compound	m.p. (°C)	b.p. (°C/mmHg)	n_D^{20}	d_4^{20}	References
$(CH_3)_3Sn—P(CH_3)_2$		15/760 decomposes			(*46*)
$(CH_3)_3Sn—P[C(CH_3)_3]_2$		73/0.1			(*45a, 59a*)
$(CH_3)_3Sn—P(C_6H_5)_2$		141–2/0.7(*11*) 150/0.8(*26, 27*)			(*1b, 11, 18b, 26, 27, 64*)
$(C_2H_5)_3Sn—P(C_2H_5)_2$		70/0.3(*11*) 170–2/0.7(*30*)			(*11, 30, 64*)
$(C_2H_5)_3Sn—P(C_6H_5)(C_2H_5)$		99–100/0.2(*11*)			(*11, 64*)
$(C_2H_5)_3Sn—P(C_6H_5)_2$		167–8/0.7(*11*) 170/0.7(*26, 27*) 170–2/0.7(*30*)			(*11, 13, 26, 27, 30, 64*)
$(n\text{-}C_3H_7)_3Sn—P(C_6H_5)_2$		176–7/0.6(*11*)			(*11, 64*)
$(n\text{-}C_3H_7)_3Sn—P(C_6H_5)(C_2H_5)$		125–6/0.3(*11*)			(*11, 64*)
$(C_4H_9)_3Sn—P(C_4H_9)_2$		122–4/0.15	1.5045	1.04	(*36*)
$(n\text{-}C_4H_9)_3Sn—P(C_6H_5)_2$	60(*53*) 90–6(*56*)	192/0.6(*11*) 160–3/0.01(*24*)			(*11, 24, 44, 53, 56, 64*)
$(C_4H_9)_3Sn—P(—OC_4H_9)_2$					(*P2*)
$(C_6H_5)_3Sn—P(C_6H_5)_2$	103–5(*11*) 126(*53*) 127–30(*56, 57*)				(*11, 19, 32, 50–53, 56, 57, 64, 66, 66a*)

(*continued*)

TABLE 2 (*continued*)

ORGANOTIN PHOSPHINES

Compound	m.p. (°C)	b.p. (°C/mmHg)	n_D^{20}	d_4^{20}	References
$[(CH_3)_3Sn]_2PCH_3$		89–90/3(*8*, *P4*) 95–7/4(*8*, *46*, *P5*)	1.5778(*8*, *P4*) 1.5768(*P5*)	1.5601(*8*, *P4*) 1.5599(*P5*)	(*8*, *10*, *46*, *46a*, *P4*, *P5*)
$[(CH_3)_3Sn]_2PC_6H_5$	37–8(*48*) 35(*60*)	131–5/0.01(*48*)			(*48*, *60*)
$[(C_2H_5)_3Sn]_2PCH_3$		143–5/3(*8*, *P4*) 135–40/1(*8*)	1.5649(*8*) 1.5621(*8*, *P4*)	1.3725(*8*, *P4*)	(*8*, *P4*, *P5*)
$[(C_2H_5)_3Sn]_2PC_6H_5$		150–1/0.3			(*11*)
$[(n\text{-}C_4H_9)_3Sn]_2PC_6H_5$		178–81/0.01			(*24*)
$[(C_6H_5)_3Sn]_2PC_6H_5$	146–50(*57*) 150(*66*)				(*19*, *32*, *50*, *57*, *64–66*, *66a*)
$[(CH_3)_3Sn]_3P$		136–7/3(*9*)	1.5970(*9*)	1.6769(*9*)	(*9*, *19*, *22*, *66a*, *67*)
$[(C_6H_5)_3Sn]_3P$	197–201(*57*) 201(*66*)				(*19*, *33*, *50*, *51*, *57*, *64*, *66*, *66a*)
$(CH_3)_2Sn[{-}P(C_6H_5)_2]_2$	110–4(*56*)				(*56*, *64*)
$(n\text{-}C_4H_9)_2Sn[{-}P(C_6H_5)_2]_2$	98–102(*56*)				(*56*, *64*)
$(C_6H_5)_2Sn[{-}P(C_6H_5)_2]_2$	144–7(*11*) 78–80(*56*)				(*11*, *56*, *64*)
$C_6H_5Sn[{-}P(C_6H_5)_2]_3$	115–7(*56*)				(*56*, *64*)
$Sn[{-}P(C_6H_5)_2]_4$	106–7(*56*)				(*51*, *56*, *64*)
$(C_2H_5)_3Sn{-}P(CH_3){-}S{-}C_2H_5$		86–8/2	1.5485	1.2498	(*10*)
$(C_6H_5)_3Sn{-}P(C_6H_5){-}Ge(C_6H_5)_3$	115–9				(*65*, *66a*)

$(C_6H_5)_3Sn-P(C_6H_5)-Pb(C_6H_5)_3$	110 (decomposes)		*(65, 66a)*
$[(C_6H_5)_3Sn]_2P-Ge(C_6H_5)_3$	160		*(65, 66a)*
$[(C_6H_5)_3Sn]_2P-Pb(C_6H_5)_3$	171–2 (decomposes)		*(65, 66a)*
$(C_6H_5)_3Sn-P(C_6H_5)-Li$			*(65, 66a)*
$[(C_6H_5)_3Sn]_2P-Li$			*(65, 66a)*
$[(C_6H_5)_3Sn]_2P-P[Sn(C_6H_5)_3]_2$	95–110*(55)*		*(55, 64)*
$(C_6H_5)_2Sn \leftarrow P(C_6H_5)_3$			*(54, 64)*
$\left[(C_6H_5)_3Sn-\overset{\oplus}{P}(CH_3)(C_6H_5)_2\right]I^{\ominus}$	−40 (decomposes)		*(56)*
$(CH_3)_3Sn-P(=N-Si(CH_3)_3)(t\text{-}C_4H_9)_2$	~120/0.1 (sublimes)		*(42)*
$(CH_3)_3Sn-P(=N-Ge(CH_3)_3)(t\text{-}C_4H_9)_2$	~120/0.1 (sublimes)		*(42)*
$(n\text{-}C_4H_9)_3Sn-P(=O)(-OCH_3)_2$		1.4834(29°C)	*(P3)*
$(n\text{-}C_4H_9)_3Sn-P(=O)(-OC_2H_5)_2$		1.4820(30°C)	*(P3)*

(continued)

TABLE 2 (*continued*)

ORGANOTIN PHOSPHINES

Compound	m.p. (°C)	b.p. (°C/mmHg)	n_D^{20}	d_4^{20}	References
$(n\text{-}C_4H_9)_3Sn{-}P(=O)(C_6H_5)_2$	217–9				(*44*)
$(C_6H_5)_3Sn{-}P(=O)(-OCH_3)_2$	126–7				(*P3*)
$(C_6H_5)_3Sn{-}P(=O)(-OC_2H_5)_2$	128.5–29				(*P3*)
$(n\text{-}C_4H_9)_2Sn[-P(=O)(C_6H_5)_2]_2$	>350				(*44*)
$[(CH_3)_2Sn{-}PC_6H_5]_3$	134				(*48, 48b*)
$[(C_4H_9)_2Sn{-}PC_6H_5]_3$	Oil				(*48, 48b*)
$[(C_6H_5)_2Sn{-}PC_6H_5]_3$	64(*53*) 55–60(*55, 57*) 155–60(*48b*)				(*32, 48, 53, 55, 57, 64, 48b*)
$[(C_6H_5)_2Sn{-}P{-}Sn(C_6H_5)_3]_3$	99(*53*) 98–101(*57*)				(*32, 53, 57, 64*)
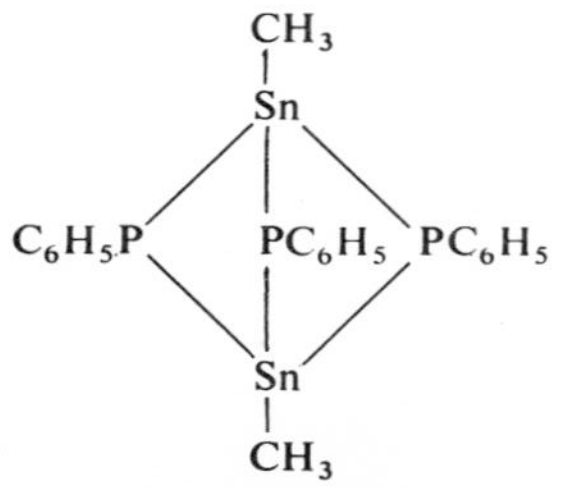 $CH_3Sn(PC_6H_5)_3SnCH_3$	152 (decomposes)				(*47*)

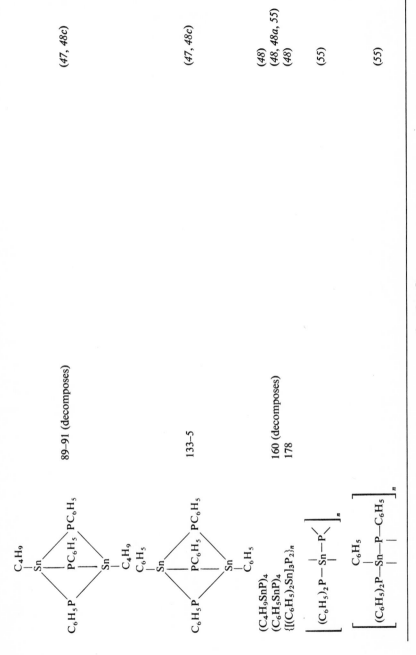

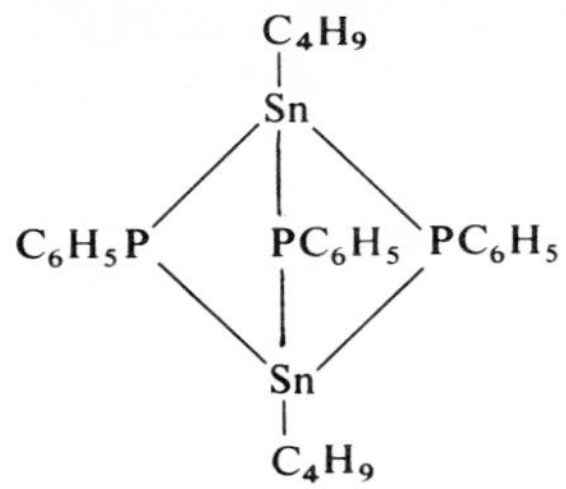 $C_4H_9Sn(PC_6H_5)_3SnC_4H_9$ (cage: two C_4H_9–Sn joined by three PC_6H_5 bridges)	89–91 (decomposes)	(*47*, *48c*)
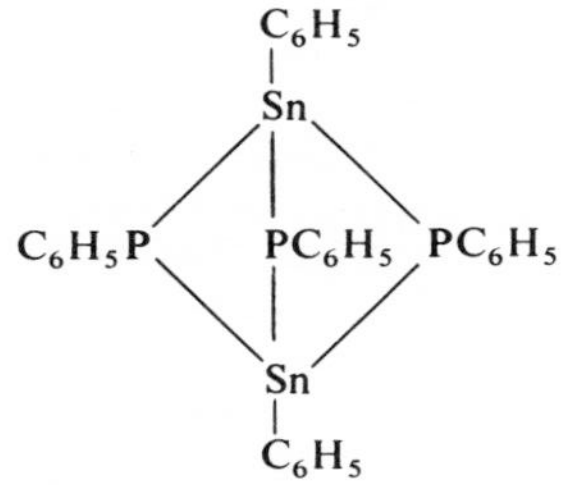 $C_6H_5Sn(PC_6H_5)_3SnC_6H_5$ (cage: two C_6H_5–Sn joined by three PC_6H_5 bridges)	133–5	(*47*, *48c*)
$(C_4H_9SnP)_4$		(*48*)
$(C_6H_5SnP)_4$	160 (decomposes)	(*48*, *48a*, *55*)
$\{[(C_6H_5)_2Sn]_3P_2\}_n$	178	(*48*)
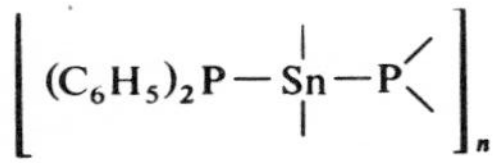 $[(C_6H_5)_2P—Sn—P\langle]_n$		(*55*)
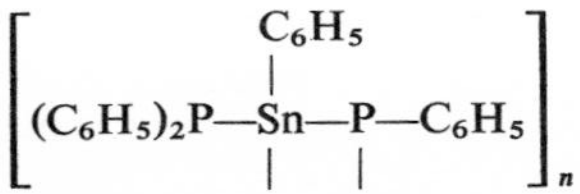 $[(C_6H_5)_2P—Sn(C_6H_5)—P—C_6H_5]_n$		(*55*)

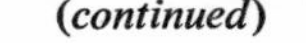

(*continued*)

TABLE 2 (*continued*)

ORGANOTIN PHOSPHINES

Compound	m.p. (°C)	b.p. (°C/mmHg)	n_D^{20}	d_4^{20}	References
$\left[\left(C_6H_5-\overset{\vert}{P}-\overset{C_6H_5}{\underset{\vert}{\overset{\vert}{Sn}}}-\right)_2PC_6H_5\right]_n$					(*55*)
$\left[\left(C_6H_5-\overset{\vert}{P}-\overset{C_6H_5}{\underset{\vert}{\overset{\vert}{Sn}}}-\right)_3P\right]_n$					(*55*)
$(CH_3)_3Sn[(CH_3]_3C)_2PMo(CO)_5$	98 (decomposes)				(*45a, 59a*)
$(CH_3)_3Sn[(CH_3]_3C)_2PFe(CO)_4$	165–70 (decomposes)				(*59a, 69c*)
$(CH_3)_3Sn[(CH_3]_3C)_2PNi(CO)_3$	130–40 (decomposes)				(*59a, 69a*)
$(CH_3)_3Sn[C_6H_5]_2PNi(CO)_3$	60 (decomposes)				(*1b, 45a, 69a*)
$[(CH_3)_3SnP(C_6H_5)_2]_2Ni(CO)_2$					(*1a, 1c*)
$(CH_3)_2Sn[-P(C_6H_5)_2]_2Ni(CO)_2$					(*1a, 1b*)
$[(CH_3)_3Sn]_2CH_3PNi(CO)_3$					(*46a*)
$[(CH_3)_3Sn]_2CH_3PCr(CO)_5$	75–8				(*46a*)
$[(CH_3)_3Sn]_2CH_3PMo(CO)_5$	95–8				(*46a*)
$[(CH_3)_3Sn]_2C_6H_9PNi(CO)_3$	50 (decomposes)				(*45a, 69a*)
$[(CH_3)_3Sn]_2C_6H_5PCr(CO)_5$	120 (decomposes)				(*45a*)
$[(CH_3)_3Sn]_3PCr(CO)_5$	190 (decomposes)				(*68*)
$[(CH_3)_3Sn]_3PMo(CO)_5$	180 (decomposes)				(*45a*)
$[(CH_3)_3Sn]_3PW(CO)_5$	180 (decomposes)				(*45a*)
$[(CH_3)_3Sn]_3PFe(CO)_4$	142 (decomposes)				(*69, 69c*)
$[(CH_3)_3Sn]PCo(NO)(CO)_2$	190 (decomposes)				(*68, 69b*)
$[(CH_3)_3Sn]_3PNi(CO)_3$	90 (decomposes)				(*67, 69a*)
$[(CH_3)_3Sn]_3PMn(CO)_2\pi C_5H_5$	140 (decomposes)				(*45a*)

$[(NO)_2(CO)Fe \leftarrow P(C_6H_5)_2]_4Sn$		(*18*)
$[(NO)(CO)_2Co \leftarrow P(C_6H_5)_2]_4Sn$		(*18*)
$\left[(CO)_4Cr \begin{matrix} \swarrow P(C_6H_5)_2- \\ \nwarrow P(C_6H_5)_2- \end{matrix}\right]_2 Sn$	170 (decomposes	(*18*)
$\left[(CO)_4Mo \begin{matrix} \swarrow P(C_6H_5)_2- \\ \nwarrow P(C_6H_5)_2- \end{matrix}\right]_2 Sn$	160 (decomposes)	(*18*)
$\left[(CO)_4W \begin{matrix} \swarrow P(C_6H_5)_2- \\ \nwarrow P(C_6H_5)_2- \end{matrix}\right]_2 Sn$	160 (decomposes) 220	(*18*)
$\left[(NO)_2Fe \begin{matrix} \swarrow P(C_6H_5)_2- \\ \nwarrow P(C_6H_5)_2- \end{matrix}\right]_2 Sn$	312 (decomposes)	(*18*)

(*continued*)

TABLE 2 (*continued*)

ORGANOTIN PHOSPHINES

Compound	m.p. (°C)	b.p. (°C/mmHg)	n_D^{20}	d_4^{20}	References
$[(NO)(CO)Co(\leftarrow P(C_6H_5)_2-)_2]_2Sn$	305 (decomposes)				(*18*)
$[(CO)_2Ni(\leftarrow P(C_6H_5)_2-)_2]_2Sn$	160 (decomposes)				(*18*)

TABLE 3

INFRARED AND RAMAN FREQUENCIES OF ORGANOTIN PHOSPHINES (cm^{-1})

Compound	Identification	References
$(Me_3Sn)_3P$	ir: $\nu_{as}PSn_3$ 351, ν_sPSn_3 284, $\delta_{as}PSn_3$ 125, δ_sPSn_3 66; Raman: $\nu_{as}PSn_3$ 359, ν_sPSn_3 284, $\delta_{as}PSn_3$ 123, δ_sPSn_3 62	(*19, 66a*)
	Raman: $\nu_{as}PSn_3$ 351, ν_sPSn_3 290	(*22*)
$(Ph_3Sn)_3P$	ir: $\nu_{as}PSn_3$ 347, ν_sPSn_3 296, $\delta_{as}PSn_3$ 125, δ_sPSn_3 88; Raman: $\nu_{as}PSn_3$ 350, ν_sPSn_3 294, $\delta_{as}PSn_3$ 123, δ_sPSn_3 88	(*19, 45, 66, 66a*)
$(Ph_3Sn)_2PGePh_3$	ir: $\nu_{as}PSn_2Ge$ 349, ν_sPSn_2Ge 297	(*65, 66a*)
$(Ph_3Sn)_2PPbPh_3$	ir: $\nu_{as}PSn_2Pb$ 345 ν_sPSn_2Pb 297	(*65, 66a*)
$(Me_3Sn)_2PMe$	ir: $\nu_{as}Sn_2P$ 349, ν_sSn_2P 300	(46a)
$(Me_3Sn)_2PPh$	ir: $\nu_{as}Sn_2P$ 350, ν_sSn_2P 327	(69*a*)
$(Et_3Sn)_2PPh$	No identification of Sn—P vibrations	(*11*)
$(Ph_3Sn)_2PPh$	ir: $\nu_{as}PSn_2$ 368, ν_sPSn_2 327, δPSn_2 168 Raman: $\nu_{as}PSn_2$ 369, ν_sPSn_2 326, δPSn_2 167	(*19, 45, 66, 66a*)
$Ph_3SnP(Ph)GePh_3$	ir: νSnP 356	(*65, 66a*)
$Ph_3SnP(Ph)PbPh_3$	ir: νSnP 342	(*65, 66a*)
$Me_3SnP(t\text{-}Bu)_2$	ir: no $\nu Sn—P$ observable!	(*59a*)
Me_3SnPPh_2	ir: νSnP 359	(*69a*)
Et_3SnPEt_2	No identification of Sn—P vibrations	(*11*)
Ph_3SnPPh_2	ir: νSnP 351; Raman: νSnP 356	(*11, 19, 45, 50, 66a*)
$Sn(PPh_2)_4$	ir: νSnP_4 264	(*18*)
$(Me_2SnPPh)_3$	ir: νSnP 388, 365, 318, 310, 284	(*48b*)
$(Bu_2SnPPh)_3$	ir: νSnP 388, 347, 307, 298, 289	(*48b*)
$(Ph_2SnPPh)_3$	ir: νSnP 372, 361, 335, 308, 297	(*48b*)
$(PhSnP)_4$	ir: νSnP 328	(*48a*)
$(Me_3Sn)_3PNi(CO)_3$	ir: νSnP 345	(*67, 69a*)
$(Me_3Sn)_3PCo(CO)_2NO$	ir: $\nu_{as}SnP_3$ 352, ν_sSnP_3 345	(*68, 69b*)
$(Me_3Sn)_3PFe(CO)_4$	ir: ν_ePSn_3 361	(*69c*)
$(Me_3Sn)_3PCr(CO)_5$	ir: νSnP 347	(*68*)
$Sn[(PPh_2)_2Fe(NO)_2]_2$	ir: νSnP 388, 319	(*18*)
$Sn[(PPh_2)_2Co(CO)(NO)]_2$	ir: νSnP, CoP 367, 332, 307, 286	(*18*)
$Sn[(PPh_2)_2Ni(CO)_2]_2$	ir: νSnP, NiP 376, 315	(*18*)
$Sn[(PPh_2)_2Cr(CO)_4]_2$	ir: νSnP, CrP 379, 362, 310	(*18*)
$Sn[(PPh_2)_2Mo(CO)_4]_2$	ir: νSnP, MoP 376, 367	(*18*)
$Sn[(PPh_2)_2W(CO)_4]_2$	ir: νSnP, WP 374, 304	(*18*)
$(Me_3Sn)_2MePNi(CO)_3$	ir: $\nu_{as}Sn_2P$ 342, ν_sSn_2P 296	(*46a*)
$(Me_3Sn)_2MePCr(CO)_5$	ir: $\nu_{as}Sn_2P$ 360, ν_sSn_2P 308	(*46a*)
$(Me_3Sn)_2MePMo(CO)_5$	ir: $\nu_{as}Sn_2P$ 358, νSn_2P 304	(*46a*)
$Me_3SnPPh_2Ni(CO)_3$	ir: νSnP 350	(*69a*)
$(Me_3Sn)_2PhPNi(CO)_3$	ir: νSnP 357	(*69a*)

$Me = CH_3$, $Et = C_2H_5$, $Bu = C_4H_9$, $Ph = C_6H_5$.

In the uv-spectrum of $NO(CO)_2CoP[Sn(CH_3)_3]_3$ two bands could be observed at 397 and 294 mμ(*69b*). The small dipole moment of 1.05 D for triethylstannyldiphenylphosphine shows that the free electron-pair of the phosphorus partly overlaps with the tin atoms. The lowering of the dipole moment to 0.96 D when going from triethylstannyldiphenylphosphine to tributylstannyldiphenylphosphine confirms the polarization of the tin-phosphorus bond, in which the tin atom represents the positive pole and the phosphorus atom the negative pole of the dipole (*11*).

The only measured ^{1}H-nmr spectra so far are those of the trimeric dimethylstannylphenylphosphine (*67*) (triplet for the methyl-protons with $J^1H—C—Sn—^{31}P = 4.4$ Hz, $J^1H—C—^{117}Sn = 46.5$ Hz, $J^1H—C—^{119}Sn = 51.5$ Hz), of bis(trimethylstannyl)-methylphosphine (*46a*) ($J^1H—C—Sn—^{31}P = 1.90$ Hz), of trimethylstannyl-di(*tert*-butyl)phosphine (*59a*) ($J^1H—C—Sn—^{31}P = 1.5$ Hz), of trimethylstannyl-diphenylphosphine (*69a*) ($J^1H—C—Sn—^{31}P = 2.15$ Hz), of bis(trimethylstannyl)-phenylphosphine (*69a*) ($J^1H—C—Sn—^{31}P = 2.05$ Hz), and those of tris(trimethylstannyl)phosphine (*66a, 67, 69a*) and its nickel tricarbonyl (*46a, 59a, 67, 69a*), chromium pentacarbonyl (*46a, 68*), molybdenum pentacarbonyl (*46a, 59a*), nitrosylcobalt dicarbonyl (*68, 69b*), and iron tetracarbonyl complexes (*59a, 69, 69c*). The doublet resonance signal of the methyl protons of tris(trimethylstannyl)phosphine shows a $^1H—C—Sn—^{31}P$ coupling constant of 1.95 Hz. In its transition metal complexes, this coupling constant shows a considerable increase (Ni = 3.35 Hz, Cr = 3.30 Hz, Co = 3.60 Hz, Fe = 3.60 Hz).

The measured δ-values of the ^{31}P-nmr spectra for tris(trimethylstannyl)-phosphine (+330 ppm), tris(triphenylstannyl)phosphine (+323 ppm), bis(triphenylstannyl)phenylphosphine (+163 ppm), trimethylstannyl di(*tert*-butyl)phosphine (−20.3 ppm), and triphenylstannyl-diphenylphosphine (+56 ppm) (*19, 66a*) are unexpectedly high; phosphoric acid (85%) was used as an external standard. From the comparison of these values with the δ-values of analogous germanium (*19*) and silicon compounds (*20*) different interpretations concerning the bonding in these molecules have been given. Fluck et al. (*20, 41a*) assume pure $p\sigma(P)—(sp)^3\sigma$(Si, Ge, Sn)-bonds between the group IV elements and the phosphorus atom. But Engelhardt et al. (*19*) conclude that there is a significant π-participation in the Sn—P bond in spite of the pyramidal structure of these compounds.

The ^{31}P-nmr spectra have also been recorded for a number of transition metal complexes of these organotin phosphines (*59a, 69a,c*). The coupling constant $J^{31}P—^{119}Sn$ was determined by double resonance experiments for trimethylstannyl-diphenylphosphine as +598 Hz(*18b*).

III. Organotin–Arsenic Compounds

A. Preparation

(Diphenylbismuthyl)diisopropylstannyldi(hydroxytolyl)arsine, the first organometallic compound with a covalent tin-arsenic bond, was described in a patent in 1936 (*P1*). In 1964 Jones and Lappert stated (*26, 27*) that trimethylstannyldimethylamine is split by diphenylarsine with formation of trimethylstannyldiphenylarsine and Schumann and Schmidt (*63*) isolated tris(triphenyl stannyl)arsine from the reaction of triphenyltin lithium with arsenic trichloride:

$$(CH_3)_3Sn{-}N(CH_3)_2 + (C_6H_5)_2AsH \longrightarrow (CH_3)_3Sn{-}As(C_6H_5)_2 + (CH_3)_2NH$$

$$3\,(C_6H_5)_3SnLi + AsCl_3 \longrightarrow [(C_6H_5)_3Sn]_3As + 3\,LiCl$$

Campbell et al. (*12*) as well as Schumann et al. (*45, 58, 61a*) succeeded shortly afterward in synthesizing a larger number of such organotin arsines. Thus, in liquid ammonia, organotin halides or tin tetrachloride and sodium diphenylarside form organotin diphenylarsines, in which up to four diphenylarsine groups are bonded to tin (*12, 58*):

$$R_{4-n}SnCl_n + n\,NaAs(C_6H_5)_2 \longrightarrow n\,NaCl + R_{4-n}Sn[As(C_6H_5)_2]_n$$

$$(R = C_2H_5, C_3H_7, C_4H_9, C_6H_5;\quad n = 1, 2, 3, 4)$$

For the preparation of a second series of compounds, in which up to three triphenyltin units are bonded to arsenic, the reaction of triphenyltin lithium with phenylarsenic chlorides or arsenic trichloride in tetrahydrofuran is suited (*45, 58*):

$$n\,(C_6H_5)_3SnLi + Cl_nAs(C_6H_5)_{3-n} \longrightarrow n\,LiCl + [(C_6H_5)_3Sn]_nAs(C_6H_5)_{3-n}$$

$$(n = 1, 2, 3)$$

To eliminate interfering side reactions it is absolutely necessary to drop triphenyltin lithium into the solution of the arsenic chlorides so slowly that an excess of the very reactive tin-lithium compound is never present. Even with these precautions, the yield of organotin arsines is very low. The main product is hexaphenylditin, for apparently, condensation and metal exchange reactions take place:

$$6\,(C_6H_5)_3SnLi + 2\,AsCl_3 \longrightarrow 3\,(C_6H_5)_3Sn{-}Sn(C_6H_5)_3 + 2\,As + 6\,LiCl$$

This condensation reaction does not play such an important part in the synthesis of bis(triphenylstannyl)phenylarsine and it cannot be observed at all when triphenylstannyldiphenylarsine is prepared.

Reactions of organotin halides in benzene solution (using triethylamine as a hydrogen chloride acceptor) with the corresponding arsines have resulted in the formation of triphenylstannyldiphenylarsine, trimethylstannyldiphenylarsine, trimethylstannyldimethylarsine, bis(triphenylstannyl)phenylarsine, bis(trimethylstannyl)phenylarsine, bis(trimethylstannyl)methylarsine, tris(triphenylstannyl)arsine, and tris(trimethylstannyl)arsine, respectively (*61a*):

$$R_3SnCl + R_2AsH \xrightarrow[-(C_2H_5)_3N\cdot HCl]{+(C_2H_5)_3N} R_3Sn{-}AsR_2$$

$$2\,R_3SnCl + RAsH_2 \xrightarrow[-2(C_2H_5)_3N\cdot HCl]{+2(C_2H_5)_3N} (R_3Sn)_2AsR$$

$$3\,R_3SnCl + AsH_3 \xrightarrow[-3(C_2H_5)_3N\cdot HCl]{+3(C_2H_5)_3N} (R_3Sn)_3As$$

$$(R = CH_3\,,\, C_6H_5)$$

The reaction of organotin halides with potassium diphenylarside in liquid ammonia is suited (*61a*) for the preparation of diphenylstannyl-bis(diphenylarsine), phenylstannyl-tris(diphenylarsine) and tetrakis(diphenylarsino)tin. Methylstannyl and methylarsine derivatives of this kind cannot be synthesized by this method. Apparently they are unstable and rearrange with migration of methyl groups to form tetrakis(diphenylarsino)tin and trimethylstannyldiphenylarsine or tetramethyltin and tetramethyldiarsine, respectively (*61*). Recently it could be demonstrated that it is possible to synthesize organotin arsines even in oxygen-free aqueous solution (*1d*), and the synthesis of the same compound from trimethylsilyldimethylarsine and trimethyltin chloride (*1e*).

B. Chemical Properties and Reactions

Alkyltin arsines are colorless liquids which distill undecomposed. They can be easily purified by fractional distillation at lower pressure. Phenyltin arsines on the contrary are colorless or slightly yellow colored solids that dissolve easily and without decomposition in organic solvents like tetrahydrofuran, diethyl ether or aromatic hydrocarbons. Their thermal stability corresponds to that of their homologous organotin phosphines.

All organotin arsines are sensitive to oxygen (*12, 58,61b*). When exposed to air they are oxidized more or less rapidly to organotin arsinates:

$$\geq Sn{-}\bar{As}\lt + O_2 \longrightarrow \geq Sn{-}O{-}As(=O)\lt$$

The oxidation products can be isolated and characterized (*12, 58, 61b*). In a benzene-alcohol solution the tin-arsenic bond of methylstannylarsine derivatives is split by hydrogen peroxide. The isolated cleavage products diphenylarsinic acid, phenylarsonic acid, arsenic acid and trimethyltin hydroxide can be used to prove the structure of the original organotin arsine.

Halogens and covalent halides as well as H_2S and thiols (*1e*) have invariably caused fission of the Sn—As bond.

$$(CH_3)_3Sn{-}As(CH_3)_2 + Br_2 \longrightarrow (CH_3)_3SnBr + (CH_3)_2AsBr$$

$$2(CH_3)_3Sn{-}As(CH_3)_2 + H_2S \longrightarrow [(CH_3)_3Sn]_2S + 2(CH_3)_2AsH$$

$$(CH_3)_3Sn{-}As(CH_3)_2 + RSH \longrightarrow (CH_3)_3SnSR + (CH_3)_2AsH$$

Polar reagents, like methyl iodide, split the tin-arsenic bond much slower than the tin-phosphorus bond and form an organotin iodide and a tetra-organoarsonium iodide (*12, 61, 61b*). Carbon disulfide and trimethylstannyl-dimethylarsine form an addition complex, which decomposes by heating to reform the starting compounds (*1d, 61*).

Metalloid ketimines were formed by an overall 1,4 addition of trimethyltin dimethylarsine to 1,1-bistrifluoromethyl-2,2-dicyanoethylene (*1c*).

$$(CH_3)_3Sn{-}As(CH_3)_2 + (F_3C)_2C{=}C(CN)_2 \longrightarrow (CH_3)_2As{-}C(CF_3)_2{-}C(CN){=}C{=}N{-}Sn(CH_3)_3$$

In the reaction of triphenylstannyldiphenylarsine and bis(triphenylstannyl) phenylarsine with phenyl azide in benzene solution the following organotin-substituted arsinimines are produced (*61c*):

$$[(C_6H_5)_3Sn]_nAs(C_6H_5)_{3-n} + (n+1)\,C_6H_5N_3 \longrightarrow (n+1)\,N_2 + \left[(C_6H_5)_3Sn{-}\underset{}{N}(C_6H_5){-}\right]_n As(C_6H_5)_{3-n}{=}NC_6H_5 \qquad (n = 1, 2)$$

Tris(trimethylstannyl)arsine reacts in ether and at room temperature with diphenylphosphorus chloride and diphenylarsenic chloride with the formation of tris(diphenylphosphino)arsine and tris(diphenylarsino)arsine, respectively. In the same way it is possible to prepare bis(diphenylphosphino) phenylarsine or bis(diphenylarsino) phenylarsine by reacting diphenylphosphorus chloride or diphenylarsenic chloride with bis(trimethylstannyl) phenylarsine; or to prepare diphenylphosphinodiphenylarsine by reacting diphenylphosphorus chloride with trimethylstannyldiphenylarsine (*62, 62a*):

$$[(CH_3)_3Sn]_3As + 3\,(C_6H_5)_2XCl \longrightarrow [(C_6H_5)_2X]_3As + 3\,(CH_3)_3SnCl$$
$$[(CH_3)_3Sn]_2AsC_6H_5 + 2\,(C_6H_5)_2XCl \longrightarrow [(C_6H_5)_2X]_2AsC_6H_5 + 2\,(CH_3)_3SnCl$$
$$(CH_3)_3Sn{-}As(C_6H_5)_2 + (C_6H_5)_2PCl \longrightarrow (C_6H_5)_2As{-}P(C_6H_5)_2 + (CH_3)_3SnCl$$
$$(X = P, As)$$

In the same way as with organotin phosphines, organotin arsines form complexes with transition-metal carbonyl, e.g., nickel tetracarbonyl or chromium hexacarbonyl (*1a,b*).

C. Physical Properties

The infrared and Raman spectra of some organotin arsines have been measured (Table 4). The fact that there appear two tin-arsenic valence vibrations in the spectra of the tris(triorganotin) arsines confirms the assumption that these compounds are of C_{3v} symmetry and therefore of a pyramidal structure.

The chemical shifts as well as the coupling constants $J^1HC^{117}Sn$, $J^1HC^{119}Sn$, $J^1HCAs^{117}Sn$ and $J^1HCAs^{119}Sn$ of some methylstannylarsines have been determined (*1, 1d, 61, 61a*) (Table 5).

A summary of the organotin-arsines prepared up to now along with their physical properties is given in Table 6.

TABLE 4

Infrared Spectra of Organotin Arsines (cm^{-1})

Compound	Identification	References
$(Me_3Sn)_3As$	ir: $\nu_{as}AsSn_3$ 233, ν_sAsSn_3 211	(*61a*)
	Raman: $\nu_{as}AsSn_3$ 233, ν_sAsSn_3 209	(*22*)
$(Ph_3Sn)_3As$	ir: $\nu_{as}AsSn_3$ 244, ν_sAsSn_3 211	(*61a*)
$(Me_3Sn)_2AsMe$	ir: $\nu_{as}AsSn_2$ 234, ν_sAsSn_2 217	(*61a*)
$(Me_3Sn)_2AsPh$	ir: $\nu_{as}AsSn_2$ 237, ν_sAsSn_2 190	(*61a*)
$(Ph_3Sn)_2AsPh$	ir: $\nu_{as}AsSn_2$ 236, ν_sAsSn_2 213	(*61a*)
$Me_3SnAsMe_2$	ir: νAsSn 199	(*61a*)
$Me_3SnAsPh_2$	ir: νAsSn 188	(*61a*)
$Ph_3SnAsMe_2$	ir: νAsSn 180	(*61a*)
$Ph_3SnAsPh_2$	ir: νAsSn 203	(*61a*)
$Ph_2Sn(AsPh_2)_2$	ir: $\nu_{as}SnAs_2$ 261, ν_sSnAs_2 228	(*61a*)
$PhSn(AsPh_2)_3$	ir: $\nu_{as}SnAs_3$ 262, ν_sSnAs_3 228	(*61a*)
$Sn(AsPh_2)_4$	ir: ν_dSnAs_4 260	(*61a*)

$Me = CH_3$, $Ph = C_6H_5$.

TABLE 5

^{1}H-NMR DATA OF ORGANOTIN ARSINES GIVEN IN HZ

Compound	$J^1H—C—^{117}Sn$	$J^1H—C—^{119}Sn$	$J^1H—C—As—^{117}Sn$	$J^1H—C—As—^{119}Sn$	References
$(CH_3)_3Sn—As(CH_3)_2$	47.0	49.5	51.0	53.5	*(1)*
	48.0	50.5	51.5		*(61a)*
$(CH_3)_3Sn—As(C_6H_5)_2$	49.0	51.0			*(61a)*
$(C_6H_5)_3Sn—As(CH_3)_2$			56.0	58.5	*(61a)*
$[(CH_3)_3Sn]_2AsCH_3$	48.5	51.5	46.0	48.0	*(61a)*
$[(CH_3)_3Sn]_2AsC_6H_5$	49.0	52.0			*(61a)*
$[(CH_3)_3Sn]_3As$	47.5	50.0			*(61a)*

TABLE 6

ORGANOTIN ARSINES

Compound	m.p. (°C)	b.p. (°C/mm Hg)	n_D^{20}	d_4^{20}	References
$(CH_3)_3Sn—As(CH_3)_2$		170–2/760(*61a*) 52–3/7.5–8(*1d*)	1.5483 (19°C)(*1d*)	1.57(*1d*)	(*1,1b,d,e, 61a*)
$(CH_3)_3Sn—As(C_6H_5)_2$		136/0.05(*26, 27*) 150–2/1(*61a*)	1.6438(*27*)	1.4682(*27*)	(*26, 27, 61a, 64*)
$(C_2H_5)_3Sn—As(C_6H_5)_2$		140–3/0.15(*12*)			(*12, 13, 64*)
$(n\text{-}C_3H_7)_3Sn—As(C_6H_5)_2$		159–61/0.2(*12*)			(*12, 64*)
$(n\text{-}C_4H_9)_3Sn—As(C_6H_5)_2$		163–4/0.09(*12*)			(*12, 64*)
$(C_6H_5)_3Sn—As(CH_3)_2$	90 (decomposes)				(*61a*)
$(C_6H_5)_3Sn—As(C_6H_5)_2$	117–9(*58*) 132–5(*61a*)				(*58, 61a, 64*)
$[(CH_3)_3Sn]_2AsCH_3$		105–8/1			(*61a*)
$[(CH_3)_3Sn]_2AsC_6H_5$		145–8/0.001			(*61a*)
$[(C_6H_5)_3Sn]_2AsC_6H_5$	112–5(*58*) 133–5(*61a*)				(*32, 58, 61a, 64*)
$[(CH_3)_3Sn]_3As$		99–100/1(*61a*)			(*22, 61a*)
$[(C_6H_5)_3Sn]_3As$	212–6(*58, 63*) 205–8(*61a*)				(*58, 61a, 63, 64*)
$(CH_3)_2Sn[—As(CH_3)_2]_2$		55/0.01	1.67		(*1b*)
$(C_6H_5)_2Sn[—As(C_6H_5)_2]_2$	80 (decomposes)(*58*) 130–3(*61a*)				(*58, 61a, 64*)
$C_6H_5Sn[—As(C_6H_5)_2]_3$	85 (decomposes)(*58*) 84–6(*61a*)				(*58, 61a, 64*)

$Sn[-As(C_6H_5)_2]_4$	68–70(*58*) 130–3(*61a*)	(*58*, *61a*, *64*)
$(C_6H_5)_2Bi-Sn(i\text{-}C_3H_7)_2-As\left(\text{3-methyl-4-hydroxycyclohexyl}\right)_2$		(*P1*)
$(CH_3)_3Sn-As(CH_3)_2 \rightarrow Ni(CO)_3$		(*1a,b*)
$(CH_3)_2Sn[-As(CH_3)_2]_2 \rightarrow Ni(CO)_2$		(*1a,b*)
$(CH_3)_2Sn[-As(CH_3)_2]_2 \rightarrow Cr(CO)_4$		(*1a,b*)

IV. Organotin-Antimony Compounds

As expected, organotin stibines are very similar to the analogous organotin arsines. Therefore, shortly after the first work with organotin-arsenic compounds had been done, the homologous antimony derivatives were examined (*P1*). Here, too, a favorable way of synthesis was the reaction of organotin halides with sodium diphenylstibide in liquid ammonia (*12, 45, 59*) or the reaction of triphenyltin lithium with diphenylantimony chloride, phenylantimony dichloride or antimony trichloride in tetrahydrofuran (*45, 59, 63*):

$$R_3SnCl + NaSb(C_6H_5)_2 \longrightarrow R_3Sn{-}Sb(C_6H_5)_2 + NaCl$$
$$(R = C_2H_5, C_3H_7, C_4H_9)$$

$$(C_6H_5)_{4-n}SnCl_n + n\ NaSb(C_6H_5)_2 \longrightarrow (C_6H_5)_{4-n}Sn[Sb(C_6H_5)_2]_n + n\ NaCl$$
$$(n = 1, 2, 3, 4)$$

$$n(C_6H_5)_3SnLi + (C_6H_5)_{3-n}SbCl_n \longrightarrow [(C_6H_5)_3Sn]_nSb(C_6H_5)_{3-n} + n\ LiCl$$
$$(n = 1, 2, 3)$$

Analogous to the reactions with the corresponding arsenic compounds, here too, a condensation of triphenyltin lithium in the presence of antimony chloride occurs to form hexaphenylditin. This side reaction decreases the yield of tris(triphenylstannyl)stibine considerably.

Tris(trimethylstannyl)stibine is prepared in high yields by reacting trimethyltin chloride with lithium stibide in diethyl ether (*2, 3*):

$$3\ (CH_3)_3SnCl + Li_3Sb \longrightarrow [(CH_3)_3Sn]_3Sb + 3\ LiCl$$

The analogous ethyl and phenyl compounds are prepared by cleaving off ethane from triethyltin hydride or triphenyltin hydride, and triethylantimony (*13, 71, 72*):

$$3\ R_3SnH + (C_2H_5)_3Sb \longrightarrow [R_3Sn]_3Sb + 3\ C_2H_6$$
$$(R = C_2H_5, C_6H_5)$$

Tris(triethylstannyl)stibine can also be prepared by displacement of triethylgermane in the reaction of tris(triethylgermanyl)stibine with triethyltin hydride (*71, 72*):

$$[(C_2H_5)_3Ge]_3Sb + 3\ (C_2H_5)_3SnH \longrightarrow [(C_2H_5)_3Sn]_3Sb + 3\ (C_2H_5)_3GeH$$

Finally, trimethylstannyldibutylstibine is a side product in the hydrostannation of ethynyl dibutylstibine with trimethyltin hydride (*34*), which presumably follows a radical mechanism:

$$(C_4H_9)_2Sb{-}C{\equiv}CH + (CH_3)_3SnH \longrightarrow (C_4H_9)_2Sb^{\cdot} + (CH_3)_3Sn^{\cdot} + HC{\equiv}C^{\cdot} + H^{\cdot}$$
$$(C_4H_9)_2Sb^{\cdot} + (CH_3)_3Sn^{\cdot} \longrightarrow (C_4H_9)_2Sb{-}Sn(CH_3)_3$$
$$HC{\equiv}C^{\cdot} + H^{\cdot} \longrightarrow HC{\equiv}CH \quad \text{or} \quad HC{\equiv}C{-}C{\equiv}CH + H_2$$

Organotin stibines are colorless liquids or crystalline solids. They dissolve well and are monomeric in aromatic hydrocarbons and tetrahydrofuran. They can be recrystallized from aliphatic hydrocarbons if absolutely no oxygen is present.

Oxygen-free water cannot attack organotin stibines, probably because they do not get moistened enough. But all these compounds are more or less sensitive to oxidation. The amount of this sensitivity is determined to a large extent by the shielding of the free electron-pair at the antimony atoms. For instance, the highly symmetric tris(triphenylstannyl)stibine is stable for many weeks when exposed to air, whereas all other compounds are oxidized, at least at their surface, immediately. The way of oxidation is formally the same as that of homologous phosphorus and arsenic compounds and leads to the addition of two oxygen atoms for each tin-antimony bond. But the oxidation products are not defined (*12, 59*). Apparently antimony-carbon bonds are also split in the oxidation. In the cleavage reaction of tributylstannyldiphenylstibine with methyl iodide the only reaction product that could be isolated was tributyltin iodide (*12*).

Tris(triethylstannyl)stibine is split by bromine, forming triethyltin bromide and $SbBr_3$ (*73*).

Small amounts of aluminum tribromide catalyze the thermal decomposition of tris(triethylstannyl) stibine to tetraethyltin, tin, and antimony (*71, 72*):

$$4\,[(C_2H_5)_3Sn]_3Sb \xrightarrow{AlBr_3} 9\,(C_2H_5)_4Sn + 3\,Sn + 4\,Sb$$

In the cleavage reaction of tris(triethylstannyl)stibine with benzoyl peroxide one obtains antimony and triethyltin benzoate (*71, 72*):

$$2\,[(C_2H_5)_3Sn]_3Sb + 3\,(C_6H_5COO)_2 \longrightarrow 2\,Sb + 6\,(C_2H_5)_3Sn{-}O\overset{\overset{\displaystyle O}{\|}}{C}{-}C_6H_5$$

In the reaction of tris(triethylstannyl) stibine with benzyl bromide or cyclopentyl bromide at 100–150°C one obtains triethyltin bromide and the corresponding organoantimony compounds (*70c, 71, 72*):

$$[(C_2H_5)_3Sn]_3Sb + 3\,RBr \longrightarrow 3\,(C_2H_5)_3SnBr + SbR_3$$
$$(R = C_6H_5CH_2,\ \text{cyclo-}C_5H_9)$$

Tris(triethylgermyl)stibine reacts with lithium in THF yielding a red mixture, which, when treated with triethyltin hydride, formed triethylgermyltriethylstannane and a crude mixture containing some compounds with $(C_2H_5)_3Ge$ and $(C_2H_5)_3Sn$ bond to Sb. The same reaction was carried out with tris(triethylsilyl)stibine. Products with the formulas $[(C_2H_5)_3Si]_{3-n}Sb[Sn(C_2H_5)_3]_n$ and $[(C_2H_5)_3Ge]_{3-n}Sb[Sn(C_2H_5)_3]_n$ have been suggested (*70c*).

TABLE 7

Organotin Stibines

Compound	m.p. (°C)	b.p. (°C/mm Hg)	n_D^{20}	d_4^{20}	References
$(CH_3)_3Sn—Sb(C_4H_9)_2$		126/0.5	1.5500		(*34*)
$(C_2H_5)_3Sn—Sb(C_6H_5)_2$		144–6/0.18(*12*)			(*12, 64*)
$(n\text{-}C_3H_7)_3Sn—Sb(C_6H_5)_2$		168–70/0.13(*12*)			(*12, 64*)
$(n\text{-}C_4H_9)_3Sn—Sb(C_6H_5)_2$		179–80/0.15(*12*)			(*12, 64*)
$(C_6H_5)_3Sn—Sb(C_6H_5)_2$	116(*59*)				(*34, 59*)
$[(C_6H_5)_3Sn]_2SbC_6H_5$	120 (decomposes)				(*59*)
$[(CH_3)_3Sn]_3Sb$	39(*2, 3*)				(*2, 3, 22*)
$[(C_2H_5)_3Sn]_3Sb$		174–7/1.5(*71*)	1.6062(*72*)	1.615(*72*)	(*17, 71, 72*)
		220 (decomposes) (*71*)		1.615(*71*)	
$[(C_6H_5)_3Sn]_3Sb$	215(*59*)				(*13, 59, 63, 64*)
	214–5(*63*)				
	153 (decomposes) (*13*)				
$(C_6H_5)_2Sn[—Sb(C_6H_5)_2]_2$	150				(*59*)
$C_6H_5Sn[—Sb(C_6H_5)_2]_3$	90				(*59*)
$Sn[—Sb(C_6H_5)_2]_4$	75				(*59*)
$(C_3H_7)_3Sn—Sb\left(C_6H_9(CH_3)(OH)\right)_2$					(*P1*)

In the ^{1}H-nmr spectrum of tris(triethylstannyl) stibine the signals of the ethyl-protons were found to be at $\tau_{CH_2} = 9.02$ and $\tau_{CH_3} = 8.79$ (*17*). The physical data of the organotin stibines known so far are collected in Table 7.

V. Organotin-Bismuth Compounds

Compared to the relatively large number of known organotin-phosphorus, -arsenic, and -antimony compounds, there are only four organotin-bismuth compounds listed in Table 8. The first publication in this field was a patent in 1936. It describes the use of triethylstannyldiphenylbismuthine and di-(hydroxytolyl)stibyl-diphenylstannyl-diphenylbismuthine as antioxidants in lubricants. But no details about the synthesis or properties of these compounds were given (*P1*). In 1964, Schumann and Schmidt (*45*, *63*), by reacting triphenyltin lithium with bismuth trichloride in tetrahydrofuran, obtained small amounts of tris(triphenylstannyl)bismuthine. In 1967, Creemers increased the yield to more than 75% by reacting triethylbismuthine with triphenyltin hydride (*13*):

$$3\,(C_6H_5)_3SnLi + BiCl_3 \longrightarrow [(C_6H_5)_3Sn]_3Bi + 3\,LiCl$$

$$3\,(C_6H_5)_3SnH + Bi(C_2H_5)_3 \longrightarrow [(C_6H_5)_3Sn]_3Bi + 3\,C_2H_6$$

In 1965, Vyazankin et al. (*29*, *71*, *72*) prepared tris(triethylstannyl)bismuthine by reacting triethylbismuth with triethyltin hydride or by ligand exchange between tris(triethylgermanyl)bismuthine and triethyltin hydride:

$$3\,(C_2H_5)_3SnH + Bi(C_2H_5)_3 \longrightarrow [(C_2H_5)_3Sn]_3Bi + 3\,C_2H_6$$

$$3\,(C_2H_5)_3SnH + [(C_2H_5)_3Ge]_3Bi \longrightarrow [(C_2H_5)_3Sn]_3Bi + 3\,(C_2H_5)_3GeH$$

These organotin bismuthines are very sensitive towards oxidation and they decompose at temperatures of about 150°C to hexaorganoditins and bismuth (*63*, *71*). This decomposition occurs at even lower temperatures if catalytic amounts of aluminum trichloride are present. But the products formed now are tetraethyltin, tin, and bismuth (*71*):

$$2\,(R_3Sn)_3Bi \xrightarrow{\Delta} 3\,R_3Sn{-}SnR_3 + 2\,Bi$$

$$4\,[(C_2H_5)_3Sn]_3Bi \xrightarrow{AlCl_3} 9\,(C_2H_5)_4Sn + 3\,Sn + 4Bi$$

Benzoyl peroxide splits the tin-bismuth bond of tris(triethylstannyl) bismuthine with the formation of bismuth and triethyltin benzoate (*29a*, *71*, *72*):

$$2\,[(C_2H_5)_3Sn]_3Bi + 3\,(C_6H_5COO)_2 \longrightarrow 2\,Bi + 6\,(C_2H_5)_3Sn{-}O\overset{\displaystyle O}{\overset{\|}{C}}{-}C_6H_5$$

TABLE 8

ORGANOTIN BISMUTHINES

Compound	m.p. (°C)	b.p. (°C/mm Hg)	n_D^{20}	d_4^{20}	References
$(C_2H_5)_3Sn—Bi(C_6H_5)_2$					(*P1*)
$[(C_2H_5)_3Sn]_3Bi$		160–70 (decomposes)(*71*)		1.743(*71*, *72*)	(*17*, *29*, *71*, *72*)
$[(C_6H_5)_3Sn]_3Bi$	138–42(*63*) (decomposes) 140(*13*) (decomposes)				(*13*, *63*, *64*)
$(C_6H_5)_2Bi—Sn(i\text{-}C_3H_7)_2—As(C_6H_9(CH_3)(OH))_2$					(*P1*)

The only spectroscopic data known so far is the chemical shift value of the ethyl-protons in the ^{1}H-nmr spectrum of tris(triethylstannyl) bismuthine. The measured chemical shift was $\tau_{CH_2,\ CH_3} = 8.86$ (*17*).

VI. Comparative Summary

An important theoretical point of view in the examination of the organotin-phosphines, -arsines, -stibines, and -bismuthines is the following consideration. The enormous stability of the silicon-oxygen bond and the silicon-nitrogen bond is assumed to be due to $(p \rightarrow d)\pi$ multiple bond participation in these bonds. Can this assumption also be used for the bonding between the higher homologs of silicon on the one hand and the higher homologs of nitrogen on the other hand? If in the organosilyl amines the nitrogen is substituted by its next higher homolog, namely phosphorus, then because of the greater covalent radius of phosphorus the opportunity of its occupied *p*-orbitals to overlap with the empty *d*-orbitals of silicon should be less favorable. On the other hand, the opportunity of the phosphorus orbitals to overlap with orbitals of heavier homologs of silicon, like for instance tin, should again be favored.

The course of the oxidation reactions and the unsuccessful attempts to synthesize organotin phosphine oxides or organotin phosphonium salts suggest that the stability of the tin-phosphorus bond is extensively dependent on a participation of $(p \rightarrow d)\pi$-bonding. If the free electron-pair at the phosphorus atom is blocked by oxygen, the tin-phosphorus bond breaks. The initially formed organotin phosphine oxide rearranges into an organotin phosphenate with regeneration of the free electron-pair at the phosphorus atom. The manner of 1,2-dipolar addition of carbon disulfide and phenylisocyanate also supports this hypothesis. While these reactions can be done with tin tetrakis(diphenylphosphine), which has four free phosphorus electron-pairs and only one central tin atom, the electrophilic attack of those 1,2-dipoles on tris(triphenylstannyl)phosphine, which has only one free electron-pair on the central phosphorus atom and three tin atoms, is no longer possible.

Vibration spectra and ^{31}P-nmr spectra of the organotin phosphines and some reactions also seem to contradict these interpretations of the tin-phosphorus bond. The spectra can only be interpreted with the assumption of a pyramidal structure for these compounds. The vibrational spectra of organogermanium- and organolead-phosphines, -arsines, -stibines, and -bismuthines point to the same conclusion. In the meantine it could also be shown that it *is* possible to block the free electron-pair of phosphorus in such organometallic phosphines without breaking the metal-phosphorus bond. This is proven by the existence of compounds like, e.g., trimethylstannyl-di-*t*-butylphosphino-trimethylsilyl-imine and different organometallic phosphine-substituted transition metal carbonyls.

Since it is not yet proved that (1) the participation of $(p \rightarrow d)\pi$ double bonds necessarily has to be connected with the molecule becoming planar, (2) in *N,P*-organometal-substituted phosphinimines the organometallic groups bonded to nitrogen have a stabilizing effect on the phosphinimine bond, and (3) in transition-metal complexes of the kind mentioned back-bonding from the central transition metal to the ligand molecule will occur, it would appear the evidences of these arguments against $(p \rightarrow d)\pi$-bonding are not at all more conclusive than those which seem to prove $(p \rightarrow d)\pi$-bonding.

The same thoughts apply to the homologous arsenic, antimony, and bismuth compounds. The research done so far shows that one obtains the most stable compounds when the covalent radii of the group IV and V atoms are nearly equal. Tris(triphenylstannyl)stibine, e.g., practically can be considered stable towards oxygen and, as can be seen in papers by Razuvaev, the heavier group IV elements can displace their lighter homologs in the compounds with antimony and bismuth. Campbell et al. also assume $(p \rightarrow d)\pi$-double bond participation for the tin-arsenic and the tin antimony bonds. That assumption is also based on the course of oxidation reactions, measurements of the dipole moment and the nonexistence of methyl iodide addition products.

Considering what is known so far, it seems to be useful to discuss the inclusion of the free electron-pair of the group V elements phosphorus, arsenic, antimony, and bismuth into the bond with the group IV elements, silicon, germanium, tin, and lead. But a direct comparison with the conditions in the Si—O—Si or Si—N bond seems to be too great a simplification.

REFERENCES

1. E. W. Abel and D. B. Brady, *J. Organometal. Chem.*, **11**, 145 (1968).

1a. E. W. Abel, J. P. Crow, and S. M. Illingworth, *Chem. Commun.*, 817 (1968).

1b. E. W. Abel, J. P. Crow, and S. M. Illingworth, *J. Chem. Soc. A*, 1631 (1969).

1c. E. W. Abel, J. P. Crow, and J. N. Wingfield, *J. Chem. Soc. D*, 967 (1969).

1d. E. W. Abel, R. Hönigschmidt-Grossich, and S. M. Illingworth, *J. Chem. Soc. A*, 2623 (1968).

1e. E. W. Abel and S. M. Illingworth, *J. Chem. Soc. A*, 1094 (1969).

1f. E. W. Abel and S. M. Illingworth, *Organometal. Chem. Rev. A*, **5**, 143 (1970).

2. E. Amberger and R. W. Salazar, *Sci. Commun.* II. *Intern. Symp. Organosilicon Chem., Prague, 1965*, 31.

3. E. Amberger and R. W. Salazar, *J. Organometal. Chem.*, **8**, 111 (1967).

4. B. A. Arbuzov and N. P. Grechkin, *Zhur. Obshchei. Khim.*, **17**, 2166 (1947); through *C.A.*, **42**, 4522 (1948).

5. B. A. Arbuzov and N. P. Grechkin, *Zhur. Obshchei. Khim.*, **20**, 107 (1950); through *C.A.*, **44**, 5832 (1950).

6. B. A. Arbuzov and N. P. Grechkin, *Izv. Akad. Nauk SSSR, Otdel. Khim. Nauk* 440 (1956); through *C.A.*, **50**, 16661 (1956).

7. B. A. Arbuzov and A. N. Pudovik, *Zhur. Obschchei. Khim.*, **17**, 2158 (1947); through *C.A.*, **42**, 4522 (1948).
8. L. D. Balashova, A. B. Bruker, and L. S. Soborovskii, *Zhur. Obshchei Khim.*, **35**, 2207 (1965).
9. A. B. Bruker, L. D. Balashova, and L. S. Soborovskii, *Dokl. Akad. Nauk SSSR, Odtel. Khim. Nauk*, **135**, 843 (1960); through *C.A.*, **55**, 13301 (1961).
10. A. B. Bruker, L. D. Balashova, and L. S. Soborovskii, *Zhur. Obshchei Khim.*, **36**, 75 (1966); through *C.A.*, **64**, 14211 (1966).
11. I. G. M. Campbell, G. W. A. Fowles, and L. A. Nixon, *J. Chem. Soc.*, 1389 (1964).
12. I. G. M. Campbell, G. W. A. Fowles, and L. A. Nixon, *J. Chem. Soc.*, 3026 (1964).
13. H. M. J. C. Creemers, Doctoral Dissertation, Univ. Utrecht, 1967.
14. H. M. J. C. Creemers, F. Verbeek, and J. G. Noltes, *J. Organometal. Chem.*, **8**, 469 (1967).
15. A. G. Davies, *Ann. Rep. Chem. Soc.*, **63**, 353 (1967).
16. M. Dub, *Organometallic Compounds*, 2nd ed., Vol. II, Springer Verlag, Berlin, 1967.
17. A. N. Egorochkin, N. S. Vyazankin, G. A. Razuvaev, O. A. Kruglaya, and M. N. Bochkarev, *Dokl. Akad. Nauk SSSR*, **170**, 333 (1966); through *C.A.*, **66**, 4780 (1967).
18. J. Ellermann and K. H. Dorn, *Z. Naturforsch.*, **23b**, 420 (1968).
18b. H. Elsner and H. Dreeskamp, *Ber. Bunsenges. phys. Chem.*, **73**, 619 (1969).
19. G. Engelhardt, P. Reich, and H. Schumann, *Z. Naturforsch.*, **22b**, 352 (1967).
20. E. Fluck, H. Bürger, and U. Goetze, *Z. Naturforsch.*, **22b**, 912 (1967).
21. T. A. George and M. F. Lappert, *Chem. Commun.*, 463 (1966).
21a. T. A. George and M. F. Lappert, *J. Chem. Soc. A*, 992 (1969).
21b. P. G. Harrison, *Organometal. Chem. Rev. A*, **4**, 379 (1969).
21c. P. G. Harrison, *Organometal. Chem. Rev. A*, **5**, 183 (1970).
22. R. E. Hester and K. Jones, *Chem. Commun.*, 317 (1966).
23. R. K. Ingham, S. D. Rosenberg, and H. Gilman, *Chem. Rev.*, **60**, 459 (1960).
24. K. Issleib and B. Walther, *J. Organometal. Chem.*, **10**, 177 (1967).
25. K. Itoh, S. Sakai, and Y. Ishii, *Yuki Gosei Kagaku Kyokai Shi*, **24**, 729 (1966); through *C.A.*, **65**, 16998 (1966).
26. K. Jones and M. F. Lappert, *Proc. Chem. Soc.*, 22 (1964).
27. K. Jones and M. F. Lappert, *J. Organometal. Chem.*, **3**, 295 (1965).
28. K. Jones and M. F. Lappert, *Organometal. Chem. Rev.*, **1**, 67 (1966).
28a. K. A. Kockeskhov, N. N. Zemlyanskii, N. J. Sheverdina, and E. M. Panov, in A. N. Nesmejanov and K. A. Kocheshkov *Methods of Elemento-organic Chemistry, Germanium, Tin and Lead* (A. N. Nesmejanov and K. A. Kocheshkov, eds.). Moscow, 1968.
29. O. A. Kruglaya, N. S. Vyazankin, and G. A. Razuvaev, *Zhur. Obshchei Khim.*, **35**, 394 (1965).
29a. O. A. Kruglaya, N. S. Vyazankin, and G. A. Razuvajev, *Usp. Khim. Org. Perekisnykh Soedin. Autokisleniya Dokl. Vse. Konf. 3rd, 1965*, 247 (1969); through *CA*, **72**, 43820 (1970).
30. W. Kuchen and H. Buchwald, *Chem. Ber.*, **92**, 227 (1959).
31. M. F. Lappert and B. Prokai, *Advances in Organometallic Chemistry*, Vol. V, Academic, New York, 1967.
31a. J. G. A. Luijten, *Organometal. Chem. Rev. B*, **5**, 687 (1969).
32. C. Mahr and G. Stork, *Z. Anal. Chem.*, **222**, 1 (1966).
33. L. Malatesta and A. Sacco, *Gazz. Chim. Ital.*, **80**, 658 (1950); through *C.A.*, **46**, 4473 (1952).
34. A. N. Nesmeyanov, A. E. Borisov, and N. V. Novikova, *Dokl. Akad. Nauk SSSR*, **172**, 1329 (1967); through *C.A.*, **67**, 302 (1967).

35. W. P. Neumann, *Die Organische Chemie des Zinns*, Ferd. Enke Verlag, Stuttgart, 1967.
36. W. P. Neumann, B. Schneider, and R. Sommer, *Ann. Chem.*, **692**, 1 (1966).
37. A. N. Pudovik, A. A. Muratova, and E. P. Semkina, *Zhur. Obshchei Khim.*, **33**, 3350 (1963); through *C.A.*, **60**, 4175 (1964).
37a. E. W. Randall and J. J. Zuckerman, *Chem. Commun.*, 732 (1966).
37b. G. A. Razuvajev and N. S. Vyazankin, *Pure Appl. Chem.*, **19**, 353 (1969).
38. I. Ruidisch, H. Schmidbaur, and H. Schumann, *Halogen Chemistry*, Vol. II, Academic, New York, p. 233, 1967.
39. D. Seyferth and R. B. King, *Annual Survey of Organometallic Chem.*, Vol. I, 1965.
40. D. Seyferth and R. B. King, *Annual Survey of Organometallic Chem.*, Vol. II, 1966.
41. D. Seyferth and R. B. King, *Annual Survey of Organometallic Chem.*, Vol. III, 1967.
41a. H. Siebert, J. Eints, and E. Fluck, *Z. Naturforsch.*, **23b**, 1006 (1968).
41b. O. J. Scherer, *Angew. Chem.*, **81**, 871 (1969).
42. O. J. Scherer and G. Schieder, *Chem. Ber.*, **101**, 4184 (1968).
43. O. J. Scherer and G. Schieder, *Angew. Chem.*, **80**, 83 (1968).
44. H. Schindlbauer and D. Hammer, *Monatsh. Chem.*, **94**, 644 (1963).
45. H. Schumann, Habilitationsschrift, Univ. Würzburg, 1967.
45a. H. Schumann, *Angew. Chem.*, **81**, 970 (1969).
46. H. Schumann and U. Arbenz, unpublished work, 1968.
46a. H. Schumann and U. Arbenz, *J. Organometal. Chem.*, **22**, 411 (1970).
47. H. Schumann and H. Benda, *Angew. Chem.*, in press, 1968.
48. H. Schumann and H. Benda, unpublished work, 1968.
48a. H. Schumann and H. Benda, *Angew. Chem.*, **80**, 846 (1968).
48b. H. Schumann and H. Benda, *Angew. Chem.*, **80**, 845 (1968).
48c. H. Schumann and H. Benda, *Angew. Chem.*, **81**, 1049 (1969).
49. H. Schumann and P. Jutzi, *Chem. Ber.*, **101**, 24 (1968).
50. H. Schumann, P. Jutzi, A. Roth, P. Schwabe, and E. Schauer, *J. Organometal. Chem.*, **10**, 71 (1967).
51. H. Schumann, P. Jutzi, and M. Schmidt, *Angew. Chem.*, **77**, 812 (1965).
52. H. Schumann, P. Jutzi, and M. Schmidt, *Angew. Chem.*, **77**, 912 (1965).
53. H. Schumann, H. Köpf, and M. Schmidt, *Angew. Chem.*, **75**, 672 (1963).
54. H. Schumann, H. Köpf, and M. Schmidt, *Z. Naturforsch.*, **19b**, 168 (1964).
55. H. Schumann, H. Köpf, and M. Schmidt, *Chem. Ber.*, **97**, 1458 (1964).
56. H. Schumann, H. Köpf, and M. Schmidt, *J. Organometal. Chem.*, **2**, 159 (1964).
57. H. Schumann, H. Köpf, and M. Schmidt, *Chem. Ber.*, **97**, 2395 (1964).
58. H. Schumann, Th. Östermann, and M. Schmidt, *Chem. Ber.*, **99**, 2057 (1966).
59. H. Schumann, Th. Östermann, and M. Schmidt, *J. Organometal. Chem.*, **8**, 105 (1967).
59a. H. Schumann, L. Rösch, and O. Stelzer, *J. Organometal. Chem.*, **21**, 351 (1970).
60. H. Schumann and A. Roth, *J. Organometal. Chem.*, **11**, 125 (1968).
61. H. Schumann and A. Roth, unpublished work, 1968.
61a. H. Schumann and A. Roth, *Chem. Ber.*, **102**, 3713 (1969).
61b. H. Schumann and A. Roth, *Chem. Ber.*, **102**, 3725 (1969).
61c. H. Schumann and A. Roth, *Chem. Ber.*, **102**, 3731 (1969).
62. H. Schumann, A. Roth, and O. Stelzer, *Angew. Chem.*, **80**, 240 (1968).
62a. H. Schumann, A. Roth, and O. Stelzer, *J. Organometal. Chem.*, **24**, 183 (1970).
63. H. Schumann and M. Schmidt, *Angew. Chem.*, **76**, 344 (1964).
64. H. Schumann and M. Schmidt, *Angew. Chem.*, **77**, 1049 (1965).
65. H. Schumann, P. Schwabe, and M. Schmidt, *Inorg. Nucl. Chem. Letters*, **2**, 313 (1966).
66. H. Schumann, P. Schwabe, and M. Schmidt, *Inorg. Nucl. Chem. Letters*, **2**, 309 (1966).
66a. H. Schumann, P. Schwabe, and O. Stelzer, *Chem. Ber.*, **102**, 2900 (1969).

67. H. Schumann and O. Stelzer, *Angew. Chem.*, **79**, 692 (1967).
68. H. Schumann and O. Stelzer, *Angew. Chem.*, **80**, 318 (1968).
69. H. Schumann and O. Stelzer, *J. Organometal Chem.*, **13**, P25 (1968).
69a. H. Schumann, O. Stelzer, U. Niederreuther, and L. Rösch, *Chem. Ber.*, **103**, 1383 (1970).
69b. H. Schumann, O. Stelzer, and U. Niederreuther, *Chem. Ber.*, **103**, 1391 (1970).
69c. H. Schumann, O. Stelzer, U. Niederreuther, and L. Rösch, *Chem. Ber.*, **103**, 2350 (1970).
70. H. Schumann and A. Yaghmai, unpublished work, 1968.
70a. T. Tanaka, *Organometal. Chem. Rev. A*, **5**, 1 (1970).
70b. G. J. M. Van der Kerk. J. G. A. Luijten, J. G. Noltes, and H. M. J. C. Creemers, *Chimia*, **23**, 313 (1969).
70c. N. S. Vyazankin, G. S. Kalinina, O. A. Kruglaya, and G. A. Razuvajev, *Zhur. Obshch. Khim.*, **39**, 2005 (1969); through *CA*, **72**, 31946 (1970).
70d. N. S. Vyazankin and O. A. Kruglaya, *Uspek. Khim.*, **35**, 1388 (1966).
70e. N. S. Vyazankin, G. A. Razuvaev, and O. A. Kruglaya, Organometal. *Chem. Rev. A*, **3**, 323 (1968).
71. N. S. Vyazankin, G. A. Razuvaev, O. A. Kruglaya, and G. S. Semchikova, *J. Organometal. Chem.*, **6**, 474 (1966).
72. N. S. Vyazankin, O. A. Kruglaya, G. A. Razuveav, and G. S. Semchikova, *Dokl. Akad. Nauk SSSR*, **166**, 99 (1966).
73. N. S. Vyazankin, L. P. Sanian, G. S. Kalinina, and M. N. Bochkarev, *Zhur Obshch. Khim.*, **38**, 1800 (1968).
P1. British Pat. 445,813; through *Chem. Zentr*. II, 1287 (1936).
P2. British Pat. 892,137; through *C.A.*, **58**, 3557 (1963).
P3. Japanese Pat. 4575 (1966); through *C.A.*, **65**, 2298 (1966).
P4. USSR Pat. 170,976; through *C.A.*, **63**, 9985 (1965).
P5. USSR Pat. 170,977; through *C.A.*, **63**, 9985 (1965).